FLUID MOTION

FLUID MOTION

By
A.K. Sharma
&
A.S. Sharma

DISCOVERY PUBLISHING HOUSE PVT. LTD.
NEW DELHI-110 002

First Published – 2008

Reprinted – 2025

ISBN: 978-81-8356-342-0

Fluid Motion

Published by:

DISCOVERY PUBLISHING HOUSE
4383/4B, Ansari Road, Darya Ganj
New Delhi-110 002 (India)
Phone: +91-11-23279245; 23253475; 43596065
Mobile: +91 9811179893 / +91 9871656464
E-mail: discoverybooksindia@gmail.com
orderdphbooks@gmail.com
namitwasan9@gmail.com
web: www.discoverypublishinggroup.com

Printed at:
Infinity Imaging Systems
Delhi (INDIA)

Preface

This book "Fluid Motion" has been design to cover the syllabus of Mathematics required for the B.Sc. (Hon.), M.Sc. and Engineering students of the Indian Universities. The subject of Fluid Motion deals with the laws Governing the pressure and flow of fluids and the application of these laws to engineering practice. The term fluid includes liquid and gases. The term fluid is applied to all the substance which offer no resistance to change of shape. The subject matter has been arranged as to provide a clear and integrated approach to the subject–care has been taken to make the treatment of the subject simple and accessible to the average students.

Suggestion for the improvement of the book will always be most welcome.

Authors

CONTENTS

1

Equation of Continuity

INTRODUCTION

A flow filed is the description of the flow of a fluid by representing all fundamental flow properties as a functions of position in space and time. We shall discuss only the kinematics of the flow field *i.e.,* a description of the motion and the mass distribution.

The macroscopic treatment of a fluid enable us to use the concept of local velocity of the fluid. We shall now establish the relation analytically how the whole field of flow may be specified as an aggregate of such local velocities. There are two methods for describing motion in a fluid system. The first in which the history of individual fluid particles is described *i.e.,* to consider a particle point of view, is called the *Lagrangian method* (Lagrange 1736-1813), while the second, which concentrate on fixed points of space filled by the fluid *i.e.,* to adopt a field point of view is called the *Eulerian method* (Euler 1707-1783).

LAGRANGIAN METHOD

A description of fluid motion based on *studying the history of an individual fluid particle* is called Lagrangian. The flow quantities are represented as functions of time and of the choice of a material element of fluid. It corresponds to individual time rate of change. In this method, *any particle of the fluid* is selected and pursued on its onward course observing the changes in path, velocity, density or any other property associated with it at each point and at each instant. The selected fluid particle is identified in such a way that its linear extension is not involved.

Let ξ (a, b, c) be the initial position vector of the particle relative to a fixed origin of some chosen time t_0 (let $t_0 = 0$). Let r (x, y, z) the position of the same particle at a later time t, then at any subsequent time (t > 0) the particle's position is completely determined if the position vector r is given as a function of time t and the initial position ξ *i.e.*,

$$r = f(\xi, t)$$

where $r = ix + jy + k_z$, $\xi = ia + jb + kc$.

Here the scalar parameters a, b, c and time t are the independent variables. The components of velocity and acceleration of a fluid particle at any instant can be obtained as dr/dt and d^2r/dt^2.

The Lagrangian type of specification is useful in some special references such as certain one-dimensional (involving one space coordinate) problems but it leads to rather cubersomc analysis because of its non-linear nature. It provides more informations than one requires. Generally it is not as convenient and fruitful as the Euler's description.

EULERIAN METHOD

In this method the individual fluid particles are not identified but a *particular point of the space* occupied by the fluid is studied. We observe what happens at specific points that are fixed in space with in the flow field as different particles pass through them in the course of time. Accordingly, a complete description of the flow involves an instantaneous picture of the velocities at all points in the field. It corresponds to the local time rate of change. Let r represents the position of a point in space with regard to a chosen coordinate system. The velocity q, pressure p, density ρ and so on of the fluid particles that pass through the point at different instants can be expressed in the form

$$q = F(r, t)$$

where $q = iu + jv + kw$.

Here, the velocity is the dependent variable, where as x, y, z, t are the independent ones. The coordinates (x, y, z) describe a fixed point in the flow field, relative to the fixed frame of reference. The dependent variables are the fields of velocity, pressure, and density etc.

RELATIONSHIP BETWEEN THE LAGRANGIAN AND EULERIAN METHOD

To relate these two methods we establish a relation between points in space and the particle parameters.

I. Lagrangian to Eulerian

Let Q be some quantity defined in terms of Lagrangian description

$$Q = Q\ (a, b, c, t). \qquad ...(1)$$

We shall express a, b, c in terms of the coordinates x, y, z of a point in space. In Lagrangian method, it is defined as

$$x' = f_1\ (a, b, c, t),$$
$$y = f_2\ (a, b, c, t),$$
$$z = f_3\ (a, b, c, t). \qquad ...(2)$$

By solving these relations to obtain a, b, c in terms of Eulerain variables x, y, z and t, we have

$$a = g_1\ [(x, y, z, t),$$
$$y = g_2\ (x, y, x, t),$$
$$c = g_3\ (x, y, z, t)]. \qquad ...(3)$$

From (1) and (3), we get

$$Q = Q\ [g_1\ (x, y, z, t), g_2\ (x, y, z, t), g_3\ (x, y, z, t)], \qquad ...(4)$$

which represents the Eulerian description.

II. Eulerian to Lagrangian

Let Q be some quantity defined in terms of Eulerain description, we then have

$$Q = Q\ (x, y, z, t). \qquad ...(5)$$

We shall express x, y, z in terms of the particle parameter a, b, c. Let u, v, w are the velocity components at the point x, y, z at any instant t, which is defined as

$$u = F_{.1}\ (x, y, z, t),$$
$$v = F_2\ (x, y, z, t),$$
$$w = F_3\ (x, y, z, t). \qquad ...(6)$$

Again, from the Lagrangian description, we have

$$u = \frac{\partial x}{\partial t}, v = \frac{\partial y}{\partial t}, w = \frac{\partial z}{\partial t},$$

where x, y, z are functions of the variables a, b, c and t.

From (6) and (7), the velocity components of a fluid element is given by

$$\frac{\partial x}{\partial t} = F_1\ (x, y, z, t),$$

$$\frac{\partial y}{\partial t} = F_2(x, y, z, t),$$

$$\frac{\partial z}{\partial t} = F_3(x, y, z, t) \qquad ...(8)$$

which represents the first order linear differential equation. By integrating, we have

$$x = f_1(x_0, y_0, z_0, t),$$
$$y = f_2(x_0, y_0, z_0, t),$$
$$z = f_3(x_0, y_0, z_0, t), \qquad ...(9)$$

where x_0, y_0, z_0 are the initial values of x, y, z at an initial instant $t = t_0$ assumed to be constants of integration. Choosing the particle parameters a, b, c equal to x_0, y_0, z_0 respectively. Thus, we have

$$x = f_1(a, b, c, t),$$
$$y = f_2(a, b, c, t),$$
$$z = f_3(a, b, c, t). \qquad ...(10)$$

From the relations (5) and (10), we have

$$Q = Q\,[f_1(a, b, c, t), f_2(a, b, c, t), f_3(a, b, c, t)]$$

DEFINITIONS

(i) *Steady and Unsteady Flows :* Steady flow occurs when at various points of the flow field the conditions and properties associated with the fluid flow remain unaltered with time *i.e.*, independent of time at all points. Mathematically, it can be expressed as

$$\frac{\partial A}{\partial t} = 0,$$

where A represents the characteristic of the fluid, *e.g.*, velocity, density, temperature and pressure etc. Thus in steady motion time drops out of the independent variables and the various field quantities become functions of the space coordinates. For example, *Water being pumped through a fixed system at a constant rate represents steady flow.*

The flow is said to be *unsteady* when conditions at' any point change with regard to the time. For example, *Water being pumped through a fixed system at an increasing rate represents unsteady flow.*

(ii) *Uniform and Non-uniform Flows :* If at every point the velocity vector is identical in magnitude and direction at any given instant, or, the conditions and properties are independent of the coordinate of the direction in which the fluid is moving then the motion is said to be *uniform*. If the flow characteristics, at any given time t, change with distance, it is said to be *non-uniform flow*. For example, a *liquid being pumped through a long straight pipe has uniform flow where as a liquid flowing through a curved pipe has non-uniform flow.*

(iii) The motion of fluid at any point in space can be specified with respect to a coordinate system. *One dimensional* flow neglects variations or changes in velocity, pressure etc., transverse to the main flow direction. The flow characteristics vary only in the direction of flow.

The flow is said to be *two-dimensional* when the velocity vector q is everywhere at right angles to a certain direction and independent of displacements parallel to that direction. The components of velocity q (u, v, w) in a two-dimensional flow are (u, v, 0), where u and v are independent of z. The flow conditions are identical in planes perpendicular to the Z-axis.

In *three-dimensional* flow the velocity components (u, v, w) in mutually perpendicular direction are the functions of space coordinates and time x, y, z and t. For example, *the flow of a liquid towards a circular outlet in a tank is three dimensional.*

(iv) A flow field is said to be *axi-symmetric* when the velocity components (u, v, w) with regard to cylinderical coordinates (x, r, ϕ) are all independent of the azimuthal angle ϕ. The velocity component q_ϕ in the direction of increasing ϕ is called *swirl* component of velocity.

(v) *Line of Flow :* A *line of flow* is a line whose direction coincides with the direction of the resultant velocity of the fluid.

(vi) *Stream Line :* A *stream line* is a continuous line of flow drawn in the fluid so that the tangent at every point of it at any instant of time coincides with the direction of the motion of the fluid at that point. The component of velocity at right angles to the streamline is always zero. It follows that there is no flow across the streamline.

Consider ds be an element of the streamline passing through any point P (r) at an instant of time. Let q be the velocity at that point at the same instant. The direction of the tangent and direction of velocity are parallel.

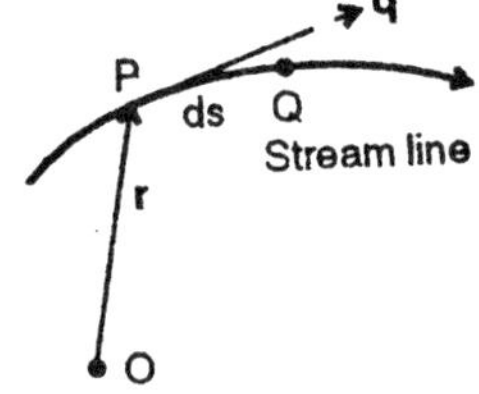

Fig. 1.1

i.e., $ds \times q = 0$

$\Rightarrow \quad (idx + jdy + kdz) \times (iu + jv + kw) = 0$

$\Rightarrow \quad (wdy - vdz)\, i + (udz - wdx)\, j + (vdx - udy)\, k = 0$

The vector equation is equivalent to three scalar equations

wdy – vdz = 0, udz – wdx = 0, vdx – udy = 0

which can be represented as

$$\frac{dx}{u} = \frac{dy}{v} = \frac{dz}{w},$$

known as the differential equation for a *streamline* where the velocity components u, v, w are the functions of x, y, z and t. When the flow is steady, the streamlines have the same form at all times. When the flow is unsteady the streamlines change from instant to instant.

(vii) *Pathline* : The curve described in space by moving fluid element is known its *pathline* or trajectory *i.e.*, a *pathline* is a line traced by a particle in the fluid. The pathline shows the direction of the velocity of the fluid particle at any instant of time. Such a line is obtained by giving the position of an element as a function of time. The pathline are given by

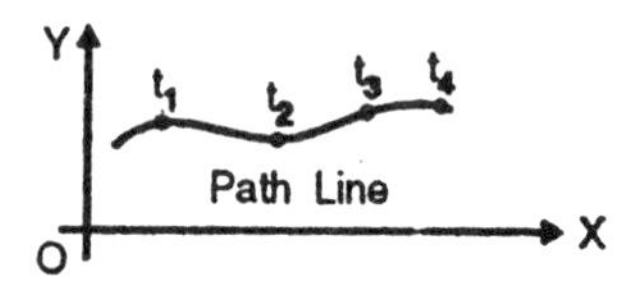

Fig. 1.2

$$q = \frac{dr}{dt}$$

i.e., $$\frac{dx}{dt} = u(x, y, z, t)$$

$$\frac{dy}{dt} v(x, y, z, t),$$

$$\frac{dy}{dt}v(x, y, z, t),$$

In general, the pathlines vary with each fluid particle. It represents the direction of velocity of a single particle of fluid at various time. For steady motion, the streamlines coincide with the paths of the fluid particles, this is not so for unsteady motion.

(viii) *Stream Surface* : A *stream surface* is a surface made by the streamlines passing through an arbitrary line in the fluid region at any instant of time.

Fig. 1.3

(ix) *Stream Tube* : A *stream* tube is obtained by drawing stream lines through every point of a closed curve in the fluid.

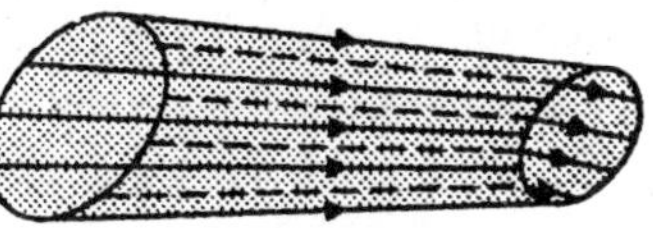

Fig. 1.4

(x) *Streak Lines* : A streak line is a line on which lie all those fluid elements that at some earlier instant passed through a particular point in space. It is a line making the position of a set of fluid particles that had passed through a fixed point in the flow field. A streakline is defined as the locus of different particles passing through fixed point. A streakline connects the locations of the particles at one instant moving with the fluid which passed through a particular point. When a dye or smoke, is injected into a fluid, the resulting trails are called streaklines.

When the flow is steady the streaklines, streamlines and pathlines coincide.

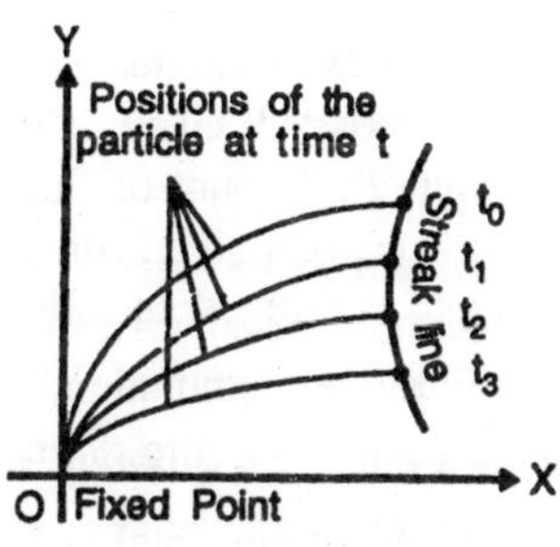

Fig. 1.5

Consider a fluid particle (x_0, y_0, z_0) passes a fixed point r_1 (x_1, y_1, z_1) in the course of time. By Lagrangian description of fluid flow, we have

$$x_1 = f_1(x_0, y_0, z_0, t),$$
$$x_2 = f_2(x_0, y_0, z_0, t),$$
$$x_3 = f_3(x_0, y_0, z_0, t).$$

Solving the equations for x_0, y_0, z_0, we have

$$x_0 = F_1(x_1, y_1, z_1, t),$$
$$y_0 = F_2(x_1, y_1, z_1, t),$$
$$z_0 = F_3(x_1, y_1, z_1, t)$$

Since a streakline is the locus of the positions (x, y, z) of the particles which have passed through the fixed point (x_1, y_1, z_1), therefore the equation of the streakline at an instant of time is given by

$$x = G_1(x_0, y_0, z_0, t),$$
$$y = G_2(x_0, y_0, z_0, t),$$
$$z = G_3(x_0, y_0, z_0, t)$$

Hence the streakline passing through the fixed point (x_1, y_1, z_1) at time t is given by

$$x = G_1(F_1, F_2, F_3, t),$$
$$y = G_2(F_1, F_2, F_3, t),$$
$$z = G_3\ F_1, F_2, F_3, t).$$

EQUATION OF CONTINUITY

Physical quantiries are said to be conserved when they do not change with regard to time during a process. The mthematical expression of the law of conservation of mass is known as the *Equation of Continuiry.*

By continuity we mean physical continuity. The fluid always remain a continuum *i.e.*, as a continuously distributed matter. When a region of fluid contain neither sources nor sinks (*i.e.*, there is no creation or annihilartion of the fluid) then the amount of fluid within the region is conserved in accordance with the principle of conservation of matter. The general conservation principle is defined as follow :

In – Out + Source – Sink = Accumulation,

where each term represents a rate for a differential element of volume.

Consider a fluid element of infinitesimal volume δv and density ρ which is situated at a point ρ at any instant t.

The mass of the element is rdv. Throughout the motion the mass of any element of fluid must be conserved, hence the mass of any fluid element remains unchanged as it moves about. This shows that the material derivative of $\rho\delta v$ vanisnes, *i.e.*,

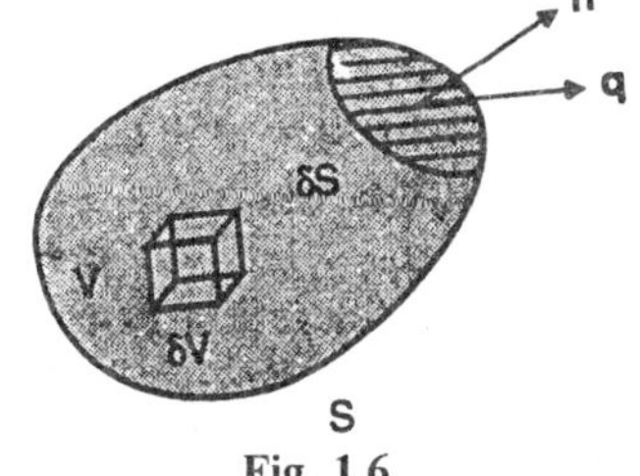

Fig. 1.6

$$\frac{D}{Dt}(\rho v) = 0, \quad ...(1)$$

which is the equiation of continuity in the simplest form.

Consider a closed surface S in a fluid medium containing a volume V fixed in space. Let n is the unit outward drawn normal at a surface element δS then the normal component of measured outward form V wil be = n . q.

Rate of mass flow across δS per unit mass = $\rho(n \cdot q)\, \delta S$

Total rate of mass flow out of V across dS = f s$\rho(n \cdot q)\, dS$.

Tolat rate of mass flow into V

$= -\int_s n \cdot (\rho q)\, dS$

$= -\int_v \nabla \cdot (\rho q)\, dV$ (By Gauss throerem) ...(2)

Also rate of increase of mass within V

$$= \frac{\partial}{\partial t}\left(\int V \rho dV\right) = \int_v \int \frac{\partial \rho}{\partial t} dV \quad ...(3)$$

By the principle of continuity, we have

$$\int_v \frac{\partial \rho}{\partial t} dV = -\int_v \nabla.(\rho q) dV$$

$$\text{or} \quad \int_v \left[\frac{\partial \rho}{\partial t} + \nabla.(\rho q)\right] dV = 0. \quad ...(4)$$

The relation (4) is valid for all arbitrary volume V, provided that it lies entirely in the fluid. Thus, the integrand must be identically zero everywhere in the fluid. Hence

$$\frac{\partial \rho}{\partial t} + \nabla.(\rho q) = 0, \quad ...(5)$$

at all points in the fluid. The relation (5) is known as the *equation of continuity or equation of the conservation of mass* wich is true at any point of a fluid free from sources and sinks. This is also called a *kinematical relation* (a relation between the velocity and density fields only).

Equation (5) may be written in a different form as

$$\frac{\partial \rho}{\partial t} + q(\nabla \rho) + \rho \nabla . q = 0$$

$$\Rightarrow \left(\frac{\partial}{\partial t} + q.\nabla \right) \rho + \rho \nabla . q = 0$$

or $\dfrac{D\rho}{Dt} + \rho \nabla . q = 0,$

since $\dfrac{D}{Dt} \equiv \dfrac{\rho}{\rho t} + q.\nabla$ ...(6)

($D\rho/Dt$) is the substantial derivative of density *i.e.*, the time derivative for a path following the fluid motion.

For a steady flow of fluid, the pattern of flow does not vary with regard to time then the relation (5) reduces to the form

$$\nabla . (\rho q) = 0 \qquad ...(7)$$

For a non-homogeneous incompressible fluid, the density of the fluid particle is invariable with time t *i.e.*, ρ remains constant throughout the entire region

$$\frac{D\rho}{Dt} = 0 \Rightarrow \nabla . q = 0 \Rightarrow div\, q = 0 \qquad ...(8)$$

The quantity Δ.q gives the rate of volume expansion of a fluid element. It may be called *dilatation or expansion.* A vector q having zero divergence is said to be *solenoidal.*

EQUATION OF CONTINUITY (Stream Tube Concept)

The continuity equation on the basis of stream tube concept can be greatly simplified. Consider a stream tube.

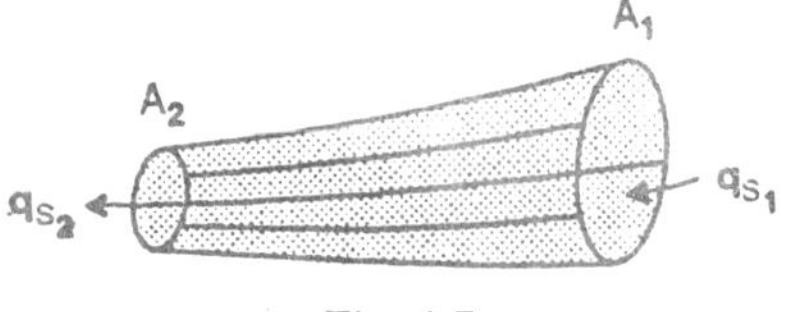

Fig. 1.7

Since the flow is always in the tangential direction and there is no component of velocity in the normal direction, both the velocity q_s and the cross-sectional area A will depend on the distance s along the axis δs of stream tube,

$$\frac{\partial}{\partial t}(\rho A \delta s) + \frac{\partial}{\partial s}(\rho q_s A)\delta s = 0$$

$$\text{or} \quad \frac{\partial}{\partial t} + \frac{1}{A}\frac{\partial}{\partial s}(\rho q_s A) = 0 \qquad ...(1)$$

If the fluid is incompressible, $\frac{\partial \rho}{\partial t} = 0$, so (1) becomes

$$\frac{\partial}{\partial s}(q_s A) = 0 \Rightarrow q_s A = \text{constant}. \qquad ...(2)$$

But $q_s A = q$ which is the rate of flow of fluid on discharge.

Hence $q = qs_1 A_1 = qs_2 A_2 = qs_3 A_3 = ...,$

where A_1, A_2, A_3... are the elementary areas along stream tube and qs_1, qs_2, qs_3... are the velocities at the centres of areas A_1, A_2, A_3....

If the fluid is compressible and flow is steady, then

$$\frac{\partial}{\partial s}(\rho q_s A) = 0 \Rightarrow \rho q_s A = \text{constant}.$$

i.e. $\rho_1 qs_1 A_1 = \rho_2 qs_2 A_2 = \rho_3 qs_3 A_3 = ...$

EQUATION OF CONTINUITY (Spnerical Polar Coordinates)

Let ρ be the density of the fluid at the point O (r, θ, ϕ). Construct a parallelopiped with O as centre, the lenght of whose edges are dr, $rd\theta$ and $r \sin\theta \, d\phi$. Let q_r, q_θ, and q_ϕ be the components of the velocity in the direction of the elements respectively.

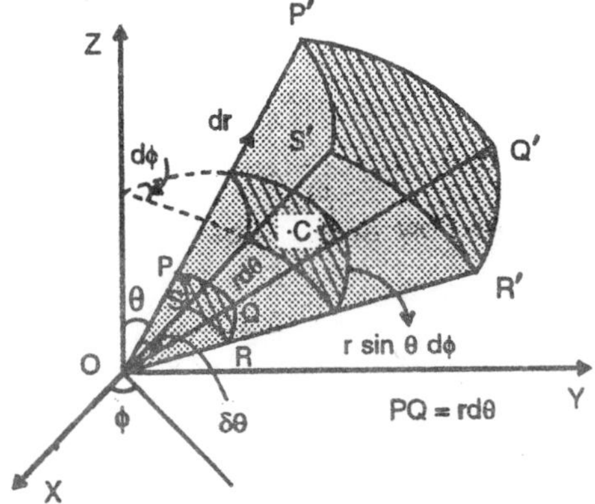

Fig. 1.8

Mass in the fluid that passes through the face PQRS.

$= \rho\ (r\delta\theta\ .\ r\sin\theta\ \delta\phi)\ q_r = f\ (r,\ \theta,\ \phi)$ (say) ...(1)

Mass of the fluid that passes through the opposite face P'Q'R'S'

$$= f(r+\delta r,\theta,\phi) = f(r,\theta,\phi) - \delta r\frac{\partial}{\partial r}f(r,\theta,\phi)+...$$

Excess of mass of flow-in over flow out through the faces PQRS and P'Q'R'S'

$$= f(r,\theta,\phi) - f(r,\theta,\phi) - \delta r\frac{\partial}{\partial r}f(r,\theta,\phi)...$$

$$= -\delta r\frac{\partial}{\partial r}f(r,\theta,\phi)$$

$$= -\delta r\frac{\partial}{\partial r}[\rho r\delta\theta . r\sin\theta\,\delta\phi]q_r \text{ per unit time}$$

$$= -\frac{\partial}{\partial r}[\rho r^2\theta_r]r^2\,\delta r\,\delta\theta\,\delta\phi. \qquad ...(2)$$

Similarly excess of mass of flow-in over flow out through the faces PSS'P' and QRR'Q'

$$= -r\delta\theta\frac{\partial}{r\partial\theta}[\rho\delta r . \rho\sin\theta\,\delta\phi]q\theta \text{ per unit time.}$$

$$= -\frac{\partial}{r\partial\theta}[\rho q\theta\sin\theta]r^2\,\delta r\,\delta\theta\delta\phi. \qquad ...(3)$$

and excess of mass flow-in over flow out throgh the faces PQQ'P and SRR'S'

$$= -r\sin\theta\,\delta\phi\frac{\partial}{r\sin\theta\,\partial\phi}[\rho\delta r . r\delta\theta]q_\phi \text{ per unit time.}$$

$$= -\frac{\partial}{r\sin\theta\,\partial\phi}[rq\phi]r^2\sin\theta\,\delta r\,\delta\theta\delta\phi. \qquad ...(4)$$

Total excess of flow-in over flow out from all the faces

$$= -\frac{\partial}{\partial r}[\rho r^2 q_r]\sin\theta\,\delta r\,\delta\theta\,\delta\phi - \frac{\partial}{r\partial\theta}[\rho\sin\theta\, q\theta]r^2\,\delta r\,\delta\theta\,\delta\phi$$

$$-\frac{\partial}{r\sin\theta\,\partial\phi}[\rho q\phi]r^2\sin\theta\,\delta r\,\delta\theta\,\delta\phi. \qquad ...(5)$$

Total mass inside the parallelopiped

$= \rho\ \delta r.\ \rho\ \delta\theta\ .\ r\sin\theta\ \delta\phi$

Rate of increase in the mass inside the parallelopiped

$$\frac{\partial}{\partial t}[\rho \delta r . r\delta\theta . r\sin\theta\delta\phi]$$

$$\frac{\partial \rho}{\partial t}.r^2 \sin\theta\,\delta r\,\delta\theta\,\delta\phi. \qquad ...(6)$$

By the principle of continuity, we have

$$\frac{\partial \rho}{\partial t}.r^2 \sin\theta\,\delta r\,\delta\phi = \frac{\partial}{\partial r}[\rho r^2 q_r]\sin\theta\,\delta r\,\delta\theta\,\delta\phi$$

$$-\frac{\partial}{r\partial\theta}[\rho\sin\theta q_\theta]r^2\,\delta r\,\delta\theta\,\delta\phi - \frac{\partial}{r\sin\theta\partial\phi}[\rho q\phi]r^2\sin\theta\,\delta r\,\delta\theta\,\delta\phi$$

or $r^2\sin\theta\frac{\partial\rho}{\partial t} + \sin q\frac{\partial}{\partial r}[\rho r^2 q_r] + r^2\frac{\partial}{r\partial\theta}[\rho\sin\theta q_\theta]$

$$+r^2\sin\theta\frac{\partial}{r\sin\theta\partial\phi}\rho q_\phi] = 0$$

or $\frac{\partial\rho}{\partial t} + \frac{1}{r^2}\frac{\partial}{\partial r}(\rho r^2 q_r) + \frac{1}{r\sin\theta}\frac{\partial}{\partial\theta}(\rho q_\theta\sin\theta) + \frac{1}{r\sin\theta}\frac{\partial}{\partial\phi}(\rho q_\phi) = 0,$

which is known as the equation of continuity in spherical polar coordinates.

EQUATION OF CONTINUITY (Cartesian Coordinates)

Consider ρ be the density of the fluid at the point P (x, y, z) and u, v, w be the velocity components parallel to the coordinate axes. Construct a small parallelopiped with edges of length δx, δy, δz parallel to their respective axes.

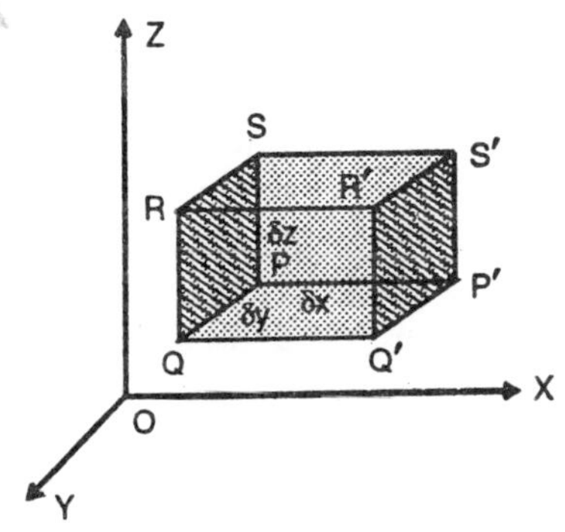

Fig. 1.9

Let any closed surface be drawn in the fluid then the increase in the mass of fluid within the surface in any time δt is equal to the excess of mass that flows in over the mass that flows out in the same time δt.

Mass of the fluid that passes in through the face PQRS

$$= (\rho\ \delta y\ \delta z\ u) \text{ per unit mass}$$

$$= f\,(x + \delta x,\ y,\ z)\ (\text{say}) \qquad ...(1)$$

Mass of the fluid that passes out through the opposite face P'Q'R'S'

$$= f(x, y, z) + \delta x.\frac{\partial}{\partial x} f(x, y, z) + ... \qquad ...(2)$$

The exccss of mass that flows-in over the mass that flows out in the direction of X-axis

= Mass the enters through the face PQRS

– Mass that leaves through the face P'Q'R'S'.

$$= f(x, y, z) - f(x, y, z) - \partial x.\frac{\partial}{\partial x} f(x, y, z)...$$

$$= -\delta x.\frac{\partial}{\partial x} f(x, y z),$$

$$= -\delta x.\frac{\partial}{\partial x}(\rho u\, \delta y\, \delta z) = -\frac{\partial}{\partial x}(\rho u)\, \delta x\, \delta y\, \delta z. \qquad ...(3)$$

Similarly the excess of flow in over flow out through the faces QQ' RR' and PP' S'S

$$= -\delta y.\frac{\partial}{\partial y}(\rho v\, \delta x\, \delta z) = -\frac{\partial}{\partial y}(\rho v)\, \delta y\, \delta z. \qquad ...(4)$$

and the excess of flow in over flow out through the faces PP' Q'Q and SS' RR'

$$= -\delta z.\frac{\partial}{\partial y}(\rho w\, \delta x\, \delta y) = -\frac{\partial}{\partial z}(\rho w)\, \delta x\, \delta y\, \delta z. \qquad ...(5)$$

The excess of flow- in over flow out from all the faces

$$= -\left[\frac{\partial}{\partial x}(\rho u) + \frac{\partial}{\partial y}\frac{\partial}{\partial z}(\rho w)\right] \delta x\, \delta y\, \delta z. \qquad ...(6)$$

Total mass inside the parallelopiped

$$= \rho\ \delta x\ \delta y\ \delta z$$

Rate of increase in the mass inside the parallelopiped

$$= \frac{\partial}{\partial t}(\rho\,\delta x\,\delta y\,\delta z) \Rightarrow \frac{\partial \rho}{\partial t}\delta x\,\delta y\,\delta z. \qquad ...(7)$$

By the principle of continuity, we have

Rate of mass accumulation = Rate of mass in – Rate of mass out,

$$\frac{\partial}{\partial t}\delta x\,\delta z\,\delta z = -\left[\frac{\partial}{\partial x}(\rho u) + \frac{\partial}{\partial z}(\rho w)\right]\delta x\,\delta y\,\delta z$$

$$\text{or} \quad \frac{\partial \rho}{\partial t} + \frac{\partial}{\partial x}(\rho u) + \frac{\partial}{\partial y}(\rho v) + \frac{\partial}{\partial z}\rho w = 0$$

$$\text{or} \quad \left(\frac{\partial \rho}{\partial t} + u\frac{\partial \rho}{\partial x} + v\frac{\partial \rho}{\partial y} + w\frac{\partial \rho}{\partial z}\right) + \rho\left(\frac{\partial u}{\partial y} + \frac{\partial v}{\partial y} + \frac{\partial w}{\partial z}\right) = 0$$

$$\text{or} \quad \frac{D\rho}{Dt} + \rho\left(\frac{\partial u}{\partial x} + \frac{\partial v}{\partial y} + \frac{\partial w}{\partial z}\right) = 0$$

which is known as the equation of continuity in cartesian coordinates.

EQUATION OF CONTINUITY (Lagrangian Method)

Let A be the region occupied by a fluidat the time t = 0 and B be the region occupied by the same fluid at any instant of time t. Consider (a, b, c) be the initial coordinates of a particle P enclosed in this element and ρ_0 be its density.

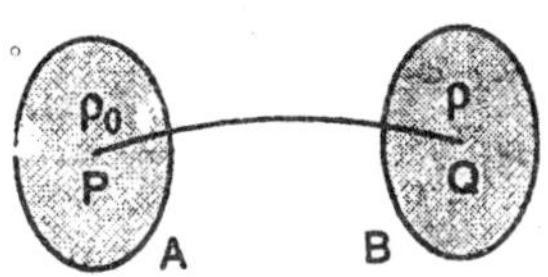

Fig. 1.10

Mass of the fluid element enclosing the point P at t = 0 is

$= \rho_0\ \delta a\ \delta b\ \delta c.$

Mass of the fluid element encolsing the subsequent point Q at any time t is

$= \rho\ \delta x\ \delta y\ \delta z,$

where ρ is the density of the fluid at Q.

There total mass inside the region A is equal to the total mass inside the reglon B, so

$$\iiint_A \rho_0\, da\, db\, dc = \iiint_B \rho\, dx\, dy\, dz$$

or $\iiint_A \rho_0 \, da\, db\, dc = \iiint_A \rho \frac{\partial(x,y,z)}{\partial(a,b,c)} da\, db\, dc$

$$\left[\text{since } \delta x\, \delta y\, \delta z = \frac{\partial(x,y,z)}{\partial(a,b,c)} \delta a\, \delta b\, \delta c\right]$$

or $\iiint_A \left\{\rho_0 - \rho \frac{\partial(x,y,z)}{\partial(a,b,c)}\right\} da\, db\, dc = 0$

The region Ais arbitrary, hence

$$\rho_o - \rho \frac{\partial(x,y,z)}{\partial(a,b,c)} = 0$$

$$\Rightarrow \rho_0 = \rho \frac{\partial(x,y,z)}{\partial(a,b,c)}$$

$$\Rightarrow \rho_0 = \rho J,$$

$$\Rightarrow \rho_0 = \rho \frac{\partial(x,y,z)}{\partial(a,b,c)} \Rightarrow \rho_0 = \rho J,$$

where the Jacobian J of the transformation of variables is expressed as

$$J = \frac{\partial(x,y,z)}{\partial(a,b,c)} = \begin{vmatrix} \partial x/\partial a & \partial x/\partial b & \partial x/\partial c \\ \partial y/\partial a & \partial y/\partial b & \partial y/\partial c \\ \partial z/\partial a & \partial z/\partial b & \partial z/\partial c \end{vmatrix}$$

known as the equation of continuity in Lagrangian form.

EQUIVALENCE OF THE TWO FORMS OF THE EQUATION OF CONTINUITY (Lagrangian and Eulerian Form)

The equation of continuity in the Lagrangian form is given by

$\rho_0 = \rho J, \text{where } J = \partial(x,y,z)/\partial(a,b,c).$ ...(1)

Differentiating (1) with regard to t, we have

$$\frac{d}{dt}(\rho J) = \frac{d}{dt}(\rho_0)$$

$$\Rightarrow \rho \frac{dJ}{dt} + J \frac{d\rho}{dt} = 0.$$

These time rates are the variations due to the motion of a particle or the variability of x, y, z. Here we shall change the variables from the Lagrangian form to Eulerian form. We know that

$$\frac{\partial u}{\partial a}=\frac{\partial}{\partial a}\left(\frac{\partial x}{\partial t}\right)=\frac{d}{dt}\left(\frac{\partial x}{\partial a}\right) etc.;$$

$$u=\frac{dx}{dt}, v=\frac{dy}{dt}, w=\frac{dz}{dt}$$

Again $\dfrac{d\rho}{dt}=\dfrac{\partial\rho}{\partial t}+u\dfrac{\partial\rho}{\partial x}+v\dfrac{\partial\rho}{\partial y}+w\dfrac{\partial\rho}{\partial z}$

Since $J=\begin{vmatrix}\partial x/\partial a & \partial y/\partial a & \partial z/\partial a\\ \partial x/\partial b & \partial y/\partial b & \partial z/\partial b\\ \partial x/\partial c & \partial y/\partial c & \partial z/\partial c\end{vmatrix}$

or $$\frac{dJ}{dt}=\begin{vmatrix}\partial u/\partial a & \partial y/\partial a & \partial z/\partial a\\ \partial u/\partial b & \partial y/\partial b & \partial z/\partial b\\ \partial u/\partial c & \partial y/\partial c & \partial z/\partial c\end{vmatrix}+\begin{vmatrix}\partial x/\partial a & \partial v/\partial a & \partial z/\partial a\\ \partial x/\partial b & \partial v/\partial b & \partial z/\partial b\\ \partial x/\partial c & \partial v/\partial c & \partial z/\partial c\end{vmatrix}$$

$$+\begin{vmatrix}\partial x/\partial a & \partial y/\partial a & \partial w/\partial a\\ \partial x/\partial b & \partial y/\partial b & \partial w/\partial b\\ \partial x/\partial c & \partial y/\partial c & \partial w/\partial c\end{vmatrix}$$

or $$\frac{dJ}{dt}=\frac{\partial(u,y,z)}{\partial(a,b,c)}+\frac{\partial(x,v,z)}{\partial(a,b,c)}+\frac{\partial(x,y,w)}{\partial(a,b,c)} \qquad ...(3)$$

But $\dfrac{\partial u}{\partial a}=\dfrac{\partial u}{\partial x}\cdot\dfrac{\partial x}{\partial a}+\dfrac{\partial u}{\partial y}\cdot\dfrac{\partial y}{\partial a}+\dfrac{\partial u}{\partial z}\cdot\dfrac{\partial z}{\partial a}$ etc.

Again $\dfrac{\partial(u,y,z)}{\partial(a,b,c)}=\begin{vmatrix}\partial u/\partial a & \partial y/\partial a & \partial z/\partial a\\ \partial u/\partial b & \partial y/\partial b & \partial z/\partial b\\ \partial u/\partial c & \partial y/\partial c & \partial z/\partial c\end{vmatrix}$

$$=\begin{vmatrix}\frac{\partial u}{\partial x}\cdot\frac{\partial x}{\partial a}+\frac{\partial u}{\partial y}\cdot\frac{\partial y}{\partial a}+\frac{\partial u}{\partial z}\cdot\frac{\partial z}{\partial a} & \frac{\partial y}{\partial a} & \frac{\partial z}{\partial a}\\ \frac{\partial u}{\partial x}\cdot\frac{\partial x}{\partial b}+\frac{\partial u}{\partial y}\cdot\frac{\partial y}{\partial b}+\frac{\partial u}{\partial z}\cdot\frac{\partial z}{\partial b} & \frac{\partial y}{\partial b} & \frac{\partial z}{\partial b}\\ \frac{\partial u}{\partial x}\cdot\frac{\partial x}{\partial c}+\frac{\partial u}{\partial y}\cdot\frac{\partial y}{\partial c}+\frac{\partial u}{\partial z}\cdot\frac{\partial z}{\partial c} & \frac{\partial y}{\partial c} & \frac{\partial z}{\partial c}\end{vmatrix}$$

$$= \begin{vmatrix} \frac{\partial u}{\partial x}\cdot\frac{\partial x}{\partial a} & \frac{\partial y}{\partial a} & \frac{\partial z}{\partial a} \\ \frac{\partial u}{\partial x}\cdot\frac{\partial x}{\partial b} & \frac{\partial y}{\partial b} & \frac{\partial z}{\partial b} \\ \frac{\partial u}{\partial x}\cdot\frac{\partial x}{\partial c} & \frac{\partial y}{\partial c} & \frac{\partial z}{\partial c} \end{vmatrix} + = \begin{vmatrix} \frac{\partial u}{\partial y}\cdot\frac{\partial y}{\partial a} & \frac{\partial y}{\partial a} & \frac{\partial z}{\partial a} \\ \frac{\partial u}{\partial y}\cdot\frac{\partial y}{\partial b} & \frac{\partial y}{\partial b} & \frac{\partial z}{\partial b} \\ \frac{\partial u}{\partial y}\cdot\frac{\partial y}{\partial c} & \frac{\partial y}{\partial c} & \frac{\partial z}{\partial c} \end{vmatrix} + \begin{vmatrix} \frac{\partial u}{\partial z}\cdot\frac{\partial z}{\partial a} & \frac{\partial y}{\partial a} & \frac{\partial z}{\partial a} \\ \frac{\partial u}{\partial z}\cdot\frac{\partial z}{\partial b} & \frac{\partial y}{\partial b} & \frac{\partial z}{\partial b} \\ \frac{\partial u}{\partial z}\cdot\frac{\partial z}{\partial c} & \frac{\partial y}{\partial c} & \frac{\partial z}{\partial c} \end{vmatrix}$$

$$= \frac{\partial u}{\partial x} \begin{vmatrix} \frac{\partial x}{\partial a} & \frac{\partial y}{\partial a} & \frac{\partial z}{\partial a} \\ \frac{\partial x}{\partial b} & \frac{\partial y}{\partial b} & \frac{\partial z}{\partial b} \\ \frac{\partial x}{\partial c} & \frac{\partial y}{\partial c} & \frac{\partial z}{\partial c} \end{vmatrix}$$

(the remaining two determinants vanish because they contain two identical coloumns.)

$$\Rightarrow \frac{\partial(u,y,z)}{\partial(a,b,c)} = \frac{\partial u}{\partial x} J.$$

Similarly $\dfrac{\partial(x,v,z)}{\partial(a,b,c)} = \dfrac{\partial v}{\partial y} J$

and $$\frac{\partial(x,y,w)}{\partial(a,b,c)} = \frac{\partial w}{\partial z} J. \quad ...(4)$$

By using the relation (4) into (3), we have

$$\frac{dJ}{dt} = J\left(\frac{\partial u}{\partial x} + \frac{\partial v}{\partial y} + \frac{\partial w}{\partial z}\right).$$

Substituing the value of $\dfrac{dJ}{dt}$ in (2), we have

$$J\frac{d\rho}{dt} = \rho J\left(\frac{\partial u}{\partial x} + \frac{\partial v}{\partial y} + \frac{\partial w}{\partial z}\right) = 0,$$

or $$\frac{d\rho}{dt} = \rho\left(\frac{\partial u}{\partial x} + \frac{\partial v}{\partial y} + \frac{\partial w}{\partial z}\right) = 0, J \neq 0.$$

or $$\frac{d\rho}{dt}\rho(\nabla.q) = 0$$

represents equation of continuity in *Eulerian form.*

Again from the equation (2), we have

$$\frac{d\rho}{\partial t}+\rho\left(\frac{1}{J}\frac{dJ}{dt}\right)=0 \Rightarrow J\frac{d\rho}{dt}+\rho\frac{dJ}{dt}=0$$

$$\Rightarrow \frac{d}{dt}(\rho J)=0 \Rightarrow \rho J=\rho_0,$$

represents the equation of continuity in Lagrangian description.

EQUATION OF CONTINUITY (Cylindrical Polar Coordinates)

Let ρ be the density of the fluid at the point P (r, θ, z). Construct a curvilinear parallelopiped with P as centre, the lenght of whose edges are dr, r$d\theta$, and dz. Let q_r, q_θ and q_z be the velocity components in the direction of the elements respectively.

Mass of the fluid that passes through the face ABCD

$= \rho\ (r\delta\theta\ \delta z)\ q_r$ per unit time

$f = (r, \theta, z)$ (say)

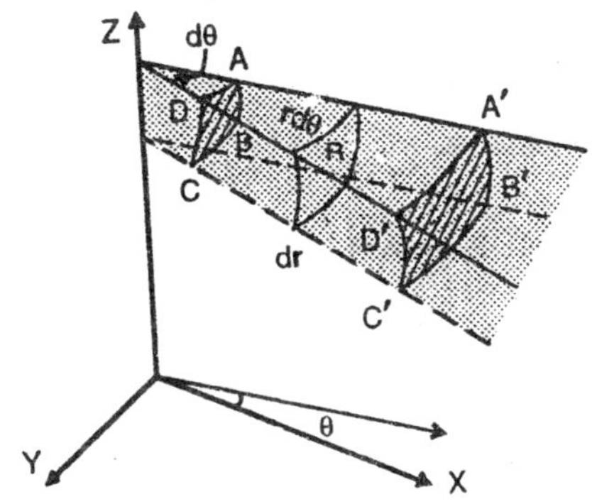

Fig. 1.11

Mass of the fluid that passes through the opposite face A'B'C'D'

$f\ (r + \delta r, \theta, z)$

$$= f(r,\theta,z)+\delta r.\frac{\partial}{\partial r}f(r,\theta,z)+...$$

Excess of mass of flow-in over flow out through the faces ABCD and A'B'C'D'.

$$= f(r,\theta,z)-f(r,\theta,z)-\delta r.\frac{\partial}{\partial r}f(r,\theta,z)-...$$

$$= -\delta r.\frac{\partial}{\partial r}f(r,\theta,z) \text{ per unit time}$$

$$= -\delta r.\frac{\partial}{\partial r}[\rho r\,\delta\theta\,\delta z\,q_r]$$

$$\Rightarrow\ -\frac{\partial}{\partial r}[\rho r\,q_r]\delta r\,\delta\theta\,\delta z. \qquad ...(1)$$

Similarly the excess of mass of flow-in over flow out through the other faces are

$$-\frac{\partial}{r\partial\theta}[\rho q_\theta]r\,\delta r\,\delta\theta\,\delta z$$

and $-\frac{\partial}{\partial z}[\rho q_z]r\,\delta r\,\delta\theta\,\delta z.$ ...(2, 3)

Total excess of flow-in over flow out from all the faces

$$= -\left\{\frac{\partial}{\partial r}[\rho_r\,q_r]\delta r\,\delta\theta\,\delta z + \frac{\partial}{r\partial\theta}[\rho q_\theta]r\,\delta r\delta\theta\,\delta z + \frac{\partial}{\partial z}[\rho q_z]r\,\delta r\,\delta\theta\,\delta z\right\} \qquad ...(4)$$

Total mass inside the parallelopiped

$= \rho\ \delta r\ .\ r\ \delta\theta\ .\ \delta z$

Rateof increase in the mass inside the parallelopiped

$$\frac{\partial}{\partial t}[\rho\,\delta r.r\,\delta\theta.\delta z] = \frac{\partial\rho}{\partial t}.r\,\delta r\delta\theta\,\delta z. \qquad ...(5)$$

By the principle of continuity, we have

$$\frac{\partial}{\partial t}(r\,dr\,dq\,dz) = -\left\{\frac{\partial}{\partial r}[\rho r\,q_r] + \frac{\partial}{r\partial\theta}[\rho q_\theta] + r\frac{\partial}{\partial z}[\rho q_z]\right\}\delta r\,\delta\theta\,\delta z,$$

or $r\frac{\partial\rho}{\partial t} + \frac{\partial}{\partial r}[\rho r q_r] + r\frac{\partial}{r\partial\theta}[\rho q_\theta] + r\frac{\partial}{\partial z}[\rho q_z] = 0,$

or $\frac{\partial r}{\partial t} + \frac{1}{r}\frac{\partial}{\partial r}[\rho r q_r] + \frac{\partial}{\partial z}[\rho q_z] = 0,$

which is known asthe equation of continuity in cylindrical polar coordinates.

SOLVED EXMPLES

Example 1: *The velocity components for a two- dimensional flow system can be given in the Eulerian system by*

$$u = 2x + 2y + 3t,\ v = x + y + \frac{1}{2}t.$$

Find the displacement of a fluid particle in the Lagrangian system.

Solution: The velocities may be expressed in terms of the displacements as follows

$$\frac{dx}{dt} = 2x + 2y + 3t,$$

$$v = \frac{dy}{dt} x + y + \frac{1}{2}t.$$

$$\frac{dx}{dt} - 2x - 2y = 3t,$$

$$\frac{dy}{dt} - x - y = \frac{1}{2}t.$$

The solution of the simultaneous differential equations can be determined by the operator method as follows:

$$(D - 2)\,x - 2y = 3t,$$

$$-x + (D - 1)\,y = \frac{1}{2}t. \qquad \text{...(1,2)}$$

Eliminating x from (1) and (2), we have

$$D\,(D - 3)\,y = 2t + \frac{1}{2},$$

whose solution is given by

$$y = A + Be^{3t} - \left(\frac{7}{18}\right) t - \left(\frac{1}{3}\right) t_2. \qquad \text{...(3)}$$

Substituting the value of y in the equation (2), we have

$$x = -A + 2Be^{3t} + \left(\frac{1}{3}\right) t_2 - \left(\frac{7}{18}\right). \qquad \text{...(4)}$$

The arbitrary constants A and B are determined by using the initial conditions: $x\ x_0$, $y = y_0$ at $t = t_0 = 0$ in (3) and (4), we have

$$y_0 = A + B,\ x_0 = -A + 2B - \left(\frac{7}{18}\right).$$

Thus $A = -\frac{1}{3}\left[x_0 - 2y_0 + (7/18)\right],$

$$B = \frac{1}{3}\left[x_0 + y_0 + (7/18)\right] \qquad \text{...(5)}$$

Using (5) into (3) and (4), we get

$$y = -\frac{1}{3}\left[x_0 - 2y_0 + \left(\frac{7}{18}\right)\right] + \frac{1}{3}[x_0 + y_0 + \frac{7}{18}]e^{3t} - \left(\frac{7}{18}\right)t - \left(\frac{1}{3}\right)t^3,$$

$$x = \frac{1}{3}\left[x_0 - 2y_0 + \left(\frac{7}{18}\right)\right] + \frac{2}{3}\left[x_0 + y_0 + \left(\frac{7}{18}\right)\right]$$

$$e^{3t} - \left(\frac{1}{3}\right)t^2 - \left(\frac{7t}{9}\right) - \left(\frac{7}{18}\right). \qquad ...(6)$$

This determines the displacement of fluid particle in the Lagrangian system.

Examle 2: *For a two-dimensional flow the velocities at a point in a fluid may be expressed in the Eulerian coordinates by*

$$u = x + y + 2t$$

and $$v = 2y + t.$$

Determine the Lagrange coordinates as functions of the initial positions x_0 *and* y_0 *and the time* t.

Solution: Equating the given relations, we have

$$u = \frac{dx}{dt} x + y + 2t$$

$$\textit{and } v = \frac{dy}{dt} = 2y + t$$

The solution of the differential equations can be determined by the operator method as follows

$$(D - 1)\, x - y = 2t,$$

$$(D - 2)\, y = t. \qquad ...(1, 2)$$

The solution for y, is given as follows:

$$y = Be^{2t} - \frac{1}{4}\,(2t + 1). \qquad ...(3)$$

The solution for x can be determined by substituting the relation (3) into equation (1),

$$(D - 1)\, x = Be^{2t} + \frac{1}{4}\,(6t - 1).$$

or, $$x = Ae^{t} + Be^{2t} + \frac{1}{4}(6t + 5), \qquad ...(4)$$

Initial $x = x_0,\ y = y_0$ at $t = t_0 = 0$.

Then $\quad B = y_0 + \frac{1}{4},$

and $\quad A = x_0 - y_0 + 1.$

Hence the solution (3) and (4) can be written in the from

$$x = F_1(x_0, y_0, t), \; y = F_2(x_0, y_0, t), \qquad ...(5)$$

Where

$$F_1(x_0, y_0, t) = (x_0 - y_0 + 1)\, e^t + (y_0 + \left(y_0 + \frac{1}{4}\right) e^{2t} - (6t + 5),$$

$$F_2(x_0, y_0, t) = \left(y_0 + \frac{1}{4}\right) e^{2t} - \frac{1}{4}(2t + 1).$$

This determines the Lagrange coordinates as a function of the initial positions x_0, y_0 and the time t.

Example 3: *The velocity q in a three-dimensional flow field for an incompressible flids is given by*

$$q = 2xi - yj - zk$$

Determine the equations of the streamlines passing through the point (1, 1, 1).

Solution: The equations of stream lines are given by

$$\frac{dx}{u} = \frac{dy}{v} = \frac{dz}{w}$$

$$\Rightarrow \frac{dx}{2x} = \frac{dy}{-y} = \frac{dz}{-z}$$

(i) (ii) (iii)

From, (i) and (ii), we have

$$\frac{dx}{2x} = \frac{dy}{-y} \Rightarrow \frac{dx}{x} + \frac{2dy}{y} = 0$$

By integrating, we obtain

$$\log x + 2 \log y = \log A$$

or $xy^2 = A$, where A is an integration constant.

Form (i) and (iii), we have

$$\frac{dx}{2x} = \frac{dz}{-z} \Rightarrow \frac{dx}{x} + \frac{2dz}{z} = 0$$

By integrating, we have

$xz^2 = B$, where B is an integration constant.

At the point (1, 1, 1) $A = 1 = B$

Hence the required streamlines are

$xy^2 = 1$ and $xz^2 = 1$.

Example 4: *Determine the streamlines and the path of the particles*

$$u = \frac{x}{(1+t)}, v = \frac{y}{(1+t)}, w = \frac{z}{(1+t)}.$$

Solution: The equations of the streamlines are given by

$$\frac{dx}{u} = \frac{dy}{v} = \frac{dz}{w}$$

or $$\frac{dx}{x/(1+t)} = \frac{dy}{y/(1+t)} = \frac{dz}{z/(1+t)}$$

or $$\frac{dx}{x} = \frac{dy}{v} = \frac{dz}{w}$$

(i) (ii) (iii)

By integrating (i) and (ii), we have

log x = log A, A is an integration constant.

$\Rightarrow$ x = Ay ...(1)

By integrating (i) and (iii), we have

log x = log z + log B, B is an integration constant.

$\Rightarrow$ x = Bz. ...(2)

Hence the streamlines are given by the intersection of (1) and (2).

$$q = \frac{dr}{dt}$$

or $$\frac{dx}{dt} = \frac{x}{1+t}, \frac{dy}{dt} = \frac{y}{1+t}, \frac{dz}{dt} = \frac{z}{1+t}$$

$$\Rightarrow \frac{dx}{x} = \frac{dt}{1+t}, \frac{dy}{y} = \frac{dt}{1+t}, \frac{dz}{z} = \frac{dt}{1+t}$$

By integrating, we have

log x = log(1 + t) + log A,

or log y = log(1 + t) + log B,

or log z = log(1+ t) +log C

$\Rightarrow$ x = A(1 + t),

y = B(1 + t),

z = C(1 + t),

which gives the required path of the particles.

Example 5: *The velocities at a point in a fluid in the Eulerian system are given by u = (x + y + z) + t, v = 2 (x + y + z) + t, w = 3 (x + y + z) +t. Find the displacement of a fluid particle in the Lagrangian system. Also determine the velocity of the fluid particle at* (x_0, y_0, z_0).

Solution: The velocity components may be expressed in terms of the displacement as

$$u=\frac{dx}{dt}=x+y+z+t,$$

$$w=\frac{dz}{dt}=3(x+y+z)+t. \qquad ...(1, 2, 3)$$

$$w=\frac{dz}{dt}=3(x+y+z)+t.$$

The differential equations can be written in form of operator as

(D–1) x – y = z + t

–2x + (D – 2) y = 2z + t

–3t –3y +(D – 3) z = t. ...(4, 5, 6)

Multiplying (4) by (D – 2) and adding to (5), we have

[(D – 1) (D – 2) –2] x = (D – 2) z + 2z +(D – 2)t

+ t(D^2–3D) x = Dz + 1 – t. ...(7)

Multiplying (4) by (2) and (5) by (D – 1) and adding, we have

[(D – 1) (D – 2)–2] y = (D – 1) (2z + t)

+ 2z + 2t(D^2 – 3D)y = 2Dz + 1 + t ...(8)

Multiplying (6) by (D^2 – 3D), we have

(D^2 – 3D) (D – 3)z = 3(D^2 – 3D)x + 3(D^2 – 3D)y + (D^2 – 3D)t

From (7) and (8), we have

(D^2 – 3D) (D – 3)z = 3(Dz + 1 – t) + 3(2Dz + 1 + t)

+ (D^2 – 3D)t D^2(D – 6)z = 3 ...(9)

The solution of the differential equation (9) is given by

$$z = A + Bt + Ce^{6t} - \frac{1}{4}t^2. \qquad ...(10)$$

From the equations (5) and (6), we have

$(D - 2)y - 2z = 2x + t,$

$-3y + (D - 3)z = 3x + t.$

Solving the equations, we have

$(D^2 - 5D)y = 2Dx + 1 - t$

$(D^2 - 5D)z = 3Dx + 1 + t. \qquad ...(11, 12)$

From (1), we have

$(D - 1)x = y + z + t$

or $(D - 1)(D^2 - 5D)x = (D^2 - 5D)y + (D^2 - 5D)z + (D^2 - 5D)t$

or $(D - 1)(D^2 - 5D)x = 2Dx + 1 - t + 3Dx + 1 + t - 5$

or $(D^3 - 6D^2)x = -3 \qquad ...(13)$

The solution of the differential equation becomes

$$x = A_1 + B_1 t + C_1 e^{6t} + \frac{1}{4}t^2. \qquad ...(14)$$

Proceeding in the same manner, we have

$y = A_2 + B_2 t + C_2 e^{6t} \qquad ...(15)$

Thus the equations (10), (14) and (15) determine the displacement of a fluid particle.

Let $x = x_0,\ y = y_0,\ z = z_0$

when $t = t_0 = 0$

The relations (14), (15) and (10) give

$x_0 = A_1 + C_1,\ y_0 = A_2 + C_2,\ z_0 = A + C$

Thus $x = x_0 - C_1 + B_1 t + C_1 e^{6t} + \frac{1}{4}t_2,$

$y = y_0 - C_2 + B_2 t + C_2 e^{6t},$

$z = z_0 - C + Bt + Ce^{6t} - \frac{1}{4}t^2 \qquad ...(16, 17, 18)$

Substituting thse values in (1), (2) and (3), we obtain the following identities

$$B_1 + 6C_2{}^{e6}t + \frac{1}{2}t = x_0 + y_0 + z_0 - (C_1 + C_2 + C) + (B_1 + B_2 + B)t + (C_1 + C_2 + C)e^{6t} + t,$$

$$B_2 + 6C_2e^{6t} = 2(x_0 + y_0 + z_0) - 2(C_1+C_2+C) + 2(B_1+B_2 + B)t + 2(C_1 + C_2 + C)e^{6t} + t,\ B + 6Ce^{6t} - \frac{1}{2}t = 3(x_0 + y_0 + z_0) - 3(C_1 + C_2 + C)$$

$$+ 3(B_1 +B_2 + B)t + 3(C_1 + C_2 + C)e^{6t} + t \quad ...(19, 20, 21)$$

Equating the coeffcients of t, e^{6t} and the constant term, we have

$$\left.\begin{aligned} x_0 + y_0 + z_0 - (C_1 + C_2 + C) &= B_1 \\ C_1 + C_2 + C &= 6C_1 \\ B_1 + B_2 + B + 1 &= \frac{1}{2} \end{aligned}\right\} \quad ...(22)$$

$$\begin{aligned} &2(x_0 + y_0 + z_0) - 2(C_1 + C_2 + C) = B_2 \\ &2(C_1 + C_2 + C) = 6C_2 \quad ...(23) \\ &2(B_1 + B_2 + B) + 1 = 0 \end{aligned}$$

$$\begin{aligned} &3(x_0 + y_0 + z_0) - 3C_1 + C_2 + C) = B \\ &3(C_1+C_2 + C) = 6C \quad ...(24) \\ &3(B_1 + B_2 + B) + 1 = -\frac{1}{2} \end{aligned}$$

From these three sets, we obtain

$$C_1 = \frac{1}{6}\left(x_0 + y_0 + z_0 + \frac{1}{12}\right),$$

$$C_2 = \frac{1}{3}\left(x_0 + y_0 + z_0 + \frac{1}{12}\right),$$

$$C = \frac{1}{2}\left(x_0 + y_0 + z_0 + \frac{1}{12}\right)$$

Also $B_1 = -\frac{1}{12}$,

$$B_2 = -\frac{1}{6},$$

$$B = -\frac{1}{4}$$

Substituting these values in therelations (16), (17), and (18) an simplifying, we have

$$x = \frac{5}{6}x_0 - \frac{1}{6}z_0 + \frac{1}{6}\left(x_0 + y_0 + z_0 + \frac{1}{12}\right)e^{6t} - \frac{1}{12}t + \frac{1}{4}t^2 - \frac{1}{72},$$

$$y = -\frac{1}{3}x_0 + \frac{2}{3}y_0 - \frac{1}{3}z_0 + \frac{1}{3}\left(x_0 + y_0 + z_0 + \frac{1}{12}\right)e^{6t} - \frac{1}{6}t - \frac{1}{36},$$

$$z = -\frac{1}{2}x_0 - \frac{1}{2}y_0 + \frac{1}{2}z_0 + \frac{1}{2}\left(x_0 + y_0 + z_0 + \frac{1}{12}\right)e^{6t} - \frac{1}{4}t - \frac{1}{4}t^2 - \frac{1}{24},$$

Determines the displacement of a flid particle in Largragian description, we have

$$u_1 = \frac{\partial x}{\partial t} = \left(x_0 + y_0 + z_0 + \frac{1}{12}\right)e^{6t} - \frac{1}{12} + \frac{1}{2}t,$$

$$v_1 = \frac{\partial y}{\partial t} = 2\left(x_0 + y_0 + z_0 + \frac{1}{12}\right)e^{6t} - \frac{1}{6},$$

$$w_1 = \frac{\partial z}{\partial t} = 3\left(x_0 + y_0 + z_0 + \frac{1}{12}\right) - \frac{1}{4} - \frac{1}{2}t.$$

Thus the velocity of the flid **particle** is given by

$$q_1 = u_1 i + v_1 j + w_1 k.$$

Example 6: *Determine the acceleration of a fluid particle of fixed identity for the velocity field,*

$$q = iA\,x^2y + jB\,y^2zt + kCzt^2.$$

Solution: Since $q = iu + jv + kw = iA\,x^2y + jB\,y^2zt + k\,Czt^2$.

$$\Rightarrow\ u = A\,x^2y,$$

$$v = B\,y^2zt,$$

$$w = C\,zt^2.$$

The acceleration f of the fluid particle is given as

$$f = \frac{\partial q}{\partial t} + u\frac{\partial q}{\partial y} + w\frac{\partial q}{\partial z}$$

$$\Rightarrow\ f = jB\,y^2z + k^2Czt + A\,x^2y(i\,2A\,xy) + B\,y^2zt(iA\,x^2 + j2B\,yzt) + C\,zt^2(j\,B\,y^2t + kCt^2)$$

$$f = A(2A\,x^3y^2 + B\,x^2y^2zt)i + B\,y^2z + (2B\,y^3z^2t^2 + C\,y^2zt^3)\,j + C(2zt + Czt^4)k.$$

which determine the acceleration of a fluid particle.

Example 7: *Find the equation of the streamlines for the flow*

$$q = -\, i(3y^2) - j(6x)$$

at the point (1, 1).

Solution: The equations of streamline are given by

$$\frac{dx}{u} = \frac{dy}{v}$$

Here q = – i(3y²) – j(6x)

$\Rightarrow$ $u = -\,3y^2,\ v = -\,6x$

or $\dfrac{dx}{-3y^2} = \dfrac{dy}{-6x}$

$\Rightarrow \dfrac{2dx}{y^2} = \dfrac{dy}{x}$

or $2x\ dx = y^2\ dy$

By integrating, we have

$x^2 = \dfrac{1}{3}y^3 + c$, where c is an integration costant.

At the point (1, 1), $c = \dfrac{2}{3} \Rightarrow 3x^2 = y^3 + 2,$

which determines the equation of the streamlines for the flow field.

Example 8: *The velocity components in a two-dimensional flow field for an-incompressible fluid are given by*

$$u = e^x \cosh y$$

and $v = -\, e^x \sinh y.$

Determine the equation of the streamlines for this flow.

Solution: The equation of the streamlines are given by

or $\dfrac{dx}{u} = \dfrac{dy}{v}$

$\Rightarrow \dfrac{dx}{e^x \cosh y} = \dfrac{dy}{e^x \sinh y}$

or $dx + \coth y\ dy = 0$

By integrating, we have

x + log sinhy = log c $\Rightarrow$ sinhy = ce^{-x}

where log c is an integration constant.

Example 9: *The velocity field at a point in fluid is given as*

$$q = \left(\frac{x}{t}, t, 0\right)$$

Obtain path lines and streak lines.

Solution: Here $q = \left(\frac{x}{t}, y, 0\right)$

The differential equations of path lines are given by

$$q = \frac{dr}{dt} = \frac{dx}{dt}i = \frac{dy}{dt}j + \frac{dz}{dt}k = \frac{x}{t}i + yj$$

$$\Rightarrow \qquad \frac{dx}{dt} = \frac{x}{t}, \frac{dy}{dt} = y, \frac{dz}{dt} = 0. \qquad ...(1, 2, 3)$$

By integrating (1), we have

$$\frac{dx}{dt} = \frac{x}{t}$$

$\Rightarrow$ log x = log t + log A

$\Rightarrow$ x = At. ...(4)

Let (x_0, y_0, z_0) be the cordinates of the chosen fluid particle at time $t = t_0$, then

$$x_0 = At_0 \Rightarrow A = \frac{x_0}{t_0}.$$

Form (4), we have $x = \frac{x_0}{t_0}t.$

By integrating (2), we have

$$x = \frac{x_0}{t_0}t.$$

or log y = t + log B

$\Rightarrow$ y = Be^t ...(5)

Aty = y_0, t = t0

$\Rightarrow$ B = $y_0e{-}t_0$

By integrating (3), we have

$$\frac{dz}{dt} = 0 \Rightarrow z = \text{c } i.e. \text{z isindependent of t} \Rightarrow \text{z} = \text{z}_0.$$

Hence the path lines are given by

$$x = \left(\frac{x_0}{t_0}\right)t, y = y_0^{e^{t-t_0}}, z = z_0. \quad ...(6)$$

Let the fluid particle (x_0, y_0, z_0) passes through a fixed point (x_1, y_1, z_1) at an instant of time t = T, where $t_0 \leq T \leq t$. Then the relation (6) reduces to

$$x_1 = \left(\frac{x_0}{t_0}\right)T, y_1 = y_0^{e^{T-t_0}}, z_1 = z_0$$

$$\text{or } x_0 = \left(\frac{x_1}{T}\right)t_0, y_0 = y_1^{e^{t_0-T}}, z_0 = z_1 \quad ...(7)$$

where T is the parameter. Substituting the relation (7) into (6), we have

$$x = \left(\frac{x_1}{T}\right)t, y = y_1^{e^{t-T}}, z = z_1$$

which gives the equaiton of streaklines passing through the point (x_1, y_1, z_1,).

Example 10: *The velocity distribution of a certain two-dimensional flow is given by*

$$u = Ay + B$$

$$and\ v = Ct,$$

where A, B, C are constants. Obtain the equation of the motion of fluid particles in Lagrngian method.

Solution: Let r (x, y) be the position of a given particle at any instant of time t. The path lines for the fluid particle are given by

$$q = \frac{dr}{dt} \Rightarrow \frac{dx}{dt} = u = Ay + B,$$

$$and \frac{dy}{dt} = v = Ct \quad ...(1, 2)$$

From (2), we have

$$y = \frac{1}{2}Ct^2 + D,$$

where D is an integration constant.
Initially $y = y_0$, $t = 0$; $D = y_0$

$$\Rightarrow \qquad y = \frac{1}{2}Ct^2 + y_0 \qquad ...(3)$$

Form (1) and (3), we have

$$\frac{dx}{dt} = A\left(\frac{1}{2}Ct^2 + y_0\right) + B.$$

$$\text{or } x = A\left(\frac{1}{6}Ct^2 + y_0 t\right) + Bt + E,$$

where E is in integration constant.

Initially $x = x_0$, $t = {}_0$; $E = x_0$.

$$\text{Therefore } x = A\left(\frac{1}{6}Ct^3 + y_0 t\right) + Bt + x_0. \qquad ...(4)$$

The equations (3) and (4) represent the path lines of fluid etements.

Example 11: *Determine the equations of the streamlines at the point in an incompressible fluid having spherical polar coordinates (r, θ, ϕ). The velocity components are*

$[2r^{-3} \cos \theta,$

$= r^{-3} \sin \theta, 0]$.

Solution: From the definition of the stream lines, we have

$$\frac{dr}{2r^{-3}\cos\theta} = \frac{rd\theta}{r^{-3}\sin\theta} = \frac{r\sin\theta d\phi}{0}.$$

By integrating, the equations of the stream lines are obtained as follows :

ϕ = constant,

$$\text{and } \frac{dr}{2r^{-3}\cos\theta} = \frac{rd\theta}{r^{-3}\sin\theta},$$

$\Rightarrow r = A \sin^2\theta$.

The equation ϕ = constant represents that the streamlines lie in the planes which pass through the axis of symmetry.

Example 12: *The velocity momponents of a flow in sylindrical polar coordinates are $(A r^2 z \cos\theta, B rz \sin\theta, C z^2 t)$. Determine the components of the acceleration of a fluid particle.*

Solution: Let q_r, q_θ and q_z be the components of velocity in cylindrical polar coordinates (r, θ,).

$q_r = A\, r^2 z \cos\theta,$

$q_\theta = B\, r_z \sin\theta,$

$q_z = C\, z^2 t$ ....(1)

Let f_r, f_θ, f_z be the components of acceleration then

$$f_r = \frac{\partial q_r}{\partial t} + q_r \frac{\partial q_r}{\partial r} + \frac{q_\theta}{r}\frac{\partial q_r}{\partial \theta} + q_z \frac{\partial q_r}{\partial z} - \frac{q\theta^2}{r}$$

$$f_\theta = \frac{\partial q_\theta}{\partial t} + q_r \frac{\partial q_\theta}{\partial r} + \frac{q_\theta}{r}\frac{\partial q_\theta}{\partial \theta} + q_z \frac{\partial q_\theta}{\partial z} + \frac{q_1 q\theta}{r}$$

$$f_z = \frac{\partial q_z}{\partial t} + q_r \frac{\partial q_z}{\partial r} + \frac{q_\theta}{r}\frac{\partial q_z}{\partial \theta} + q_z \frac{\partial q_z}{\partial z}. \qquad ...(2, 3, 4)$$

From (1) and (2, 3, 4), we have

$f_r = A\, r^2 z \cos\theta\, (2A\, rz \cos\theta) + B\, z \sin\theta\, (-A\, r^2 z \sin\theta) +$

$C\, z^2 t(A\, r \cos\theta) - B^2 r z^2 \sin^2\theta,$

or $f_r = rz^2[2A^2 r^2 \cos^2\theta - B(A\, r + B)\sin^2\theta + AC\, rt \cos\theta].$

Similarly

$f_\theta = (Ar^2 z \cos\theta)\, (Bz \sin\theta)\, (Bz \sin\theta)\, (Brz \cos\theta) + C\, z^2 t(Br \sin\theta)$

$+ AB\, r^2 z^2 \sin\theta \cos\theta,$

or $f_\theta = rz^2 \sin\theta[B(2AR + B)\cos\theta + BCt],$

and $f_z = C\, z^2 + C\, z^2 t.\ 2C\, zt,$

or $f_z = C\, z^2(1 + 2C\, zt^2).$

Example 13: *The velocity components in spherical polar coordinates (r, θ, ϕ) of a flow are*

$$q_r = \left(\frac{r^2}{t^2}\right)\sin\phi, q_\theta = \left(\frac{r}{t}\right)\cot\theta \cos ec\phi,$$

$$q_\phi = \left(\frac{r}{t}\right)\sin\theta\cos\phi.$$

Determine the componets of acceleration of a fluid particle.

Solution: Let q_r, q_θ and θ_ϕ be the components of velocity in spherical polar coordinates (r, θ, ϕ), then

$$q_r = \left(\frac{r^2}{t^2}\right)\sin\phi, q_\theta \left(\frac{r}{t}\right)\cot\theta\cos ec\phi, q_\phi = \left(\frac{r}{t}\right)\sin\theta\cos\phi \qquad ...(1)$$

Let f_r, f_θ, f_ϕ be the components of acceleration, then

$$f_r = \frac{\partial q_r}{\partial t} + q_r\frac{\partial q_r}{\partial r} + \frac{q_\theta}{r}\frac{\partial q_r}{\partial\theta} + \frac{q_\phi}{r\sin\theta}\frac{\partial q_r}{\partial\phi} - \frac{q^2\theta + q^2\phi}{r}.$$

$$f_\theta = \frac{\partial q_\theta}{\partial t} + q_r\frac{\partial q_\theta}{\partial r} + \frac{q_\theta}{r}\frac{\partial q_\theta}{\partial\theta} + \frac{q_\phi}{r\sin\theta}\frac{\partial q_\theta}{\partial\phi} - \frac{q^2\phi\cot\theta}{r}.$$

$$f_\theta = \frac{\partial q_\phi}{\partial t} + q_r\frac{\partial q_\phi}{\partial r} + \frac{q_\theta}{r}\frac{\partial q_\phi}{\partial\theta} + \frac{q_\phi}{r\sin\theta}\frac{\partial q_\phi}{\partial\phi} + \frac{q_\theta q_\phi\cot\theta}{r}.$$

...(2,3,4)

From (1)and (2, 3, 4), we have

$$f_r = -\frac{2r^2}{t^3}\sin\phi + \left(\frac{r^2\sin\phi}{t^2}\right)\left(\frac{2r}{t^2}\sin\phi\right) + \left(\frac{\cos\phi}{t}\right)\left(\frac{r^2\cos\phi}{t^2}\right)$$

$$-\frac{r}{t^2}(\cot^2\theta\cos ec^2\phi + \sin^2\theta\cos^2\phi),$$

$$\text{or } f_\theta = -\frac{r}{t^2}\cot\theta\cos ec\phi + \left(\frac{r^2\sin\phi}{t^2}\right)\left(\frac{\cot\theta\cos ec\phi}{t}\right)$$

$$+\left(\frac{\cot\theta\cos ec\phi}{r}\right)\left(\frac{-r\cos ec^2\theta\cos ec\phi}{t}\right)$$

$$+\left(\frac{\cos\phi}{t}\right)\left(\frac{r\cot\theta\cos ec\phi\cot\phi}{t}\right) + \frac{r^2\cot\theta}{t^3} - \frac{r\sin^2\theta\cos^2\phi\cot\theta}{t^2},$$

$$\text{and } f_\phi = -\frac{r}{t^2}\sin\theta\cos\phi + \left(\frac{r^2\sin\phi}{t^2}\right)\left(\frac{\sin\theta\cos\phi}{t}\right)$$

$$+\left(\frac{\cot\theta\cos ec\phi}{t}\right)\left(\frac{r\cos\theta\cos\phi}{t}\right) + \left(\frac{\cos\phi}{t}\right)\left(-\frac{r\sin\theta\sin\phi}{t}\right)$$

$$+\frac{r^2\sin\theta\sin\phi\cos\phi}{t^3} + \frac{r\sin\theta\cot^2\theta\cot\phi}{t^2}.$$

Example 14: *Determine the acceleration of fluid particle from the flow field*

$q = i(A\,xy^2t) + (B\,xy^2t) + k(C\,xyz)$

Solution: Let f be the acceleration of a fluid particle, then

$$f = \frac{\partial q}{\partial t} + u\frac{\partial q}{\partial x} + v\frac{\partial q}{\partial y} + w\frac{\partial q}{\partial z}$$

or $f = i(A\ xy^2) + j(B\ x^2y) + A\ xy^2t[i(A\ y^2t) + j(2B\ xyt)$
$+ k(C\ yz)] + B\ x^2yt\ [i(2A\ xyt) + j(B\ x^2t)$
$+ k(Cxz)] + Cxyz[k(Cxy)]$

$f = iA[xy^2 + Axy^4t^2 + 2Bx3y^2t^2] + jB[x^2y + 2Ax^2y^3t^2 + Bx^4yt^2]$
$+ kC[Axy^3zt + Bx^3yzt + x2y^2z].$

The components of the acceleration are

$f_x = A[xy^2 + A\ xy^4t^2 + 2B\ x^3y^2t^2],$

$f_y = B[x^2y + 2A\ x^2y^3t^2 + B\ x^4yt^2],$

$f_z = C[A\ xy^3zt + Bx^3tzt + x^2y^2z].$

Example 15: *Determine the acceleration at the point (2, 1, 3) at $t = 0.5$ sec, if $u = yz + t$, $v = xz - t$ and $w = xy$.*

Solution: Let $q = iu + jv + kw$.

or $q = (yz + t)i + (xz - t)j + xyk.$

The acceleration f of a fluid particle is given by

$$f = \frac{\partial q}{\partial t} + u\frac{\partial q}{\partial x} + v\frac{\partial q}{\partial y} + w\frac{\partial q}{\partial z}$$

or $f = (i - j) + (yz + t)(zj + yk) + (xz - t)(zi + xkk) + xy(yi + xj)$

or $f = (1 + xz^2 + xy^2 - tz)i + (-1 + yz^2 + x^2y + zt)j$
$+ (y^2z + x^2z + yt - xt)k.$

At the point (2, 1, 3) and $t = 0{\cdot}5$ sec, we have

$f = 19{\cdot}5i + 13{\cdot}5j + 6{\cdot}5k.$

Thus the comoponents of acceleration of a fluid particle are

$f_x = 19{\cdot}5\text{m/sec}^2,$

$f_y = 13{\cdot}5\text{m/sec}^2,$

$f_z = 6{\cdot}5\text{m/sec}^2.$

Example 16: *The velocity vector q is given by*

$q = ix - jy.$

Determine the equation of the stream lines.

Solution: From the definition of a stream line, we have

$$q \times dr = 0$$

or $(ix - jy) \times (idx + jdy) = 0$

or $(xdy + ydx)\, k = 0$

or $\frac{dx}{x} = -\frac{dy}{y}$.

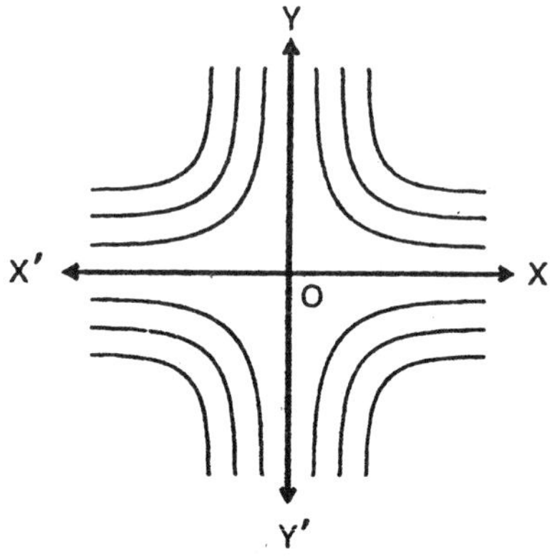

Fig. 1.12

By integrating, we obtain

$$\log x + \log y = \log c$$

or $xy = c$,

which represents the rectangular hyperbolas where c is an arbitrary constant.

Example 17: *Show that the velocity field $q_r = 0$, $q_\theta = Ar + B/r$, $q_z = 0$, satisfy the equation of motion*

$$\frac{d^2 q_\theta}{dr^2} + \frac{d}{dr}\left(\frac{q_\theta}{r}\right) = 0,$$ *where A and B are arbitrary constants.*

Solution: Here $q_r = 0, q_\theta = Ar + \frac{B}{r}, q_z = 0,$

$$\frac{dq_\theta}{dr} = A - \frac{B}{r^2}, \frac{d^2 q_\theta}{dr^2} = \frac{2B}{r^3}.$$

L.H.S $\frac{d^2 q_\theta}{dr^2} + \frac{d}{dr}\left(\frac{q_\theta}{r}\right) = \frac{2B}{r^3} + \frac{d}{dr}\left(A + \frac{B}{r^2}\right)$

$$= \frac{2B}{r^3} - \frac{2B}{r^3} = 0 = R.H.S$$ **Proved.**

Example 18: *The particles of a fluid move symmertically in space with regard to a fixed centre; prove that the equation of continuity is*

$$\frac{\partial \rho}{\partial t} + u\frac{\partial \rho}{\partial r} + \frac{\rho}{r^2} \cdot \frac{\partial}{\partial r}(r^2 u) = 0,$$

where u is the velocity at a distance r.

Solution: Let us concider a point P (r, θ, ϕ) in the fluid. Construct a parallelopiped with its edges PP' (= δr), PQ (= r δθ), and PS (= r sinθ δϕ). Let u, v, w be the components of the velocity inthe direction of the respective elements.

Let theorigin O be the fixed centre. Since the fluid particles move symmetrically in space with regard to an origin, it follows that the motion is only along the direction PP' and there is no motion along other edges r δθ and r sinθ δϕ.

Excess of mass of flow-in over flow out from the faces PQRS and P'Q'R'S' along PP' is

$$= -\delta r \frac{\partial}{\partial r}\{\rho u . r\delta\theta\, r \sin\theta\, \delta\phi\} \text{ per unit time.}$$

The excess of mass of flow-in over flow out along PQ and PS vanish as there is no motion along these directions.

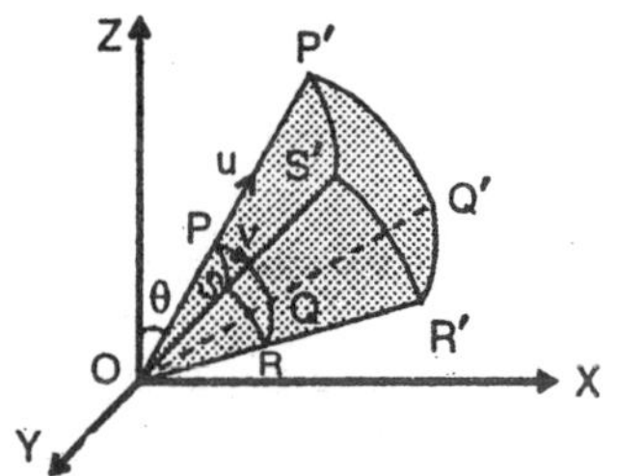

Fig. 1.13

Mass of the fluid inside the element = ρ δr. rδθ. rsinθ δϕ.

Rate of increase in the mass of the element

$$= \frac{\partial}{\partial t}(\rho\, \delta r . r\delta\theta . r \sin\theta\, \delta\phi) \text{ per unit time.}$$

From the equation of continuity, we have

$$= \frac{\partial}{\partial t}(\rho\, \delta r . r\delta\theta . r \sin\theta\, \delta\phi) = -\delta r \frac{\partial}{\partial r}(\rho u . r\delta\theta . r \sin\theta\, \delta\phi)$$

$$\Rightarrow \frac{\partial \rho}{\partial t}(\delta r . r\delta\theta . r\sin\theta\,\delta\phi)+\frac{\partial}{\partial r}(\rho u r^2)\delta r . \delta\theta \sin\theta\,\delta\phi) = 0.$$

$$\Rightarrow \frac{\partial \rho}{\partial t}+\frac{1}{r^2}\frac{\partial}{\partial r}(\rho u r^2) = 0$$

$$\Rightarrow \frac{\partial \rho}{\partial t}+u\frac{\partial \rho}{\partial r}+\frac{\rho}{r^2}\frac{\partial}{\partial r}(r^2 u) = 0.$$ **Hence Proved.**

Example 19: *A mass of fluid is in motion so that the lines of motion lie on the surface of coaxial cylinders. Show that the equation of continuity is*

$$\frac{\partial \rho}{\partial t}+\frac{1}{r^2}\frac{\partial}{\partial \theta}(\rho q_r)+\frac{\partial}{\partial z}(\rho q_z) = 0,$$

where q_r, q_z are the velocities perpendicular and parallel to z.

Solution: Let us concider a point P (r, θ, z) in the fluid. Construct a parallelopiped at P with edges PS (= rδθ), PR (= δr) and PP' (= δz). Let q_r, q_θ and and q_z be the velocity components along PR, PS, PP'. Since the lines of motion of the fluid lie on the surface of coasial cylinders so there is no motion along PR.

Excess of mass of flow-in over flow out along the direction PS

$$= -r\,\delta\theta\frac{\partial}{\partial z}(\rho q_\theta\,\delta r . \delta z)$$

and excess of mass of flow-in over flow out along the direction PP'

$$= -\delta z . \frac{\partial}{\partial z}(\rho q_z\,\delta r . r\delta z).$$

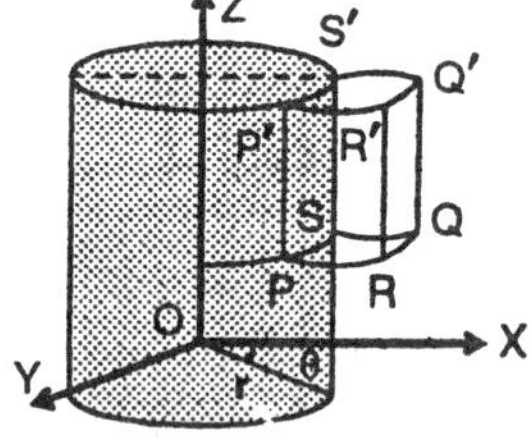

Fig. 1.14

Mass of the fluid element

= ρrδθ. δr. δz.

Rate of increase in the mass of the fluid element

$$= \frac{\partial}{\partial t}(\rho r\delta\theta . \delta r . \delta z).$$

From the equation of continuity, we have

$$\frac{\partial}{\partial t}(\rho r\delta\theta\,\delta r\,\delta z) = -r\delta\theta . \frac{\partial}{r\partial\theta}(\rho\theta_z\,\delta r\,\delta_z)-\delta_z\frac{\partial}{\partial z}(Pq_z\delta r\delta\theta)$$

$$\Rightarrow \frac{\partial \rho}{\partial t}(r\delta\theta . \delta r\,\delta z) = +\frac{\partial}{r\partial\theta}(r\delta\theta\,\delta r\,\delta z)+\frac{\partial}{\partial z}(\rho q_z)(r\delta\theta\,\delta r\,\delta z) = 0$$

$$\Rightarrow \frac{\partial \rho}{\partial t} + \frac{\partial}{r\partial \theta}(\rho q_\theta) + \frac{\partial}{\partial z}(\rho q_z) = 0.$$ **Hence Proved.**

Example 20: *If the linesof motion are curves on the surfaces of cones having their vertices at the origin and the axis of z for common axis. Prove that the equation of continuity is*

$$\frac{\partial \rho}{\partial t} + \frac{\partial}{\partial r}(\rho q_r) + \frac{2\rho q_r}{r} + \frac{\cos ec\theta}{r}\frac{\partial}{\partial \omega}(\rho q_\omega) = 0.$$

Solution: Let OZ be the common axis of Z with O as vertex. Let A(r, θ, ω) be a point on the surface of the cone and q_r, q_θ, $q\omega$ be the components of the velocity along the edges of the parallelopiped AA' (= δr), AB (= r$\delta\theta$) and AD (= rsinθ $\delta\omega$) respectively. Since the lines of motion are curves on the surfaces of cones so there will be no motion perpendicular to the surface of the cone.

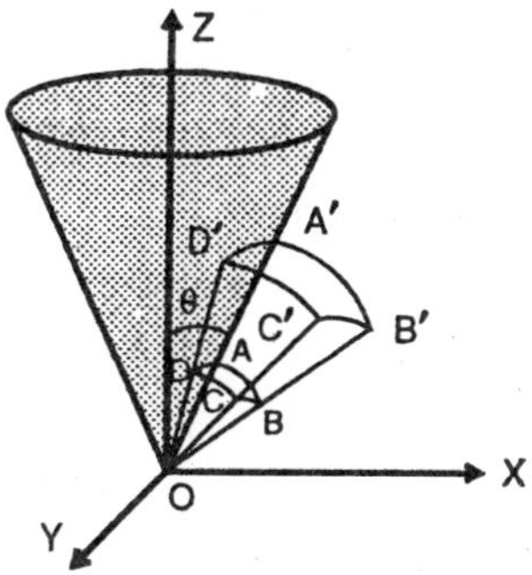

Fig. 1.15

Excess of flow- in over flow out in the direction AA' *i.e.*, from the face ABB'A' and the opposite face in time δt

$$= -\frac{\partial}{\partial r}\{\rho q_w . r\delta\theta . r\sin\theta\delta\omega\}\delta r\delta t$$

and the excess of flow-in over flow out in the direction AD *i.e.*, from the face ABB'A' and the opposite face in time δt.

$$= -\frac{\partial}{r\sin\theta\,\delta\omega}\{\rho q\omega . r\delta\theta\,\delta r\}\, r\sin\theta\,\delta\omega\,\delta t.$$

Total excess of mass flow-in over flow out in time δt

$$= -\frac{\partial}{\partial r}\{\rho q_r\, r\delta\theta . r\sin\theta\,\delta\omega\}\,\delta r\,\delta t$$

$$-\frac{\partial}{r\sin\theta\,\partial\omega}\{\rho q_\omega . r\delta\theta\,\delta r\}\, r\sin\theta\,\delta\omega\,\delta r\,\delta t \quad ...(1)$$

Rate of increase in the mass of an element in time δt

$$= \frac{\partial}{\partial t}\{\rho r\,\delta\theta . r\sin\theta\,\delta\omega . \delta r\}\,\delta t \quad ...(2)$$

From the equation of continuity, we have

$$= \frac{\partial\rho}{\partial t}\{r^2\sin\theta\,\delta r\,\delta\theta\,\delta\omega\}\,\delta t = \frac{\partial}{\partial r}\{\rho q_r . r\delta\theta . r\sin\theta\,\delta\omega\}\delta r\,\delta t$$

$$-\frac{\partial}{r\sin\theta\partial\omega}\{\rho q\omega\, r\delta\theta\,\delta r\}\, r\sin\theta\,\delta\omega\,\delta t$$

$$\Rightarrow \frac{\partial\rho}{\partial t}+\frac{1}{r^2}\frac{\partial}{\partial r}(\rho r^2 q_r)+\frac{1}{r\sin\theta}\frac{\partial}{\partial\omega}(\rho q_\omega) = 0$$

$$\text{or } \frac{\partial\rho}{\partial t}+\frac{1}{r^2}\left\{r^2\frac{\partial}{\partial r}(\rho q_r)+(\rho q_r).2r\right\}+\frac{\operatorname{cosec}\theta}{r}\frac{\partial}{\partial\omega}(\rho q_\omega) = 0$$

$$\text{or } \frac{\partial\rho}{\partial t}+\frac{\partial}{\partial r}(\rho q_r)+\frac{2\rho q_r}{r}+\frac{\operatorname{cosec}\theta}{r}\frac{\partial}{\partial\omega}(\rho q_\omega) = 0.$$ **Hence Proved.**

Example 21: *If every particle moves on the surface of the sphere, prove that the equation of continuty is*

$$\cos\theta\frac{\partial\rho}{\partial t}+\frac{\partial}{\partial\theta}(\rho\omega\cos\theta)+\frac{\partial}{\partial\phi}(\rho\omega'\cos\theta) = 0.$$

ρ being the density, θ, ϕ the latitude and longitude of any element and ω' the angular velocities of the element in latitude and longitude respectively.

Solution: Consider an elementary parallelopiped PQRS and P'Q'R'S' on the surface of the sphere, whose lenght of the edges are as follows

PP' = δr, PQ = $r\delta\theta$,

PS = $r\cos\theta\ \delta\phi$.

Since ω and ω' are the angular velocities of the element along latitutde and longitude respectively so $r\omega$ and $r\cos\theta\,\omega'$ are the velocities along PQ and PS. Since the particle moves on the surface of the sphere it follows that there will be no velocity normal to the surface of the sphere *i.e.*, the velocity along PP' vanish.

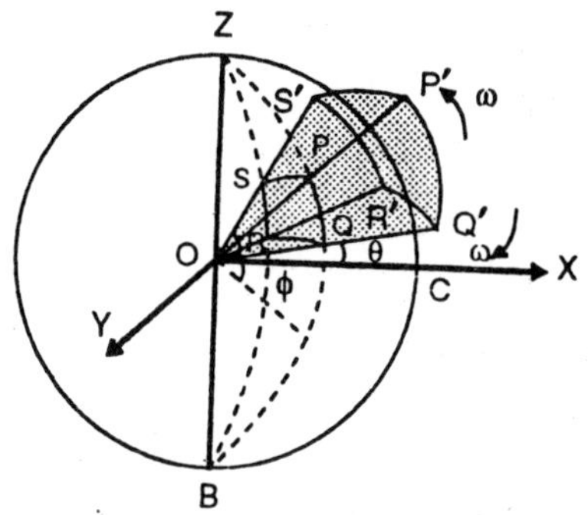

Fig. 1.16

Excess of flow-in over flow out along the faces perpendicular to PQ *i.e.*, from the face QQ' RR' and PP' S'S is

$$= -r\delta\theta.\frac{\partial}{r\partial\theta}(\rho r\omega\delta r.r\cos\theta\delta\phi) \text{ per unit time}$$

Simmilarly excess of flow-in over flow out along the face perpendicular to PS is

$$= -r\cos\theta\delta\phi.\frac{\partial}{r\cos\theta\partial\phi}(\rho r\cos\theta\,\omega'\delta r\, r\delta\theta) \text{ per unit time}$$

Total excess of flow-in over flow out through all the faces

$$= -[r\delta\theta\frac{\partial}{r\partial\theta}(\delta r\rho\omega.\rho\cos\theta\phi)$$

$$+r\cos\theta\,\delta\phi\frac{\partial}{r\cos\theta\partial\phi}(\rho r\cos\theta\omega'.\delta r.r\delta\theta)] \text{ per unit time} \quad ...(1)$$

Rate of increase in the mass per unit time is

$$= \frac{\partial}{\partial t}(\rho\delta r.r\delta\theta.r\cos\theta\,\delta\phi.) \quad ...(2)$$

The equation of continuity is given by the relation

$$\frac{\partial}{\partial t}(\rho\delta r.r\delta\theta.r\cos\theta\,\delta\phi) = -r\delta\theta.\frac{\partial}{r\partial\theta}(\rho r\omega.\delta r.r\cos\theta\delta\phi)$$

$$-r\cos\theta\delta\phi.\frac{\partial}{r\cos\theta\partial\phi}(\rho r\cos\theta\omega'.\delta r.r\delta\theta)$$

$$\Rightarrow \frac{\partial\rho}{\partial t}+(r^2\cos\theta\,\delta r\,\delta\theta\delta\phi)+r\delta\theta.\frac{\partial}{r\partial\theta}(\rho\omega\cos\theta)r^2\delta r\delta\phi$$

$$+r\cos\theta\delta\phi.\frac{\partial}{r\cos\theta\partial\phi}(\rho\omega'\cos\theta)r^2\delta r\,\delta\theta = 0$$

$$\Rightarrow \frac{\partial \rho}{\partial t} + \frac{1}{\cos\theta}\frac{\partial}{\partial \theta}(\rho\omega\cos\theta) + \frac{1}{\cos\theta}\frac{\partial}{\partial \phi}(\rho\omega'\cos\theta) = 0$$

$$\Rightarrow \cos\theta\frac{\partial \rho}{\partial t} + \frac{\partial}{\partial \theta}(\rho\omega\cos\theta) + \frac{\partial}{\partial \phi}(\rho\omega'\cos\theta) = 0.$$ **Hence proved.**

Example 22: *If the lines of motion are curves on the surfaces of spheres all touching the plane of –XY at the origin O, the equation of continuity is*

$$r\sin\theta\frac{\partial \rho}{\partial t} + \frac{\partial}{\partial \phi}(\rho v) + \sin\theta\frac{\partial}{\partial \theta}(\rho u) + 1 + 2\cos\theta) = 0,$$

where r is the radius CP of one of the spheres, θ the angle PCO, u the velocity in the plane PCO, to a fixed plane through the axis of Z.

Solution: Let C and C' be the centres of the Z-axis of the two consecutive spheres of radii r and r + δr respectively, such that CC' = δr. Join the line CP which meets the second sphere in Q. Consider P be a point on the inner sphere and PQ,PR and PS be the edgesof the elementary parallelopiped of lengths

PR = rδθ,

and PS = rsinθ δϕ,

where ϕ isthe angle which the plane PCO makes with a fixed plane through the Z-axis *i.e.*, with – XOZ plane.

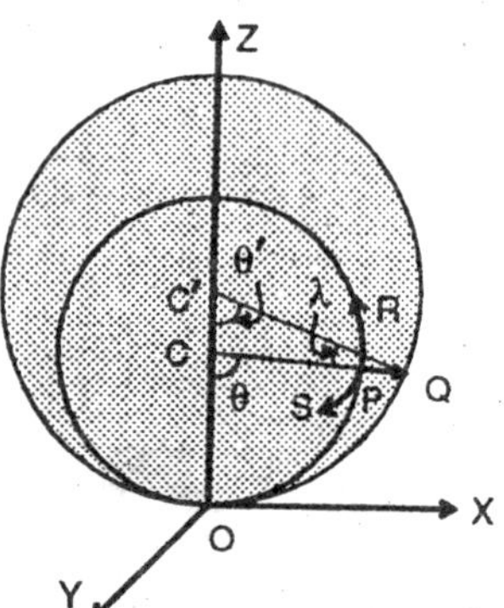

Fig. 1.17

Now we shall determine the lenght of the edge PQ.

CP = r, C' Q = r + δr,

CC' = δr, ∠ PCO = θ.

Let $\angle$ CC'Q = θ' and $\angle$ C'QC = λ.

Then θ' + λ = θ

$\Rightarrow$ θ' = θ − λ

From the Δ CQC', we have

$CQ^2 = C''Q^2 = + CC'^2 - 2C'Q.\ CC' \cos\theta'$

$$\Rightarrow (r+PQ)^2 = (r+\delta r)^2 + (\delta r)^2 - 2(r+\delta r)\delta r\cos(\theta-\lambda)$$

$$\Rightarrow r^2 + 2rPQ + PQ^2 = r^2 + 2r\delta r + \delta r^2 + \delta r^2$$

$$-2(r\delta r + \delta r^2)\cos(\theta-\lambda)$$

Neglecting the squares of the quantities PQ and δr, being small, we have

$$2r.PQ = 2r\,\delta r - 2r\,\delta r.\cos(\theta-\lambda)$$

or $PQ = \delta r\{1-\cos(\theta-\lambda)\}$

or $PQ = \delta r\{1-(\cos\theta\cos\lambda + \sin\theta\sin\lambda)\}$

or $PQ = \delta r\{1-\cos\theta - \lambda\sin\theta)\}$

or $PQ = \delta r\{1-\cos\theta\}$ (neglecting the small quantities)

Since lines of motion are curves on the surfaces of spheres, so there will be no component of velocity along PQ, u and vbe the velocity components along the edegs PR and PS in the direction of θ and ϕ increasing.

Excess of flow-in over out in the direction PR

$$= -\frac{\partial}{r\partial\theta}\{\rho u\, r\sin\theta\,\delta\phi(1-\cos\theta)\,\delta r\}\,r\,\delta\theta \text{ per unit time.}$$

Similarly the excess of flow-in over flow out in the direction PS

$$= -\frac{\partial}{r\sin\theta\partial\phi}\{\rho v\, r\,\delta\theta\,\delta r(1-\cos\theta)\,\delta r\}\,r\sin\theta\,\delta\phi \text{ per unit time}$$

Total excess of flow-in over flow out through all the faces

$$= [-\frac{\partial}{r\partial\theta}\{\rho u\, r\sin\theta\,\delta\phi(1-\cos\theta)\,\delta r\}\,r\,\delta\theta$$

$$-\frac{\partial}{r\sin\theta\partial\phi}\{\rho v\, r\,\delta\theta\,\delta r\,(1-\cos\theta)\}\,r\sin\theta\,\delta\phi] \qquad ...(1)$$

Volume of the parallelopiped

$$= (1-\cos\theta)\,\delta r . r\delta\theta\, r\sin\theta\delta\phi.$$

Rate of increase in the mass of the parallelopiped.

$$= \frac{\partial}{\partial t}\{\rho(1-\cos\theta)\,\delta r\,\delta\theta . r\sin\theta\delta\phi\} \text{ per unit time} \qquad ...(2)$$

From the equation of continuity, we have

$$\frac{\partial}{\partial t}\{\rho(1-\cos\theta)\,\delta r\, r\delta\theta . r\sin\theta\delta\phi\}$$

$$= \frac{\partial}{r\partial\theta}\{\rho u\, r\sin\theta\delta\phi(1-\cos\theta)\,\delta r\}\, r\,\partial\theta$$

$$-\frac{\partial}{r\sin\theta\partial\phi}\{\rho v\, r\,\delta\theta\,(1-\cos\theta)\,\delta r\}\, r\sin\theta\,\delta\phi$$

$$\Rightarrow \frac{\partial\rho}{r\partial\theta}\{(1-\cos\theta)\,\delta r\, r\delta\theta\, r\sin\theta\delta\phi\}$$

$$+\frac{\partial}{r\partial\theta}\{\rho u\sin\theta\,(1-\cos\theta)\}r^2\delta r\,\delta\theta\,\delta\phi$$

$$+\frac{\partial}{r\sin\theta\,\partial\phi}\{\rho v\}\, r^2(1-\cos\theta)\,\delta r\,\delta\theta\,\delta\phi = 0$$

$$\Rightarrow \frac{\partial\rho}{\partial t}+\frac{1}{r\sin\theta\,(1-\cos\theta)}\frac{\partial}{\partial\theta}\{\rho u\sin\theta\,(1-\cos\theta)\}$$

$$+\frac{1}{r\sin\theta}\frac{\partial}{\partial\phi}(\rho v) = 0$$

$$\Rightarrow \frac{\partial\rho}{\partial t}+\frac{1}{r\sin\theta\,(1-\cos\theta)}[\sin\theta\,(1-\cos\theta)\frac{\partial}{\partial\theta}(\rho u)$$

$$+\rho u\{\cos\theta\,(1-\cos\theta)+\sin^2\theta\}]+\frac{1}{r\sin\theta}\frac{\partial}{\partial\phi}(\rho v) = 0$$

$$\Rightarrow \frac{\partial\rho}{\partial t}+\frac{1}{r}\frac{\partial}{\partial\theta}(\rho u)+\frac{\rho u\,(1+\cos\theta-2\cos^2\theta)}{r\sin\theta\,(1-\cos\theta)}+\frac{1}{r\sin\theta}\frac{\partial}{\partial\phi}(\rho v) = 0$$

$$\Rightarrow \frac{\partial\rho}{\partial t}+\frac{1}{r}\frac{\partial}{\partial\theta}(\rho u)+\frac{\rho u\,(1+2\cos\theta)}{r\sin\theta}+\frac{1}{r\sin\theta}\frac{\partial}{\partial\phi}(\rho v) = 0$$

$$\Rightarrow r\sin\theta\frac{\partial\rho}{\partial t}+\frac{\partial}{\partial\phi}(\rho v)+\sin\theta\frac{\partial}{\partial\theta}(\rho u)(1+2\cos\theta) = 0.$$

Hence Proved.

INDEPENDENT VARIABLES IN LAGRANGE'S METHOD

Let a point P_0 (a, b, c) reach a point P (x, y, z) in time t. It is clear that (x, y, z) are functions of t. But since partictes which have inetially different positions occupy different positions which after the motion is allowed, so the final position P (x, y, z) depends on the initial position P_0 (a, b, c) also. Thus x, y, z are functions of a, b, c and t, which are the four independent variables in this method.

Thus
$$x = f_1 (t, a, b, c),$$
$$y = f_2 (t, a, b, c),$$
$$z = f_3 (t, a, b, c).$$

In most of the problems the motion is every where.

Continuous, *i.e.*, f_2 f_3 are continous functions of a, b, c and t. We shall assume that f_1, f_2, f_3 possess partial derivatives of first and second order with respect to a, b, c and t.

Clearly velocity and acceleration components along the axes are

$$\frac{\partial x}{\partial t}, \frac{\partial y}{\partial t}; \frac{\partial z}{\partial t}; \frac{\partial^2 x}{\partial t^2}, \frac{\partial^2 y}{\partial t^2}, \frac{\partial^2 z}{\partial t^2}$$

FURTHER ABOUT EULER'S METHOD

In this method the state of the fluid is given by the following five quantities.

Three component velocities u, v, w of the fluid at the particutar point:

(i) The pressure P there

(ii) The density P there.

DIFFERENTIATION FOLLOWING THE MOTION OF FLUID

Let f (r, t) represent a flow parameter (velocily, densitg).

The change or increment δf in f is given by

$$\delta f = f (r + \delta r, t + \delta t) - f (r, t)$$

where δr and δt are inerements of position vector and time, we write

$$\delta f = [f(r + \delta r, t + \delta t) - f(r + \delta t)] + f(r, t + \delta t) - f(r, t)]$$

$$= \delta r \cdot \nabla f (r, t + \delta t) + dt \cdot \frac{\partial}{\partial t} f (r, t).$$

$$\frac{\delta f}{\delta t} = \frac{\delta r}{\delta t}.\Delta f(r, t+\delta t) + \delta t.\frac{\delta}{\partial t} f(r,t)$$

Proceeding to limit as δt → 0, we have

$$\frac{\delta f}{\delta t} = \frac{\delta f}{\delta t} + (q.\nabla) f$$

This relation is true whether f is a sealar or a vector point function. Thus

$$\frac{\delta}{\delta t} = \frac{\delta}{\delta t} + (q.\nabla)$$

The first tern in (1) is called that local (orlemporal) rate of change in f and the second term the convective rate of change. The operator

δ/δt for this reason is termed as 'Defferentiation' following the motion of the fluid or Mobile derivative' or substantive derivative.

Acceleration

Acceleration is defined as the total derivative of velocity w, r, t time

$$a = \frac{\delta q}{\delta t}$$

$$= \frac{\delta q}{\delta t} + (q.\nabla) q.$$

In the rectangular cartesian coordinatesif u, v, w be the components of velocity the components of acceleration are

$$ax = \frac{\delta u}{\delta t} + u\frac{\delta u}{\delta x} + v\frac{\delta u}{\delta y} + w\frac{\delta u}{\delta z},$$

$$av = \frac{\delta v}{\delta t} + u\frac{\delta v}{\delta x} + v\frac{\delta v}{\delta y} + w\frac{\delta v}{\delta z},$$

$$az = \frac{\delta w}{\delta t} + u\frac{\delta w}{\delta x} + v\frac{\delta w}{\delta y} + w\frac{\delta w}{\delta z}.$$

Cor. If a motion is along the curve δ only and q be the velocity at a point P then

q = f (s, t)

and q + δq = f (s + δs, t + δt)

= f (s + qδt, t + δt) as δs = qδt

$$= f(s,t) + (q\frac{\delta f}{\delta s} + \frac{\delta f}{\delta t})\delta t + \ldots$$

$$\frac{\delta q}{\delta t} = (q\frac{\delta f}{\delta s} + \frac{\delta f}{\delta t} + \ldots$$

$$\text{acceleration} \quad = \lim_{dt \to 0} \frac{\delta q}{\delta t} = \frac{\delta t}{\delta t} + q\frac{\delta f}{\delta s}$$

INDEPENDENT VARIABLES IN EULER'S METHOD

Since in this method the fluid is studied at all its points and at every instant of time the independent variables are x, y, z and t.

Thus if u, v, w be the velocities point (x, y, z) the values of u, v, w will tell us changes at that point as t changes. And at a particular time (t constant). We find the conditions of every point of the fluid

To connect the Eulerian and Lagrangian method we have in general,

$$\text{u (in Eulerian)} = \frac{\partial x}{\partial t} \text{ (in Lagrangian)}$$

$$\text{v (in Eulerian)} = \frac{\partial y}{\partial t} \text{ (in Lagrangian)}$$

$$\text{w (in Eulerian)} = \frac{\partial z}{\partial t} \text{ (in Lagrangian)}$$

VELOCITY OF A FLUID PARTICLE

At any time t, let the particle be at P, such that

OP = r

After a time δt, let this particle reach Q, such that

OQ = r + δr.

If q is the velocity of the particle at then q is a vector given by

$$q = \lim_{\delta t \to 0} \frac{(r + \delta r) - 2}{\delta t}$$

$$= \lim_{\delta t \to 0} \frac{(r + \delta r)}{\delta t} = \frac{\delta r}{\delta t}$$

If u, v, w be the components of q along the cartosian axes and t, j, k the unit vectors along the axes, then q = ui + vj + wk

EQUATION OF CONTINUITY

The equation of continuity simply expresses the law of conservation of mass in a math ematical expresses the law of conservation of mass in a mathematical form

Consider a closed surface δ in a fluid medium enclosing a volume V fixed in space. Let P (r, t) be the density of the fluid. If n is the unit outword normal at a surace element δs, where the fluid velocity is q, then

Rate of mass flow out of ds

$$= pn.q \, ds$$

Total rate of mass flow out of $v = \frac{\phi}{s} pn.qds$

$$\int_v \nabla.(pq)dv \qquad \text{[By Gauss' Theorem]}$$

This must equal the negative of the rate of increase of mass within V.

$$\therefore \int_v \nabla.(pq)dv = -\frac{\partial}{\partial t}\int_v r\delta v = -\int_v \frac{\partial p}{\delta t} dv,$$

$$\int_v [\frac{\partial p}{\partial t} + \nabla.(pq)dv = 0$$

This relation must hold no matter how small V is and thus

$$\frac{\partial p}{\partial t} + \nabla.(pq) = 0$$

This is the equation of continuty other forms of (1) are

$$\frac{\partial p}{\partial t} + p\nabla.q + q.\nabla_p = 0$$

$$\frac{dp}{dt} + p\nabla.q = 0 \qquad \text{...(1a)}$$

or $\frac{d}{dt}(\log P) + \nabla.q = 0$ ...(1b)

where d/dt denotes defferentiation following the motion.

In case of steady flow $\partial p / \partial t = 0$ where (1) given

$$\nabla . (pq) = 0 \qquad \text{...(2)}$$

For an incompressible fluid the density of any fluid partile is invariant w, r, t time so that

$$= dp/dt = 0$$

In such a fluid called nonhomogeneous incompressible fluid. In both these eases (1a) gives

$$\nabla .q = 0. \qquad ...(3)$$

EQUATION OF CONTINUITY IN CARTESIAN CO-ORDINATES

Let there be a point P (x, y, z) in the fluid and P be the density of the fluid there. Also let u, v, w be the velocity compodents at P paralled to axes.

Now construct a small parallelopiped with edgesof lengiths δx, δy and δz parallel to Co-ordinate axes, having P at one of its angular points

Now mass of fluid that passes in thraugh the plane face PABC

= P (δy.δz) u in unit time

as δy.δz is the area of the cross-section and u is the velocity with which the fluid crosses this face

= f (x, y, z), say

Now mass of the fluid that passes out through the plane face QA'B'C'

$$= f(x + \delta x, y, z)$$

$$= f(x, y, z) + \delta x \frac{\partial}{\partial x} f(x, y, z) + ...$$

Therefore the excess of of flow-in over flow out

= mass that enter s in through P.ABC – Mass that leaves through QAB'C' – mass that leaves through QAB'C'

$$= f(x,y,z) - [f(x,y,z) + \delta x \frac{\partial}{\partial x} f(x,y,z) + ...]$$

$= - \delta x \; \partial / \partial x \; f(x, y, z)$ to the first order of approximation

$$= - \delta x \frac{\partial}{\partial x}(pu, \delta y, \delta z) \qquad \text{from (1)}$$

$$= - \delta x, \delta y, \delta z \frac{\partial (pu)}{\partial x} \qquad ...(4)$$

as x, y, z are independent variables and p, u, v, w are their functions.

Similarly the excess of flow-in over flow-out from faces

$$PQCC'.AA'B'B' = -\delta y \frac{\partial}{\partial y}(pv\,\delta x\,\delta z)$$

$$= -\delta x\,\delta y\,\delta z\frac{\partial(pv)}{\partial y}$$

and that from the faces PQAA' add CC'B'B

$$= -\delta z\frac{\partial}{\partial z}(pw\delta y\delta x)$$

$$= -\delta x\,\delta y\,\delta z\frac{\partial}{\partial z}(pw) \qquad ...(5)$$

Again the total mass in the parallelopiped = P.δx δyδz. Hence increase in the mass of the parallelopiped in unit time

$$= \frac{\partial}{\partial t}-(p\,\delta x\,\delta y\,\delta z) = \delta x\,\delta y\,\delta z = \frac{\partial p}{\partial t} \qquad ...(6)$$

Now the increase in mass = total excess of flow-in over flow-out from all the faces, *i.e.*,

$$\delta x\,\delta y\,\delta z\frac{\partial p}{\partial t} = -\delta x\,\delta y\,\delta z\left[\frac{\partial(pu)}{\partial x}+\frac{\partial(pv)}{\partial y}+\frac{\partial(pw)}{\partial z}\right],$$

$$\frac{\partial p}{\partial t}+\frac{\partial(pu)}{\partial x}+\frac{\partial(pv)}{\partial y}+\frac{\partial(pw)}{\partial z} = 0.$$

This is the equation of continuity in Cartesian co-ordinates.

Note 1: The above equation of continuity can also be written

as $$\frac{\partial p}{\partial t}+u\frac{\partial p}{\partial x}+v\frac{\partial p}{\partial y}+w\frac{\partial p}{\partial z}\left(\frac{\partial u}{\partial x}+\frac{\partial v}{\partial y}+\frac{\partial w}{\partial z}\right) = 0$$

Note 2: If the fluidis incompressible, then

$$\frac{\partial p}{\partial t}+u\frac{\partial p}{\partial x}+v\frac{\partial p}{\partial y}+w\frac{\partial p}{\partial z} = 0$$

and the equation of continuity in this case becomes

$$\frac{\partial u}{\partial x}+\frac{\partial v}{\partial y}+\frac{\partial w}{\partial z} = 0$$

EQUATION OF CONTINUITY IN POLAR CO-ORDINATES

Let P (r, θ, w) be a point in the fluid constract a parallelopiped with PQ (= δr), PR (= r δθ) and PS (= rsinθ δw) as edges.

Also let (u, v, w) as edges.

Also let u, v, w be the component velocities in the directions of the elements δr, rδθ and rsinθ δw.

Now mass of the liquid that-passes along PQ, *i.e.*, through the faces PRTS = R.rδθ rsinθ δw u per unit time

= f (r + δr, θw) say

And the mass of liquid that passes ot along PQ, *i.e.*, throgh the face QR'T'S'

Therefore excess of flow in over flow out along PQ

= Mass that enters through PRTS – mass that flows out

through QR'T'S'

$$= -\delta r\frac{\partial f(r,\theta,w)}{\partial r}$$ to the first order of approximation

$$= -dr\frac{\partial}{\partial r}[Pr\delta\theta, r\sin\theta\, Sw.u.]$$ per unit time

as f (r, θ, w) = Prδθ r sinθ δw.u

Similarly excess of flow-in over flow-out from faces PS'S'Q and RTTR'

$$= -r\delta\theta\frac{\partial}{r\partial\theta}[P.\delta r, r\sin\theta\,\delta wv]$$

$$= -r\delta\theta\,\delta w\frac{\partial}{\partial\theta}[Pv\sin\theta]$$

and that from faces PRR'Q and STT'S'

$$= -r\sin\theta\,\delta w\frac{\partial}{r\sin\partial w}[P.\delta r, r\delta\theta\, w]$$

$$= -r\delta w\,\delta r\,\delta\theta\frac{\partial}{\partial w}[Pw]$$

But the total massinside the parallepopiped

= p.δr, rδθ, rsinθ δw.

So that the change in the maas of the liquid inside the parallelopiped

$$= \frac{\partial}{\partial t}(P\,\delta r, r\delta\theta . r\sin\theta\,\delta w) \text{ per unit time}$$

$$= r^2 \sin\theta\,\delta r . \delta\theta\,\delta w \frac{\partial p}{\partial t}.$$

Now the inerease in mass = total excess of flow-in over flow-out

i.e., $r^2 \sin\theta\,\delta r . \delta\theta\,\delta w \dfrac{\partial p}{\partial t}$

$$= -\,\delta r\,\delta\theta\,\delta w \sin\theta \frac{\partial}{\partial r}(pr^2 u) - r\delta r\,\delta\theta\,\delta w \frac{\partial}{\partial\theta}(pv\sin\theta)$$

$$-r\delta r\,\delta\theta\,\delta w \frac{\partial}{\partial w}(pw).$$

or $r^2 \sin\theta \dfrac{\partial p}{\partial t} = -\sin\theta \dfrac{\partial}{\partial r}(r^2 pu) - \dfrac{r\partial}{\partial\theta}(pv\sin\theta) - r\dfrac{\partial}{\partial w}(pw)$

or $\dfrac{\partial p}{\partial t} + \dfrac{1}{r^2}\dfrac{\partial}{\partial r}(r^2 pu) + \dfrac{1}{r\sin\theta}\dfrac{\partial}{\partial\theta}(pv\sin\theta) + \dfrac{1}{r\sin\theta}\dfrac{\partial}{\partial w}(pw) = 0$

which is the equation of continuity in spherical polar Co-ordinates.

MOTION IN TWO Δ DIMENSIONS

When there is no velocity parallel to Z-axis and the motion in all planes parallel to xy-plane is the same, then we call ita motion in two dimensions. Thus if there is a motion in two climensions, u and v are the functions of x, y and t only and w = o. In such cases the motion in xy-plane is only considered. Since the whole motion is known when we known it in z = 0 plane.

LAGRANGE'S STREAM FUNCTION OR CURRENT FUNCTION

When the motion in two-climensional,

w = 0

∴ the differential equation of the stream line is

$$\frac{dx}{u} = \frac{dy}{v}$$

or dx – udy = 0 ...(1)

and the equation of continuity is

$$\frac{du}{x}+\frac{dv}{y}=0 \qquad ...(2)$$

But (2) is the condition that (1) is an exat differential equation.

Thus vdx – udy is a complete differential say $d\psi$, so that

$$v\ dx - u\ dy = d\psi$$

$$=\frac{\partial\psi}{\partial x}dx+\frac{\partial\psi}{\partial y}dy,$$

i.e. $$u=\frac{-\partial\psi}{\partial x}$$

and $$v=\frac{\partial\psi}{\partial y}.$$

The function ψ is called the stream function or the current function.

Stream lines are obtained by integrating (1),

i.e., by $d\psi = 0$

as $v\ dx - u\ dy = d\psi = 0$,

i.e., by ψ = const.

PHYSICAL MEANING OF ψ_2-Ψ_1

Let AB be curve such that the values of the current functions at A and B are $\psi_1$and ψ_2 respectively.

Take a point P on the curve, where tangent makes an angel θ, say, with x-axis and u and v are the velocities parallel to axis of xancly.

Velocity along the inward drawn normal PN = $v\cos\theta - u\sin\theta$.

Therefore flow across the element δs at P from right to left

$$= (v\cos\theta - u\sin\theta)\ \delta s$$

$$=\left(\frac{\partial\psi}{\partial x}.\frac{dx}{dx}+\frac{\partial\psi}{\partial y}.\frac{dy}{ds}\right)ds$$

as $u=\frac{\partial\psi}{\partial y}, v=\frac{\partial\psi}{\partial x}$,

$$= d\psi \cos\theta = \frac{dx}{ds}$$

and $\cos\theta = \frac{dx}{ds}$

Thus total flow across AB from right to left

$$= \int_{\psi_1}^{\psi_2} d\psi = \psi_2 - \psi_1$$

STREAM FUNCTION

The continuity equation establishes the existence of a stream function. The volume flow rate across all curves connecting the fixed reference point O (x_0, y_0) andan arbitrary point P (x, y) must be the same because of mass conservation and is, therefore a function of (x, y) indepently of curves from O to P. The velocity components are expressed as derivatives of the stream function in such a way that the equation of continuity is satisfied automatically. The introduction of stream function is useful in two ways:

(i) The equation of continuity is already taken care of.

(ii) Instead of dealing with two functions u and v it remains only one function named as stream function.

Consider a curve P in-xy plane joining the two points O and A.

Let (u, v) be the velocity components at the point P (x, y) on the curve, and Q ($x + \delta x$, $y + \delta y$) be its neighbouring point such that PQ $= \delta s$. Let the tangent at the point P makes an angle Q with OX.

The velocity component at P along inward drawn normal is

$= (v\cos\theta - u\sin\theta)$.

The volume of the fluid which crosses unit thickness PQ normal to the plane of flow per time becones.

$= (v\cos\theta - u\sin\theta)\ \delta s$.

The flux of volume is constant, let it be equal to $d\psi$, then

$$d\psi = (v\cos\theta - u\sin\theta)\ \delta s = v\ \delta x - u\ \delta y, \qquad \text{...(1)}$$

Where $\cos\theta = \delta x/\delta s$ and $\sin\theta = \delta y/\delta s$.

The total flux of volume flow Q per unit thickness per unit time across any plane curve joining O to A is

$$Q = \int^{A} d\psi = \int_{O}^{A} vdx - udy$$

where ψ_0 is a constant and the line intergal is taken along an arbitrary are joining fixed reference point O to the arbitrary point A. Q is positive when measured in the sense right to left.

Since the flux of volume across any curve joining two point is equal to the difference between the values of ψ at these two point *i.e.*,

$\psi_A - \psi_O = 0 \Rightarrow$ that ψ is constant along the curve. The family of curves ψ = const. are the stream lines in the plane z = 0. The function ψ (x, y) is called the stream function which exists for all two-dimensional flows, both rotational as well as irrotational. When the points O and A are fined the R.H.S of (1) is independent of the shape of the curve P joining them. In any source or sink free region the mass conservation equation is

$$\frac{\partial u}{\partial x} + \frac{\partial v}{\partial y} = 0.$$

This determines the recessary and suffient condition that (vdx – udy) is the exact total differential of some funtion ψ (x, y),then

$$vdx - udy = d\psi = \frac{\partial \psi}{\partial x} dx + \frac{\partial \psi}{\partial y} dy$$

$$\Rightarrow \qquad u = -\frac{\partial \psi}{\partial y}$$

and $v = \dfrac{\partial \psi}{\partial x}$, ...(3, 4)

where the function ψ (x, y) is called the stream function.

We can determine the stream functioin for any velocity field that (3) partially with regard to y both the sides, we have

$$\psi(x, y, t) = -\int u(x, y, t) dy + f(x, t), \qquad ...(5)$$

where f (x, t) is an integration constant. Differentitating (5) partially with respect to x, we have

$$\frac{\partial \psi}{\partial x} = -\frac{\partial}{\partial x}[\int u(x, y; t)\, dy] + \frac{\partial}{\partial x}\delta(x, t)$$

$$\Rightarrow \qquad \frac{\partial}{\partial x} = f(x, t) = \frac{\partial \psi}{\partial x} + \frac{\partial}{\partial x}[\int u(x, y; t)\, dy]$$

$$\Rightarrow \qquad \frac{\partial}{\partial x} = f(x, t) = v(x, y; t) + \frac{\partial}{\partial x}[\int u(x, y; t)\, dy]$$

$$\Rightarrow \qquad \frac{\partial}{\partial x} = f(x, t) = p(x, y; t) \text{ (say)} \qquad ...(6)$$

Integrating (6) partially with regard to x, we have

f (x, t) = fP (x, y; t) dx + Q (t) ...(7)

where Q (t) is an arbitrary function of t.

Again $P(x,y;t) = v(x,y;t) + \frac{\partial}{\partial x}[fu(x;y;t)\,dy]$ ...(8)

Differentitating (8) partially with regard to y, we have

$$\frac{\partial p}{\partial y} = \frac{\partial v}{\partial y} + \frac{\partial}{\partial y}\left[\frac{\partial}{\partial x} fu(x,y;t)\,dy\right]$$

$$\frac{\partial p}{\partial y} = \frac{\partial v}{\partial y} + \frac{\partial}{\partial x}\left[\frac{\partial}{\partial y} fu(x,y;t)\,dy\right]$$

$\Rightarrow \frac{\partial p}{\partial y} = \frac{\partial v}{\partial y} + \frac{\partial u}{\partial x} = 0 \Rightarrow p$ is a function of x and t only.

From the relations (5) and (7), we have

ψ (x, y; t) = – fu (x, y, t) dy + fP (x, y; t) dx + Q (t).

It follows that the stream function ψ can be determined upto an arbitrary functions of it. In an irrotational motion, the velocity potential ϕ exists such that

$$u = \frac{\partial \phi}{\partial x}$$

and $v = -\frac{\partial \phi}{\partial y}$. ...(9)

From the relation (3, 4)

and $\frac{\partial \phi}{\partial y} = -\frac{\partial \phi}{\partial x}$ ...(10, 11)

This constiute the Cauchy-Riemann equations and the velocity potential ϕ and the stream function ϕ are the real and imaginary parts of complex functions.

Differentiating (10) and (11) partially with regard to x and y respectively, we have

$$\frac{\partial^2 \phi}{\partial x^2} = \frac{\partial^2 \psi}{\partial x \partial y}$$

and $\frac{\partial^2 \phi}{\partial y^2} = -\frac{\partial^2 \psi}{\partial x \partial y}$

$$\Rightarrow \frac{\partial^2 \phi}{\partial x^2} + \frac{\partial^2 \phi}{\partial y} = 0. \qquad ...(12)$$

Similarly, differentiating (10) and (11) partially with regard to y and x respectively, we have

$$\frac{\partial^2 \phi}{\partial y \partial x} = \frac{\partial^2 \phi}{\partial y^2}$$

and $$-\frac{\partial^2 \phi}{\partial y \partial x} = \frac{\partial^2 \psi}{\partial x^2}$$

$$\Rightarrow \frac{\partial^2 \psi}{\partial x^2} + \frac{\partial^2 \psi}{\partial y^2} = 0. \qquad ...(13)$$

It follows that the velocity potential ϕ and the stream function ψ are the Harmonic functions and satisfy the laplace equation.

Again from (10) and (11), we obtian

$$\frac{\partial \phi}{\partial x}\frac{\partial \psi}{\partial x} + \frac{\partial \phi}{\partial y}\frac{\partial \psi}{\partial x} = 0. \qquad ...(14)$$

The relation (14) represents that the two system of curves of constant velocity potential and the stream function cut each other orthogonally. The curves ϕ (x, y) = const are the curves of equal velocity potential and the curves ψ (x, y) = const are the stream lines.

Let Y be the stream function at the pont P (r, θ) and (ψ + dψ) is the stream function at the neighbouring point θ (ψ δr, θ + $\delta\theta$). Let q/r and q_θ be the radial and transverse component of velocity.

$q/r = -(\partial/\partial r)$,

$q_\theta = -(\partial \phi / r\partial\theta)$.

The flux of volume flow across any curve PQ is $\delta\phi$. The flux, out of the polar triangle PNQ, across NQ is (δr qθ) and across NP is ($-$ r$\delta\theta$ qr). Therefore, for all r and θ in the absence of sources or sinks, we have

$$\partial \psi = \frac{\partial \psi}{\partial r}\delta r + \frac{\partial \psi}{\partial \theta}\delta\theta$$

$$\Rightarrow \frac{\partial \psi}{\partial r}\delta r + \frac{\partial \psi}{\partial \theta}\delta\theta = q_\theta\, \delta r - r q_r\, \delta\theta,$$

which gives $q_\theta = \dfrac{\partial \psi}{\partial r}$

and $\qquad \delta r = -\frac{1}{r}\frac{\partial \psi}{\partial \theta}.$

Let w = f (2),

where w = ϕ + i ψ , z = reθi

⇒ ϕ + i ψ = f (reθi)

Differentiating partially with regard to r and θ respectively, we have

$$\frac{\partial \phi}{\partial r} + i\frac{\partial \psi}{\partial r} = e\theta if'(re\theta i)$$

$$\frac{\partial \phi}{\partial \theta} + i\frac{\partial \psi}{\partial \theta} = rie\theta if'(re\theta i),$$

$$\Rightarrow ri\left(\frac{\partial \phi}{\partial r} + i\frac{\partial \psi}{\partial r}\right) = \frac{\partial \phi}{\partial \theta} + i\frac{\partial \psi}{\partial \theta}$$

Equating real and imaginary parts, we have

$$\frac{r\partial \phi}{\partial r} = \frac{\partial \psi}{\partial \theta},$$

$$-r\frac{\partial \psi}{\partial r} = \frac{\partial \phi}{\partial \theta}$$

$$\Rightarrow \frac{\partial \phi}{\partial r} = \frac{1}{r}\frac{\partial \psi}{\partial \theta},$$

$$\frac{\partial \psi}{\partial r} = -r\frac{\partial \psi}{\partial \theta}$$

$$\Rightarrow \frac{\partial \phi}{\partial r} = \frac{1}{r}\frac{\partial \psi}{\partial \theta}$$

and $\frac{\partial \psi}{\partial r} = -\frac{1}{r}\frac{\partial \phi}{\partial \theta}$

PHYSICAL INTERPRETATION OF STREAM FUNCTION

The flow across an element ds of OAP from left to right is P (udy – vdx) per unit time at any instant. Total net flow acrass. OAP from left to right is

$P\int_{O}^{P}(u - dy - vdx),$ per unit time at the same instant.

Similarly, total new flow across OBP from left to right is

$$\int_O^P (udy - vdx).$$

Since the paths A and B are arbitary except that they are from O to P, so at any instant for arbitrary paths between O and P mass flow across OAP must push the same amount of mass flow across OBP

$$\int_O^P (udy - vdx) = \int_O^P (udy - vdx)$$

Therefore, the line integral on R.H.S. depends only on the end points, not on the path. Let (x_0, y_0) be the coordinate of the fined reference point O, then

$$\int_{O\ (x_0, y_0)}^{P\ (x, y)} (udy - vdx) = \text{a function of } (x, y; t)$$

$$= \psi\ (x, y; t).$$

This shows that the stream function ψ (x, y; t) is the rate of volume flow left point right past any curve joining the arbitrary point (x, y). The difference of the stream function between two points (x_1, y_1) and (x_2, y_2) at any instant t is given by.

$$\psi\ (x_2, y_2; t) - \psi\ (x_1, y_1; t)$$

$$= \int_O^{(x^2, y^2)} (udy - vdx) - = \int_O^{(x, y)} (udy - vdx)$$

$$= \int_{(x_1, y_1)}^{(x^2, y^2)} (udy - vdx)$$

which represents the volumeflow from feft to right across any curve joining (x_1, y_1) and (x_2, y_2).

When the region between two paths is wholly occupiped by the incompreseible flow then the flux of volume across the closed curve farmed from any two different paths joining O and P is necessarily zero.

COMPLEX POTENTIAL

The conditions (3) are satisfied if $\phi + i\psi$ is taken as a function of z, *i.e.* x + iy.

For if $\phi + i\psi = f\ (x + iy)$,

$$\frac{\partial \phi}{\partial x} + i\frac{\partial \psi}{\partial x} = f'(x + iy), \text{ diff. partially w. r. t. y}$$

and $$\frac{\partial \phi}{\partial y} + i\frac{\partial \psi}{\partial y} = if'(x + iy), \text{ diff. partially w. r. t. y}$$

$$= i\left(\frac{\partial \phi}{\partial x} + i\frac{\partial \psi}{\partial x}\right).$$

Equating real and imoginary parts, we get

$$\frac{\partial \phi}{\partial x} = \frac{\partial \psi}{\partial y} \text{ and } \frac{\partial \phi}{\partial y} = -\frac{\partial \psi t}{\partial x} \qquad ...(1)$$

Which are the sameas given in (3) of the above article thus if we take w = ϕ + iψ, where w is a function of z = x + iy, then the function w is called the complex potintial.

By virtue of condifions (1), w is an analytic function of z = x = iy.

COMPLEX POTENTIAL AND COMPLEX VELOCITY

Since the function ϕ (x, y) and y (x, y) constitute the Cauchy-Riemann equations so they provide the hecessary and sufficient condition for the function

$$F(2) = \phi (x, y) + i\psi (x, y) \qquad ...(1)$$

to be an analytic function of the complex variable z = (x + iy).

Differentitating (1) partially with regard to x and y, we have

$$\frac{\partial \phi}{\partial x} + i\frac{\partial \psi}{\partial x} = F'(x+iy), \frac{\partial \phi}{\partial y} + i\frac{\partial \psi}{\partial y} = iF'(x+iy)$$

$$\Rightarrow \frac{\partial \phi}{\partial x} + i\frac{\partial \psi}{\partial x} = -i\left(\frac{\partial \phi}{\partial y} + i\frac{\partial \psi}{\partial y}\right)$$

$$\Rightarrow \frac{\partial \phi}{\partial x} = \frac{\partial \psi}{\partial y}, \quad \frac{\partial \phi}{\partial y} = -\frac{\partial \psi}{\partial x},$$

which represent the cauchy-riemann equations.

Thus, the real and imaginary parts of any analytic function may be regarded as the potentias function and stream function of a possible irrotational flow an inviscid, fluid in two dimensions. The complex function F, whose real and imaginary parts are the velocity polential and the stream function respectively, is termed the complex function respectively, is termed the complex potential of the liquid motion. It complex potential of the liquid by sources, sinks as vortices. Differentiating (1) Partially with regard tox, we have

$$\frac{dF}{dz}, \frac{\partial z}{\partial x} = \frac{\partial \phi}{\partial x} + i\frac{\partial \psi}{\partial x};$$

$$\frac{dF}{dz}=\frac{\partial F}{\partial z} \text{ as F is a function of z only.}$$

$$\text{or } \frac{dF}{dz}=\frac{\partial \phi}{\partial x}-i\frac{\partial \phi}{\partial y}=u+iv; \qquad \text{...(2)}$$

$$\begin{cases} z=x+iy \\ \partial z/\partial x=1 \\ \partial \psi/\partial x=-\partial \phi/\partial y \end{cases}$$

which is called the complex velocity. Let q be the magnitude of the velocity at any point then

$$q=\left|\frac{dF}{dz}\right|=\left[\left(\frac{\partial \phi}{\partial x}\right)^2+\left(\frac{\partial \phi}{\partial y}\right)^2,\right]^{1/2}=(u^2+v^2)^{1/2}$$

$$\text{or } q^2=(u+iv)(u-iv)=\frac{dF}{dz}.\frac{d\bar{F}}{dz} \qquad \text{...(3)}$$

where $\bar{F}$ is the complex conjugate. The point where velocity is zero are called stangnation points.

Thus for stangnation point (dF/dz) = 0.

UNIFORM FLOWS

(I) Let the complex potential is proportional to z with an imaginary constant of proportionality, then

w = iaz ⇒ u + iv = iaz = –1

where A is a real and positive constant.

From (1), we have

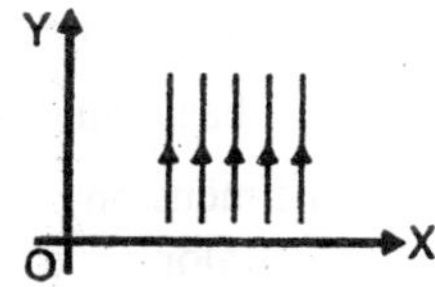

Fig. 1.18

$$\frac{dw}{dz}=-u+iv=iA$$

⇒ u = 0, v = A, which is a uniform vertical flow.

Thus the complex potential for such a flow whose magnitude of the stream is v in the positive Y-direction becomes

$$w = ivz.$$

(II) Let the complex potential is proportional to z, such that

$$w = -Ae^{-i\alpha}z, \qquad ...(1)$$

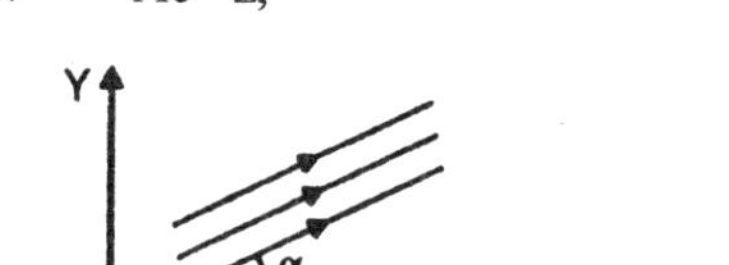

Fig. 1.19

where A and α are real constants.

From (1), we have

$$\frac{dw}{dz} = -u + iv = Ae^{-i\alpha}z,$$

$$-u + iv = -A(\cos a - i \sin \alpha)$$

$$\Rightarrow \qquad u = A \cos \alpha, \text{ and } v = A \sin \alpha,$$

which corresponds to a uniform flow inclined at an angle α to the X-axis. Thus, the complex potential for such a flow whose magnitude is v reduces to

$$w = -ve^{-i\alpha}z.$$

In particular, at $\alpha = \pi/2$, the complex potential represents a uniform vertical flow.

TWO DIMENSIONAL SOURCE AND SINK

Now, we shall introduce the basic singularities of the fluid flow.

Source : Any point in a two dimensional field where fluid is assumed to flow uniformly in all direction along radial paths is called a simple source. It consists of outward radial flow from a point. The lines of flow will be straight radial lines *i.e.*, $q_\theta = 0$. A source is a point at which the fluid is continuously created. If the total flux outwards across a small closed surface surrounding the point be (+m) then (+m) is called the *strength of the source*. Infact this is a purely abstract conception which does not occur in nature.

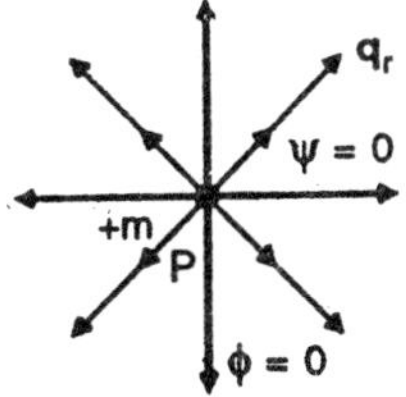

Fig. 1.20

Sink : A source of negative strength (-m) or inward radial flow is called a sink. This is a point at which the fluid continuously annihilates. Whirlpool is an example of a sink.

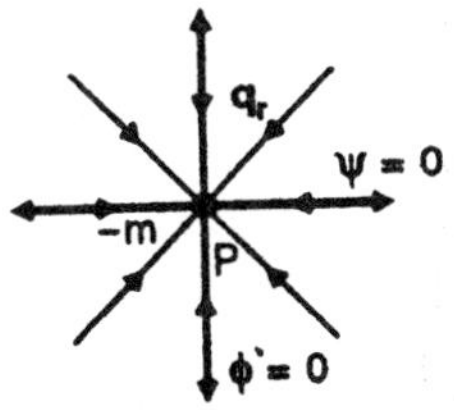

Fig. 1.21

The source and sink are the points at which the velocity potential and stream function becomes infinite.

STRENGTH

Strength of a source is defined as total volume of flow per unit time from it. If $2\pi m$ is the total volume of flow across any small circle surrounding the source then it is called the *strength of the source.*

COMPLEX POTENTIAL OF A SOURCE

Complex a source of a strength +m at an origin O. Let q_r be the radial velocity at a distance r from the source. Since the flow is radial and symmetric, the flux across a circle of radius r is $2\pi r\rho q_r$. Also a source of strength +m is such that the flow across any small curve surrounding it is $2\pi\rho m$. From conservation of mass in a steady incompressible flow, we have

$$2\pi r\rho q_r = 2\pi\rho m$$

or $$q_r = \frac{m}{r} = -\frac{\partial\phi}{\partial r} = -\frac{1}{r}\frac{\partial\psi}{\partial\theta} \qquad ...(1)$$

or $$q_\theta = 0 = -\frac{1}{r}\frac{\partial\phi}{\partial\theta} = \frac{\partial\psi}{\partial r} \qquad ...(2)$$

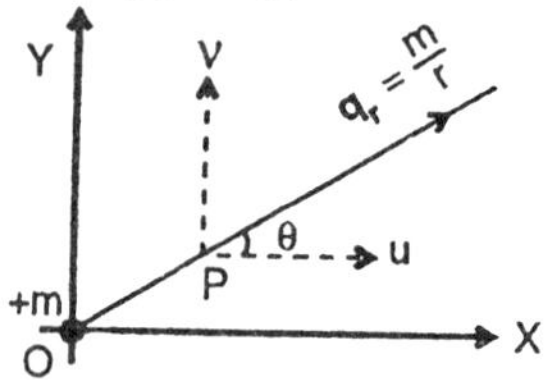

Fig. 1.22

From the equation (1), we have

$$\theta = -m \log r,$$

and $$y = -mq. \qquad ...(3)$$

Let u and v be the velocity components at the point P, the

$$u = \frac{m}{r}\cos\theta$$

and $$v = \frac{m}{r}\sin\theta, \qquad ...(4)$$

The complex potential w of the flow due to a source of strength m at the origin is given by

$$w = \phi + iy'$$

or $$\frac{\partial w}{\partial x} = \frac{\partial\phi}{\partial x} + i\frac{\partial\psi}{\partial x},\ z = x + iy, \frac{\partial z}{\partial x} = 1$$

or $$\frac{\partial w}{\partial z} = \frac{\partial\phi}{\partial x} + i\frac{\partial\phi}{\partial y} = -u + iv, \qquad ...(5)$$

From (3) and (4), we have

$$\frac{dw}{dz} = -\frac{m}{r}(\cos\theta - i\sin\theta) = -\frac{m}{re^{\theta i}} = -\frac{m}{z}$$

By integrating, we get

$$w = -m \log z. \qquad ...(6)$$

which gives the *complex potential due to a source at origin* and is regular in a domain excluding the origin. The velocity potential ϕ and the stream function ψ exist every where except at $z = 0$. The singularity at $z = 0$ is due to the source there.

Similarly, the complex potential of a source of equal strength +m situated at a point $z = z_1$, becomes

$$w = -m \log (z - z_1). \qquad ...(7)$$

Therefore, the complex potential of the source of strength m_1, m_2, m_3, ... situated at the points z_1, z_2, z_3, ... are given by

$$w = -m_1 \log (z - z_1) - m_2 \log (z - z_2) - m_3 \log (z - z_3) \quad ...(8)$$

SOLVED EXAMPLES

Example 1: *Find the stream function $\psi(x, y; t)$ for the given velocity fiel $u = Ut$, $v = x$.*

Solution: We know that

$$u = -\frac{\partial \psi}{\partial y},$$

$$u = -\frac{\partial \psi}{\partial y},$$

Hence u = Ut,

$$v = x \Rightarrow \frac{\partial \psi}{\partial y} = -u = -Ut.$$

By integrating, we have

$$\psi = -Uyt + f(x, t), \qquad ...(1)$$

where f (x, t) is an integration constant, independent of y.

Also $\dfrac{\partial \psi}{\partial x} = v = x$ ...(2)

From (1), we have $\dfrac{\partial \psi}{\partial x} = \dfrac{\partial f}{\partial x}$ ...(3)

From (2) and (3), we have

$$\frac{\partial f}{\partial x} = x \Rightarrow f = \frac{1}{2}x^2 + g(t),$$

where g(t) is an integration constant, independent of x.

From (1) adn (4), we have

$$\psi(x, y, t) = -Uyt + 1/2x^2 + g(t)$$ **Proved**

Example 2: *Show that the velocity vector q is everywhere tangent to lines in the-XY plane along which $\psi(x, y) = const.$*

Solution: We know that

$$d\psi = \frac{\partial \psi}{\partial x} dx + \frac{\partial \psi}{\partial y} dy \quad ...(1)$$

Since ψ (x, y) = const. ⇒ dψ = 0 then (1) becomes

$$\frac{\partial \psi}{\partial x} dx + \frac{\partial \psi}{\partial y} dy = 0. \quad ...(2)$$

From the definition of stream function, we have

$$u = -\frac{\partial \psi}{\partial y},$$

$$v = \frac{\partial \psi}{\partial x}$$

or $v\,dx - u\,dy = 0 \Rightarrow \frac{dx}{u} = \frac{dy}{v}.$

It follows that the velocity vector q is tangent to the lines ψ (x, y) = const. **Proved**

Example 3: *A velocity field is given by q = -xi + (y + t) j. Find the stream function and the stream lines for this field at t = 2.*

Solution: Here q = ui + vj = -xi + (y + t) j

⇒ u = -x, v = y = t.

We know that

$$u = -\frac{\partial \psi}{\partial y} = -x \text{ and } v = \frac{\partial \psi}{\partial x} = y + t \quad ...(1, 2)$$

By integrating (1) which regard to y, we have

$$\psi = xy + f(x, t) \quad ...(3)$$

where f(x, t) is an integration constant.

$$\text{or} \quad \frac{\partial \psi}{\partial x} = y + \frac{\partial f}{\partial x} \quad ...(4)$$

From (2) and (4), we have

$$y + \frac{\partial f}{\partial x} = y + t$$

$$\Rightarrow \quad f(x, t) = xt + g(t) \quad ...(5)$$

From (3) and (5), we have

$$\psi = xy + xt + g(t)$$

At $t = 2, \psi = x\ (y + 2) + g(2)$.

The streamlines are given by ψ = const., therefore

$$x\ (y - 2) = \text{const.},$$

which represent rectangular hyperbolas. **Ans.**

Example 4: *Show that $u = 2Axy$, $v = A(a^2 + x^2 - y^2)$ are the velocity components of a possible fluid motion. Determine the stream function.*

Solution: Here $u = 2Axy$, $v = A\ (a^2 + x^2 - y^2)$

This will be a possible fluid motion if it satisfies the equation of contnuity *i.e.*,

$$\frac{\partial u}{\partial x} + \frac{\partial v}{\partial y} = 0 \Rightarrow \ 2Ay - 2Ay = 0.$$

which is true. Therefore, the given velocity components constitute a possible fluid motion.

We know that $u = -\dfrac{\partial \psi}{\partial y}$ and $v = \dfrac{\partial \psi}{\partial x}$

So $$\frac{\partial \psi}{\partial y} = -\ 2Axy,$$

and $$\frac{\partial \psi}{\partial x} = A(a^2 + x^2 - y^2) \qquad ...(1)$$

By integrating, we have

$$\psi = -\ Axy^2 + f(x, t). \qquad ...(2)$$

Differentiating (2), we hvae

$$\frac{\partial \psi}{\partial x} = -\ Ay^2 + \frac{\partial f}{\partial x} \qquad ...(3)$$

From (1) and (3), we have

$$-Ay^2 + \frac{\partial f}{\partial x} = A(a^2 + x^2 - y^2)$$

$$\Rightarrow \frac{\partial f}{\partial x} = A(a^2 + x^2)$$

By integrating, we have

$$f(x,t) = A\left(a^2x + \frac{1}{3}x^3\right) + g(t)$$

Substituting the value of f (x, t) in (2), we have

$$\psi = A\left(a^2x - xy^2\ \frac{1}{3}x^3\right) + g(t)$$

which is the required stream fucntion.

Example 5: *If $\phi = A(x^2 - y^2)$ represents a possible flow phenomenon, determine the stream function.*

Solution: Here $\phi = A(x^2 - y^2)$

Since $\dfrac{\partial\phi}{\partial x} = \dfrac{\partial\psi}{\partial y} \Rightarrow \dfrac{\partial\psi}{\partial y} = 2Ax$

or $\psi = Axy + C,$

where C is an integration constant, which is the required stream fnction.

Example 6: *The velocity potentials $\phi_1 = x^2 - y^2$ and $\phi_2 = \sqrt{r}\cos(\theta/2)$ are solutions of the Laplace equation. Prove that the velocity potential $\phi_3 = (x^2 - y^2) + \sqrt{r}\cos(\theta/2)$ satisfies $\nabla^2\phi_3 = 0$.*

Solution: Here $\phi_1 = x^2 - y^2$ and $\phi_2 = \sqrt{r}\cos(\theta/2)$

The laplace's equation in cartesian coordinates and cylndrical polar coordinates are given as

$$\nabla^2\phi_1 = \frac{\partial^2\phi_1}{\partial x^2} + \frac{\partial^2\phi_1}{\partial y^2} = 2 - 2 = 0$$

and $$\nabla^2\phi_2 = \frac{\partial^2\phi_2}{\partial r^2} + \frac{1}{r^2}\frac{\partial^2\phi_2}{\partial\theta^2} = \frac{1}{r}\frac{\partial\phi_2}{\partial r}$$

or $$\nabla^2\phi_2 = -\frac{1}{4r^{3/2}}\cos(\theta/2) - \frac{1}{4r^{3/2}}\cos(\theta/2) + \frac{1}{2r^{3/2}}\cos(\theta/2) = 0$$

$\Rightarrow$ that ϕ_1 and ϕ_2 satisfy Laplace's equation.

Again $\phi_3 = (x^2 - y^2)\sqrt{r}\cos\theta/2$

or $\phi_3 = r^2\cos 2\theta + \sqrt{r}\cos\theta/2$

Since $$\nabla^2\phi_3 = \frac{\partial^2 f_3}{\partial r^2} + \frac{1}{r^2}\frac{\partial^2\phi_3}{\partial\theta^2} + \frac{1}{r}\frac{\partial\phi_3}{\partial r}$$

or $$\nabla^{2r}\phi_3 = \left(2\cos 2\theta - \frac{1}{4r^{3/2}}\cos\frac{\theta}{2}\right) - \frac{1}{r^2}$$

$$\left(4r^2\cos 2\theta + \frac{1}{4r^{3/2}}\cos\frac{\theta}{2}\right) + \frac{1}{r}\left(2r\cos 2\theta\frac{1}{2r^{1/2}}\cos\frac{\theta}{2}\right) = 0$$

$\Rightarrow$ that θ_3 also satisfies Laplace's equation. **Proved.**

Example 7: *Prove that if the speed is every where the same, the stream lines are straight.*

Solution: The equation to the streamlines are given by

$$\frac{dx}{u} = \frac{dy}{v} = \frac{dz}{w}$$

or $v\,dx - u\,dy = 0,$

$v\,dz - w\,dy = 0,$

$u\,dz - w\,dx = 0.$

By integrating, we have

$vx - uy = \text{const.},$

$vz - wy = \text{const.},$

$uz - wx = \text{const.}$

This shows that the intersection of these planes represent the straight lines.

Example 8: *Find the stream function ψ for the given velocity potential $\phi = cx$, where c is constant. Also, draw as set of streamlines and equipotential lines.*

Solution: The velocity potential $\phi = cx$ represents fluid flow because it satisfies Laplace equation $\nabla^2\phi = 0$.

Since $-\frac{\partial \phi}{\partial x} = -c = u$ and $u = -\frac{\partial \psi}{\partial y}$

Therefore $\frac{\partial f}{\partial y} = c \Rightarrow \psi = cy + f(x)$ ...(1)

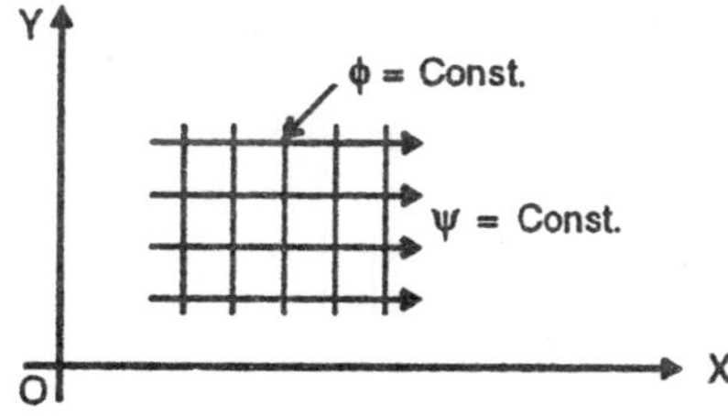

Fig. 1.23

Differentiating with regard to x, we have

$$\frac{\partial \psi}{\partial x} = f'(x)$$

But $$\frac{\partial \psi}{\partial x} = v \Rightarrow -\frac{\partial \phi}{\partial y} = v = 0.$$

Therefore f'(x) = v = 0, hence f(x) = const. ...(2)

The stream function ψ is given as

$$\psi = \text{const.} + cy.$$

which represents parallel flow in which stream lines are parallel to X axis.

The corresponding stream lines and equipotential lines are represented as follows:

EXERCISES

1. Determine the streamlines, pathlines and streaklines for the flow whose velocity field is given by

   $u = -x + t + 2,\ v = y - y + 2.$

2. Does the one-dimensional incompressssible flow given by

   $u\ (v) = Ay^2 + by + C,\ v = 0 = w,$

   where A, B, C are arbitrary constants, satisfy the equation of continuity?

3. Does the velocity distribution q = 5xi + 5yj – 10zk satisfy the law of conservation of mass for incompressible flow?

   Hint: $\nabla.q = 0 \Rightarrow \nabla.\ (5xi + 5yj - 10zk) = 0$

   $\Rightarrow \quad 5 + 5 - 10 = 0$

   $\Rightarrow$ that the given velocity distribution satisfies the law of mass continuity.

4. Does the two-dimensional incompressible flow given by

   $$u = (x, y) = A\left[\frac{x^2 + y^2 - B^2}{(x^2 + y^2 - B^2) + (4B^2y^2)}\right],$$

   $$v(x, y) = 2A\left[\frac{xy}{(x^2 + y^2 - B^2) + (4B^2y^2)}\right], w = 0,$$

   where A and B are arbitrary constant, satisfy the continuity equation.

5. Does the three-dimensional incompressible flow given by

$$u(x,y,z) = \frac{Ax}{(x^2+y^2+z^2)^{3/2}},$$

$$v = (x,y,z) = \frac{Ay}{(x^2+y^2+z^2)^{3/2}}$$

$$w(x,y,z) = \frac{Az}{(x^2+y^2+z^2)^{3/2}},$$

satisfy the equation of continuity ? Where A is an arbitrary constant

6. Does the two-dimensional incompressible flow satisfies the continuity equation?

$$q_r(r,\theta) = A\left(\frac{1}{r^2}-1\right)\cos\theta,$$

$$q_\theta(r,\theta) = A\left(\frac{1}{r^2}+1\right)\sin\theta, r>0,$$

where A is an arbitrary non-zero constant.

Hint: The equation of continuity in spherical polarcoodinates is given by

$$\frac{1}{r^2}\frac{\partial}{\partial r}(\rho r^2 q_r) + \frac{1}{r\sin\theta}\frac{\partial}{\partial\theta}(\rho q_\theta \sin\theta) = 0.$$

7. Whether the one-dimensional incompressible flow given by

$q_r = A/r^2$, $q_\theta = 0 = q_\phi$,

where A is an arbitrary constant and $r > 0$ satisfy the equation of continuity ?

8. Whether the three-dimensional incompressible flow given by

$$q_r = \frac{(3r^2-2rz^2)}{4\pi(z^2+r^2)7/2}\cos 2\theta, q_\theta = \frac{r\sin 2\theta}{2\pi(z^2+r^2)5/2},$$

$$q_z = \frac{5zr^2\cos 2\theta}{4\pi(z^2+r^2)7/2},$$

satisfy the equation of continuity ?

9. Determine the constants l, m and n in order that the velocity q = {(x + lr) i + (y + mr) j + (z + nr) k}/{r (x + r)}, where r = √(x² + y² + z²) may satisfy the equation of continuity for a liquid.

   Hint: The equation of continuity is

   $$\partial u/\partial x + \partial v/\partial y + \partial w/\partial z = 0.$$

   $$\frac{\partial u}{\partial x} = \frac{\partial}{\partial x}\left\{\frac{x+lr}{r(x+r)}\right\} = \frac{1+l(x/r)}{r(x+r)} + (x+lr)$$

   $$\times\left\{-\frac{x}{r^3(x+r)} - \frac{1}{r^2(x+r)}\right\}.$$

   $$\frac{\partial w}{\partial z} = \frac{\partial}{\partial y}\left\{\frac{y+mr}{r(x+r)}\right\} = \frac{1+m(y/r)}{r(x+r)} + (y+mr)$$

   $$\times\left\{-\frac{y}{r^3(x+r)} - \frac{y}{r^2(x+r)^2}\right\}$$

   $$\frac{\partial w}{\partial z} = \frac{\partial}{\partial z}\left\{\frac{z+nr}{r(x+r)}\right\} = \frac{1+n(z/r)}{r(x+r)} + (z+nr)$$

   $$\times\left\{-\frac{z}{r^3(x+r)} - \frac{z}{r^2(x+r)^2}\right\}$$

   Hence

   $$r^3(x+r)^2.\left(\frac{\partial u}{\partial x} + \frac{\partial v}{\partial y} + \frac{\partial w}{\partial z}\right) = r^2\{r(1-1) + x(1-1) - my - nz\}.$$

   This expression will be identically zero iff l = 1, m = 0 = n.

10. Show that the equaiton of continuity reduces to Laplace's equation when the liquid is incompressible and the motion is irrotarional.

2

Velocity Potential and Motion of Equation

EQUATION OF CONTINUITY IN CYLINDRICAL CO-ORDINATES

Let P be a point whose cylindrical co-ordinates are (r, θ, z). With P as end construct a parallelopiped with edges

$PQ = \delta r$, $PR = r\ \delta\theta$

and $PS = \delta z$.

Also let u, v, w be the velocity components along PQ, PR and PS.

The mass of the liquid that passes in along PQ, *i.e.* through the face PRS′ S

= rr dq.dz u per unit time

= f (r, q, z) say,

and the mass of the liquid that passes out along $\pi\theta$. i.e, through the face QRP′ O′

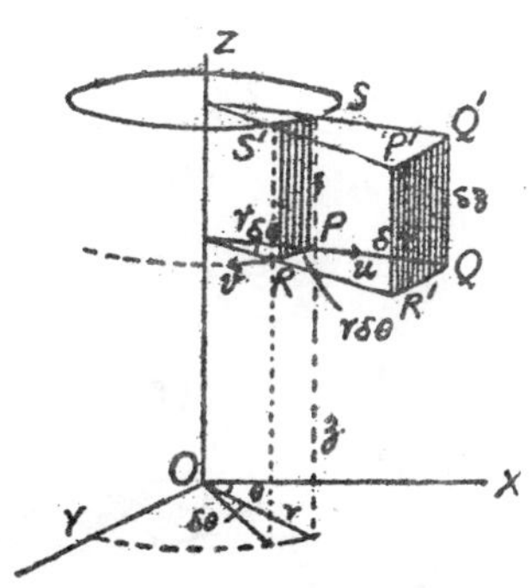

Fig. 2.1

$= f(r + \delta r, \theta, z)$

$= f(r, \theta, z) + \delta r \dfrac{\partial f(r,\theta,z)}{\delta r}$ approximately.

$\therefore$ excess of flow-in over flow- out along PQ

$= -\delta r. \dfrac{\partial f(r,\theta,z)}{\partial r}$ approximately

$= -\delta r. \dfrac{\partial}{\partial r}(\rho r\, \delta\theta\, \delta z\, u) = -\delta r\, \delta q\, \delta z \dfrac{\partial}{\partial r}(\rho r u).$

Similarly excess of flow-in over flow-out along PR

$= -r\, \delta\theta. \dfrac{\partial}{r\partial\theta}(\rho\, \delta r\, \delta z\, v) = -\delta r\, \delta\theta\, \delta z \dfrac{\partial}{\partial\theta}(\rho v)$

and excess of flow-in over flow-out along PS

$= -r\, \delta z. \dfrac{\partial}{\partial z}(\rho\, \delta r.\, r\, \delta\theta\, w) = -\delta r\, \delta\theta\, \delta z\, r \dfrac{\partial(\rho w)}{\partial z}.$

Also total mass of the fluid in the parallelopiped

so that change in the mass of the fluid in the parallelopiped

$= \dfrac{\partial}{\partial t}(\rho\, \delta r.\, r\, \delta\theta\, \delta\mu z) \dfrac{\partial\rho}{\partial t} r\, \delta r\, \delta\theta\, \delta z$ per unit time.

$\therefore$ the equation of continuity is

$$-dr\, dq\, dz \left[\frac{\partial}{\partial r}(\rho r u) + \frac{\partial}{\partial\theta}(\rho v) + r\frac{\partial}{\partial z}(\rho w)\right] = \frac{\partial\rho}{\partial t} r\, \delta r.\delta\theta\delta :$$

$$\frac{\partial\rho}{\partial t} + \frac{1}{r}\frac{\partial}{\partial r}(\rho r u) + \frac{1}{r}\frac{\partial}{\partial\theta}(\rho w) = 0.$$

ORTHOGONAL CURVILINEAR CO-ORDINATES

Through each point P (x, y, z) in cartesian frame, there pass three surfaces, one of each of the families :

$$\lambda_1 = f_1(x, y, z) = \text{constant}$$
$$\lambda_2 = f_2(x, y, z) = \text{constant} \qquad ...(1)$$
$$\lambda_3 = f_3(x, y, z) = \text{constant.}$$

Moreover if these surfaces intersect orthogonally, and $\lambda_1, \lambda_2, \lambda_3$ are continuous functions of space co-ordinates, equations (1) maybe solved as

$x = \phi_1(\lambda_1, \lambda_2, \lambda_3)$

$$y = \phi_2(\lambda_1, \lambda_2, \lambda_3) \qquad ...(2)$$

$$z = \phi_3(\lambda_1, \lambda_2, \lambda_3)$$

Now in cartesian co-ordinates, line element ds^2 is given by

$$ds^2 = dx^2 + dy^2 + dz^2. \qquad ...(3)$$

From (2)

$$dx = \frac{\partial x}{\partial \lambda_1} d\lambda_1 + \frac{\partial x}{\partial \lambda_2} d\lambda_2 + \frac{\partial x}{\partial \lambda_3} d\lambda_3,$$

$$dy = \frac{\partial y}{\partial \lambda_1} d\lambda_1 + \frac{\partial y}{\partial \lambda_2} d\lambda_2 + \frac{\partial y}{\partial \lambda_3} d\lambda_3$$

$$dz = \frac{\partial z}{\partial \lambda_1} d\lambda_1 + \frac{\partial z}{\partial \lambda_2} d\lambda_2 + \frac{\partial z}{\partial \lambda_3} d\lambda_3.$$

Using the fact that the three surfaces (1) intersect orthogonally, (3) maybe written as

$$ds^2 = h_1^2\, d\lambda_1^2 + h_2^2\, d\lambda_2^2 + h_3^2\, d\lambda_3^2$$

where h_1, h_2, h_3, called Lame's Parameters, are given by

$$h_1 = \sqrt{\left\{\left(\frac{\partial x}{\partial \lambda_1}\right)^2 + \left(\frac{\partial y}{\partial \lambda_1}\right)^2 + \left(\frac{\partial z}{\partial \lambda_1}\right)^2\right\}}$$

$$h_2 = \sqrt{\left\{\left(\frac{\partial x}{\partial \lambda_2}\right)^2 + \left(\frac{\partial y}{\partial \lambda_2}\right)^2 + \left(\frac{\partial z}{\partial \lambda_2}\right)^2\right\}}$$

$$h_3 = \sqrt{\left\{\left(\frac{\partial x}{\partial \lambda_3}\right)^2 + \left(\frac{\partial y}{\partial \lambda_3}\right)^2 + \left(\frac{\partial z}{\partial \lambda_3}\right)^2\right\}}$$

$(\lambda_1, \lambda_2, \lambda_3)$ are called the curvilinear co-ordinates of the point P (x, y, z).

EQUATION OF CONTUNUITY IN ORTHOGONAL CURVILINEAR COORDINATES

Let P $(\lambda_1, \lambda_2, \lambda_3)$ be a point in the fluid whose curvilinear co-ordinates are $(\lambda_1, \lambda_2, \lambda_3)$. At P construct a curvilinear parallelopiped with edges PQ = δs_1, PR = δs_2, PS = δs_3. Let q_1, q_2, q_3 be the components of velocity in the increasing direction of λ_1, λ_2, λ_3 respectively.

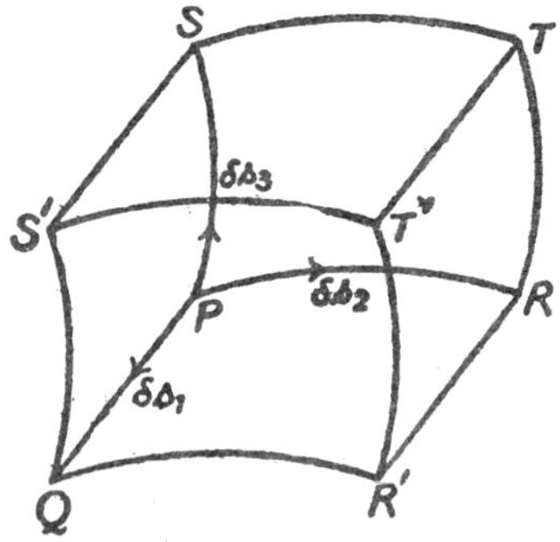

Fig. 2.2

The mass of liquid that flows in per unit time along PQ *i.e.* through the face PRTS = $\rho q_1 \; \delta s_2 \; \delta s_3$.

The mass of liquid that passes out along pq i.e. through the face QR′T′S′

$$= -\rho q_1 \; \delta s_2 \; \delta s_3 + \frac{\partial}{\partial s_1} (\rho q_1 \; \delta s_2 \; \delta s_3) \; \delta s_1.$$

$\therefore$ excess of rate of flow-in over flow-out along PQ

$$= -\frac{\partial}{\partial s_1} (\rho q_1 \; \delta s_2 \; \delta s_3) \; \delta s_1.$$

$\therefore$ rate of accumulation of liquid from the 3 pairs of opposite face

$$= -\sum \frac{\partial}{\partial s_1} (\rho q_1 \; \delta s_2 \; \delta s_3) \; \delta s_1.$$

This must equal the rate of increase of mass of fluid inside the parallelopiped.

Now total mass inside theparallelopiped

$$= \rho \; \delta s_1 \; \delta s_2 \; \delta s_3.$$

$$\therefore \; \frac{\partial}{\partial t} (\rho \; \delta s_1 \; \delta s_2 \; \delta s_3) = -\frac{\partial}{\partial s_1} (\rho q_1 \; \delta s_2 \; \delta s_3) \; \delta s_1$$

$$- \frac{\partial}{\partial s_2} (\rho q_2 \; \delta s_1 \; \delta s_3) \; \delta s_2 - \frac{\partial}{\partial s_3} (\rho q_3 \; \delta s_2 \; \delta s_1) \; \delta s_3 \; .$$

Putting $\delta s_1 = h_1 \; d\lambda_1$, $\delta s_2 = h_2 \; \delta\lambda_2$, $\delta s_3 = h_3 \; \delta\lambda_3$, we get

$$\frac{\partial \rho}{\partial t}+\frac{1}{h_1 h_2 h_3}\left[\frac{\partial}{\partial \lambda_1}(\rho q_1 h_2 h_3)+\frac{\partial}{\partial \lambda_2}(\rho q_2 h_3 h_1)+\frac{\partial}{\partial \lambda_3}(\rho q_2 h_1 h_2)\right]=0.$$

This is the equation of continuity in orthogonal curvilinear co-ordinates.

EQUATION OF CONTINUITY BY LAGRANGIAN METHOD

The equation of continuity is always found by the fact that the mass contained inside a given volume of fluid remains unaltered throughout the motion.

Let R_0 be the region occupied by a portion of a fluid at t = 0, and R the fluid region occupied by the same portion of the fluid at time t.

Let ρ_0 be density of the fluid when inside R_0 ; the mass of fluid element is p_0 $\delta a\ \delta b\ \delta c$, where a, b, c are co ordinates of P_0 enclosed in this element.

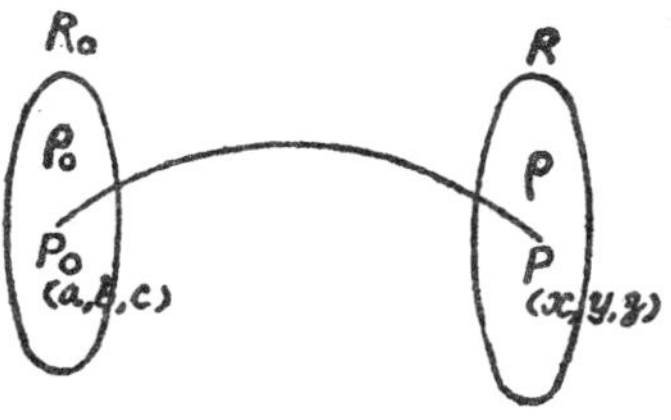

Fig. 2.3

Also the mass of the element enclosing the subsequent point P (x, y, z) is $\rho\ \delta x\ \delta y\ \delta z$, where ρ is the density of the fluid there.

Since the total mass inside R_0 is equal to the total mass inside R, we have

$$\iiint_{R_0} \rho_0 da\, db\, dc = \iiint_{R} \rho\, dx\, dy\, dz$$

$$= \iiint_{R_0} \rho \frac{\partial(x, y, z)}{\partial(a, b, z)} da\, db\, dc$$

as $\delta x\ \delta y\ \delta z = \dfrac{\partial(x, y, z)}{\partial(a, b, c)} \delta a\ \delta b\ \delta c$

i.e., $\displaystyle\iiint_{R_0}\left[\rho_0-\rho \frac{\partial(x, y, z)}{\partial(a, b, c)}\right] da\, db\, dc = 0.$

But since R_0 is arbitrary,

$$\rho_0 - \rho \frac{\partial(x, y, z)}{\partial(a, b, c)}$$

or $\rho_0 = \rho J$,

$$\text{where } J = \begin{vmatrix} \frac{\partial x}{\partial a} & \frac{\partial x}{\partial b} & \frac{\partial x}{\partial c} \\ \frac{\partial y}{\partial a} & \frac{\partial y}{\partial b} & \frac{\partial y}{\partial c} \\ \frac{\partial z}{\partial a} & \frac{\partial z}{\partial b} & \frac{\partial z}{\partial c} \end{vmatrix}$$

which is the equation of continuity in Lagrangian form.

Example 1: *By considering the constancy of mass of a finite volume of the fluid, obtain the equation of continuity.*

Solution: Do your self

Example 2: *Show that equation of continuity reduces to Laplace's equation when the liquid is incompressible and the motion is irrotational.*

EQUIVALENCE OF EQUATIONS OF CONTINUITY IN LAGRANGIAN AND EULERIAN FORMS

Lagrangian form of the equation of continuity is

$$\rho_J = \rho_0, \quad \text{where } J = \frac{\partial(x, y, z)}{\partial(a, b, c)}.$$

Differentiating it w.r.t. t, we get

$$\rho \frac{dJ}{dt} + J \frac{d\rho}{dt} = 0.$$

To connect this with the Eulerian form, we shall put

$$\frac{d}{dt}\left(\frac{\partial x}{\partial a}\right) = \frac{\partial}{\partial a}\left(\frac{dx}{dt}\right) = \frac{\partial u}{\partial a} \text{ etc.} \qquad ...(1)$$

as $u = \frac{dx}{dt}$, $v = \frac{dy}{dt}$ and $w = \frac{dz}{dt}$.

$$\text{Also } \frac{d\rho}{dt} = \frac{\partial \rho}{\partial t} + \frac{\partial \rho}{\partial x}\frac{dx}{dt} + \frac{\partial \rho}{\partial y}\frac{dy}{dt} + \frac{\partial \rho}{\partial z}.\frac{dz}{dt}$$

$$= \frac{\partial \rho}{\partial t} + u\frac{\partial \rho}{\partial x} + v\frac{\partial \rho}{\partial y} + w\frac{\partial \rho}{\partial z}. \qquad ...(2)$$

Now

$$J = \frac{\partial(x,y,z)}{\partial(a,b,c)} = \begin{vmatrix} \frac{\partial x}{\partial a} & \frac{\partial y}{\partial a} & \frac{\partial z}{\partial a} \\ \frac{\partial x}{\partial b} & \frac{\partial y}{\partial b} & \frac{\partial z}{\partial b} \\ \frac{\partial x}{\partial c} & \frac{\partial y}{\partial c} & \frac{\partial z}{\partial c} \end{vmatrix}$$

$$\therefore \frac{dJ}{dt} = \begin{vmatrix} \frac{\partial u}{\partial a} & \frac{\partial y}{\partial a} & \frac{\partial z}{\partial a} \\ \frac{\partial u}{\partial b} & \frac{\partial y}{\partial b} & \frac{\partial z}{\partial b} \\ \frac{\partial u}{\partial c} & \frac{\partial y}{\partial c} & \frac{\partial z}{\partial c} \end{vmatrix} + \begin{vmatrix} \frac{\partial x}{\partial a} & \frac{\partial v}{\partial a} & \frac{\partial z}{\partial a} \\ \frac{\partial x}{\partial b} & \frac{\partial v}{\partial b} & \frac{\partial z}{\partial b} \\ \frac{\partial x}{\partial c} & \frac{\partial v}{\partial c} & \frac{\partial z}{\partial c} \end{vmatrix} + \begin{vmatrix} \frac{\partial x}{\partial a} & \frac{\partial y}{\partial a} & \frac{\partial w}{\partial a} \\ \frac{\partial x}{\partial b} & \frac{\partial y}{\partial b} & \frac{\partial w}{\partial b} \\ \frac{\partial x}{\partial c} & \frac{\partial y}{\partial c} & \frac{\partial w}{\partial c} \end{vmatrix}$$

$$= \frac{\partial(u,y,z)}{\partial(a,b,c)} + \frac{\partial(x,v,z)}{\partial(a,b,c)} + \frac{\partial(x,y,w)}{\partial(a,b,c)}.$$

Now $\frac{\partial u}{\partial a} = \frac{\partial u}{\partial x}\frac{\partial x}{\partial a} + \frac{\partial u}{\partial y}\frac{\partial y}{\partial a} + \frac{\partial u}{\partial z}\frac{\partial z}{\partial c}$

$$\frac{\partial u}{\partial b} = \frac{\partial u}{\partial x}\frac{\partial x}{\delta b} + \frac{\partial u}{\partial y}\frac{\partial y}{\partial b} + \frac{\partial u}{\partial z}\frac{\partial z}{\partial c}$$

and $\frac{\partial u}{\partial c} = \frac{\partial u}{\partial x}\frac{\partial x}{\partial x} + \frac{\partial u}{\partial y}\frac{\partial y}{\partial c} + \frac{\partial u}{\partial z}\frac{\partial z}{\partial c}$

Eliminating $\frac{\partial u}{\partial y}$ and $\frac{\partial u}{\partial z}$ from these, we get

$$\begin{vmatrix} \frac{\partial u}{\partial x}\frac{\partial x}{\partial a} - \frac{\partial u}{\partial a} & \frac{\partial y}{\partial a} & \frac{\partial z}{\partial a} \\ \frac{\partial u}{\partial x}\frac{\partial x}{\partial b} - \frac{\partial u}{\partial b} & \frac{\partial y}{\partial b} & \frac{\partial z}{\partial b} \\ \frac{\partial u}{\partial x}\frac{\partial x}{\partial c} - \frac{\partial u}{\partial c} & \frac{\partial y}{\partial c} & \frac{\partial z}{\partial c} \end{vmatrix} = 0.$$

i.e. $\frac{\partial u}{\partial x}\frac{\partial(x,y,z)}{\partial(a,b,c)} = \frac{\partial(u,y,z)}{\partial(a,b,c)}$,

i.e. $\frac{\partial(u,y,z)}{\partial(a,b,c)} = \frac{\partial u}{\partial x}.J.$

Similarly $\dfrac{\partial(x,y,z)}{\partial(a,b,c)} = \dfrac{\partial v}{\partial y} J$ and $\dfrac{\partial(x,y,w)}{\partial(a,b,c)} = \dfrac{\partial w}{\partial x} J,$

$\therefore$ (3) gives $\dfrac{dJ}{dt} = J\left(\dfrac{\partial u}{\partial x} + \dfrac{\partial v}{\partial y} + \dfrac{\partial w}{\partial z}\right) = J\nabla.\, q.$ ...(9)

With the help of (2) and (4), becomes

$$JP\left(\frac{\partial u}{\partial x} + \frac{\partial v}{\partial y} + \frac{\partial w}{\partial z}\right) + J\frac{d\rho}{dt} = 0,$$

i.e. $\dfrac{d\rho}{dt} + \rho\left(\dfrac{\partial u}{\partial x} + \dfrac{\partial v}{\partial y} + \dfrac{\partial w}{\partial z}\right) = 0,$

or $\dfrac{d\rho}{dt} + \rho \operatorname{div} \mathbf{q} = 0$

which is the Euler's form of equation of continuity.

Example 1: *A mass of fluid moves in such a way that each particle describes a circle in one plane about a fixed axis: Show that the equation of continuity is*

$$\frac{\partial \rho}{\partial t} + \frac{\partial(\rho\omega)}{\partial \theta} = 0,$$

where ω is the angular velocity of a particle whose azimuthal angle is θ at time t.

Solution: Here the motion is in a plane. Consider a particle moving in a circle of radius r.

At this point consider an element PABQ, such that

PA = δr, PQ = $r\,\delta\theta$.

There is no motion along pA.

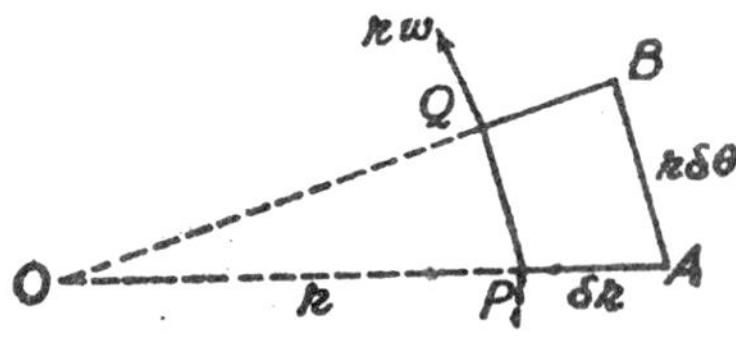

Fig. 2.4

The excess of flow - in over flow-out along PQ

$= r\delta\theta \dfrac{\partial}{r\partial\theta} [\rho r\omega\, \delta r]$ per unit time.

Also mass of the fluid inside the element

$= \rho \, dr \, r \, d\theta.$

$\therefore$ change in the mass of the element

$= \dfrac{\partial}{\partial t}(\rho \, \delta r \, r \, \delta\theta)$ in unit time.

The equation of continuity is

$$\frac{\partial}{\partial t}(\rho \, \delta r \, r \, \delta\theta) = -\, r \, \delta\theta \, \frac{\partial}{\partial \theta}[\rho r \omega \, \delta r],$$

i.e. $\dfrac{\partial \rho}{\partial t} + \dfrac{\partial(\rho\omega)}{\partial \theta} = 0,$

cancelling $\delta r \, r \, \delta\theta$ throughout.

Example 2: *The particles of a fluid move symmetrically in space with regard to a fixed centre; prove that the equation of continuity is*

$$\frac{\partial \rho}{\partial t} + u\frac{\partial \rho}{\partial r} + \frac{\rho}{r^2}\frac{\partial}{\partial r}(r^2 u) = 0.$$

where u is the velocity at distance r.

Solution: If the fixed centre be the origin, then in this example there is motion only along PQ and no motion along other edges of the element.

$\therefore$ excess of flow-in over flow-out along PQ

$= -\, \delta r \, \dfrac{\partial}{\partial r}[\rho u. \, r \, \delta\theta. \, r \sin\theta \, \delta\omega]$ per unit time.

Excess of flow-in over flow-out along PR = 0

and excess of flow- in over flow-out along PS = 0,

there being no motion of the fluid along these directions.

Now the mass of fluid inside the element

$= \rho \, \delta r \, r \, \delta\theta \, r \sin\theta \, \delta\omega.$

$\therefore$ change in the mass of the element

$= \dfrac{\partial}{\partial t}(\rho \, \delta r \, r \, \delta\theta \, r \sin\theta \, \delta\omega)$ per unit time.

Therefore the equation of continuity is

$$\frac{\partial}{\partial t}(\rho \, \delta r \, r \, \delta\theta \, r \sin\theta \, \delta\omega) = -\, dr \frac{\partial}{\partial r}(\rho u r \, \delta\theta \, r \sin\theta \, \delta\omega)$$

i.e. $\frac{\partial \rho}{\partial t} + \frac{1}{r^2}\frac{\partial}{\partial r}(\rho u r^2) = 0.$

i.e. $\frac{\partial \rho}{\partial t} + \frac{1}{r^2}\left[u r^2 \frac{\partial \rho}{\partial r} + \rho \frac{\partial (u r^2)}{\partial r}\right] = 0.$ This gives the result.

Example 3: *A mass of fluid is in motion so that the lines of motion on the surface of coaxial cylinders; show that the equation of continuity is*

$$\frac{\partial \rho}{\partial t} + \frac{1}{r}\frac{\partial (\rho v_\theta)}{\partial \theta} + \frac{\partial (\rho v_z)}{\partial z} = 0,$$

where v_θ v_z *are the velocities perpendicular and parallel to z.*

Solution: Consider a point P, whose cylindrical co-ordinates are (r, q, z).Construct an element at P with edges

PQ = r $\delta\theta$, PR = δr

and PS = δz.

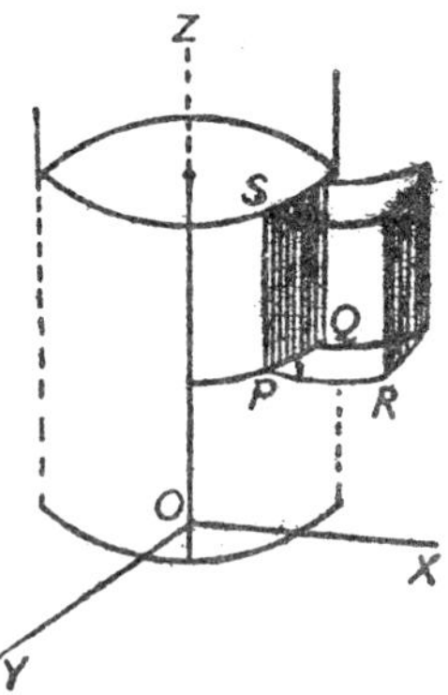

Fig. 2.5

The fluid moves on the surface of coaxial cylinders; hence there is no motion along PR.

Excess of flow-in over flow-out along PQ

$$= - r\,\delta\theta \frac{\partial}{\partial \theta}[\rho\, v_\theta\, \delta r\, \delta z]$$

and excess of flow-in over flow-out along PS

$$= - dz \frac{\partial}{\partial z}[\rho\, v_z\, \delta r.\, r\, \delta\theta].$$

Excess of flow-in over flow-out along PR = 0 there being no motion of the fluid in this direction.

Also mass of the fluid in the element

$= \rho r\ \delta\theta.\ \delta r.\ \delta z.$

$\therefore$ change in mass of the element

$$= \frac{\partial}{\partial t}(\rho r\ \delta\theta\ \delta r\ \delta z) = r\ \delta\theta\ \delta r\ \delta z \frac{\partial \rho}{\partial t}.$$

The equation of continuity is

$$(r\ \delta\theta\ \delta r\ \delta z)\ \frac{\partial \rho}{\partial t} = r\ \delta\theta \frac{\partial}{\partial \theta}[\rho v_\theta\ \delta r\ \delta z] - \delta z\ \frac{\partial}{\partial z}[\rho v_z\ \delta r.\ r\ \delta\theta]$$

i.e. $\frac{\partial \rho}{\partial t} + \frac{1}{r}\frac{\partial(\rho v_\theta)}{\partial \theta} + \frac{\partial(\rho v_z)}{\partial z} = 0,$

dividing by $r\ \delta\theta\ \delta r\ \delta z$.

Example 4: *If the lines of motion are curves on the surface of cones having their vertices at the origin and the axis of z for common axis, prove that the equation of continuity is*

$$\frac{\partial \rho}{\partial t} + \frac{\partial(\rho q_r)}{\partial r} + \frac{2\rho q_r}{r} + \frac{\operatorname{cosec}\theta}{r}\frac{\partial(\rho q_\omega)}{\partial \omega} = 0.$$

Solution: Let O, the common vertex, be the origin and OZ the common axis, the axis of z. Consider a cone OAB of semivertical angle θ. Let P (r, θ, ω) be a point on the surface of the cone, PQ, PR, PS being edges of the elementary parallelopiped.

Since the lines of motion are curves on the surfaces of cones there will be no motion perpendicular to the surface of the cone, *i.e.*, there is no velocity along the edge PR.

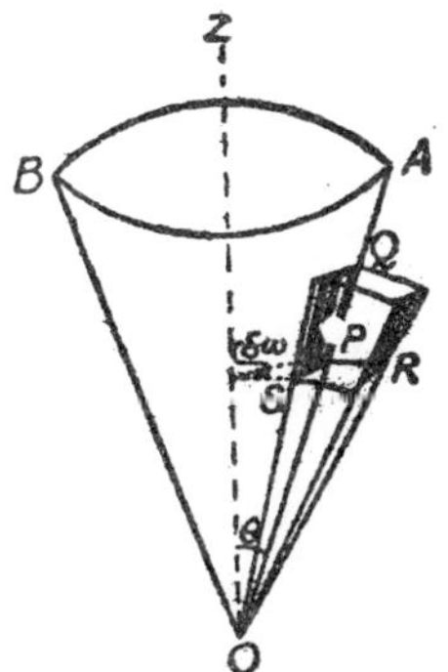

Fig. 2.6

Here $PQ = \delta r$, $PR = r\ \delta\theta$ and

$PS = r \sin\theta\ \delta\omega$.

Also volume of the elementary parallelopiped

$= r\ \delta\theta.\ r \sin\theta\ \delta\omega\ \delta r$

and q_r and q_ω are the velocity components along PQ and PS (directions in which r and ω increase).

Therefore excess of the flow- in over the flow-out in the direction PQ

$$= -\frac{\partial}{\partial r}[\rho q_{r}.\ r\ \delta\theta\ \sin\theta\ \delta\omega]\ \delta r \text{ per unit time}$$

and excess of the flow-in over the flow-out in the direction PS

$$= -\frac{\partial}{r\sin\theta\partial\omega}[\rho q_{\omega} r\ \delta\theta\ \delta r]\ r\sin\theta\ \delta\omega.$$

Also the increase in the mass of the element

$$= \frac{\partial}{\partial t}[\rho r\ \delta\theta\ r\sin\theta\ \delta\omega.\ \delta r] \text{ per unit timc.}$$

$\therefore$ the equation of continuity is given by

$$\frac{\partial}{\partial r}[\rho r\ \delta\theta.\ r\sin\theta\ \delta\omega.\ \delta r]$$

$$= \frac{\partial}{\partial r}[\rho\theta_{r}.\ r\ \delta\theta.\ r\sin\theta\ \delta\omega]\ \delta r - \frac{\partial}{r\sin\theta\partial\omega}[\rho q_{\omega} r\delta\theta.\delta r]\ r\sin\theta\delta\omega$$

The factor r sin θ δω. r dq dr can be cancelled and we get

$$\frac{\partial\rho}{\partial t} = -\frac{1}{r^2}\frac{\partial}{\partial r}[\rho q_r r^2] - \frac{1}{r\sin\theta}\frac{\partial}{\partial\omega}[\rho q_{\omega}],$$

$$\frac{\partial\rho}{\partial t} + \frac{1}{r^2}\left[r^2\frac{\partial}{\partial r}(\rho q_r) + \rho q_r\frac{\partial}{\partial r}(r^2)\right] + \frac{1}{r\sin\theta}\frac{\partial}{\partial\omega}(\rho q_{\omega}) = 0$$

$$\text{or}\quad \frac{\partial\rho}{\partial t} + \frac{\partial(\rho q_r)}{\partial r} + \frac{2\rho q_r}{r} + \frac{\operatorname{cosec}\theta}{r}\frac{\partial(\rho q_{\omega})}{\partial\omega} = 0.$$

This proves the result.

Example 5: *If the lines of motion are curves on the surfaces of spheres all touching the plane of xy at the origin O, the equation of continuity is*

$$r\sin\theta\frac{\partial\rho}{\partial t} + \frac{\partial(\rho v)}{\partial\phi} + \sin\theta\frac{\partial(\rho u)}{\partial\theta} + \rho u(1 + 2\cos\theta) = 0,$$

where r is the radius CP of one of the spheres, θ the angle PCO, u the velocity in the plane PCO, v the perpendicular velocity, and φ the inclination of the plane PCO to a fixed plane through the axis of z.

Solution: Let C and C′ be the centres of two spheres of radii r and r + δρ ρεσπεχτιϖελψ. P is a point on the smaller sphere.

Let PQ, PR and PS be the edges of the elementary parallelopiped. Now PR = r δθ,

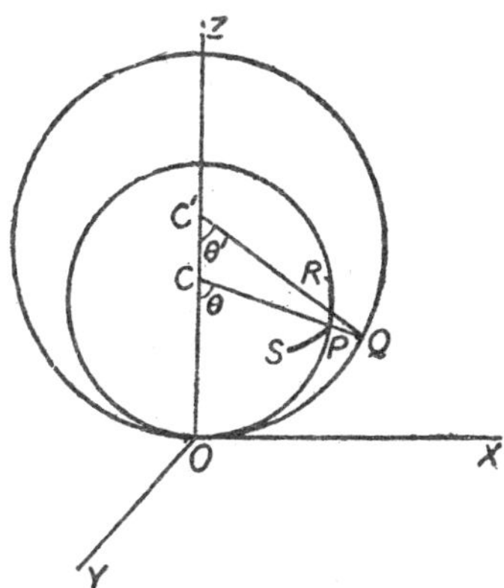

Fig. 2.7

$$PS = r \sin \theta\, \delta\phi,$$

(where ϕ is the angle that the plane PCO makes with a fixed plane through z-axis, i.e. XOZ- plane say).

To find the length PQ, WE have CP = r, C′Q = r + δr, CC′ = δr and ∠PCO = θ.

Suppose ∠CC′Q = θ′ and angle C′OC = ∈;

then $\cos \theta' = \cos (\theta - \in) = \cos \theta \cos \in + \sin \theta \sin \in$

$= \cos \theta + \in \sin \theta$ as ∈ is small.

Also $CQ^2 = C'Q^2\ CC'^2 - 2C'Q.\ CC' \cos \theta'$,

i.e. $(r + PQ)^2 = (r + \delta r)^2\ (\delta r)^2 - 2(r + \delta r)\ \delta r \cos \theta'$

or $PQ = \delta r\ (1 - \cos \theta')$.

neglecting PQ^2, δr^2, these being squares of small quantities. Thus

$PQ = \delta r\ [1 - (\cos \theta + \in \sin \theta)]$ from (1)

$= \delta r\ [1 - (\cos \theta]$ to the first order of small quantities.

After having determined the three edes of the elementary volume, we proceed as usual.

Volume of the elementary parallelopiped

$= PQ.\ PR.\ PS$

$= \delta r\ (1 - \cos \theta).\ r\, \delta\theta.\ r \sin \theta\, \delta\phi.$

Also since the lines of motion lie on the sphere, there is no motion along PQ.

The component velocities in the direction of θ and ϕ increasing are given to be u and v.

Hence the excess of flow- in over flow-out in the direction PR per unit time

$$= -\frac{\partial}{r\partial\theta}\ [\ \rho u.\ r \sin\theta\ \delta\phi.\ \delta r\ (1-\cos\theta)\ r\ \delta\theta$$

and the excess of flow-in over flow-out in the direction PS per unit time

$$= -\frac{\partial}{r\sin\theta\partial\phi}[\ \rho v.\ r\ \delta\theta.\ \delta r\ (1-\cos\theta)]\ r\sin\theta\ \delta\phi$$

and the increase in the mass of the elementary parallelopiped per unit time

$$= \frac{\partial}{\partial t}\ [\ \rho.\ \delta r\ (1-\cos\theta)\ r\ \delta\theta.\ r\sin\theta\ \delta\phi].$$

The equation of continuity therefore becomes

$$\frac{\partial}{\partial t}\ [\ \rho.\ \delta r\ (1-\cos\theta)\ r\ \delta\theta.\ r\sin\theta\ \delta\phi].$$

$$= \frac{\partial}{r\sin\theta\partial\phi}[\ \rho v.\ r\ \delta\theta.\ \delta r\ (1-\cos\theta)]\ r\sin\theta\ \delta\phi.$$

Cancelling r sin θ δϕ. r δθ δr (1– cos θ)] from both the sides, the equation of continuity becomes

$$\frac{\partial\rho}{\partial t} = -\frac{1}{\sin\theta(1-\cos\theta)}\frac{\partial}{r\partial\theta}[\rho u\sin\theta\,(1-\cos\theta)] - \frac{\partial}{r\sin\theta\partial\phi}(\rho v)$$

$$\text{or}\quad \frac{\partial\rho}{\partial t} + \frac{1}{r\sin\theta(1-\cos\theta)}\frac{\partial}{\partial\theta}(\rho u).\sin\theta\,(1-\cos\theta)$$

$$+\frac{\rho u}{r\sin\theta(1-\cos\theta)}\frac{\partial}{\partial\theta}[\sin\theta(1-\cos\theta)] + \frac{\partial}{r\sin\theta\partial\phi}(\rho v) = 0$$

$$\text{or}\quad \frac{\partial\rho}{\partial t} + \frac{1}{r}\frac{\partial}{\partial\theta}(\rho u) + \frac{\rho u\,[\cos\theta(1-\cos\theta)+\sin^2\theta]}{r\sin\theta\,(1-\cos\theta)} + \frac{\partial}{r\sin\theta\partial\phi}(\rho v) = 0$$

$$\text{or}\quad \frac{\partial\rho}{\partial t} + \frac{1}{r}\frac{\partial}{\partial\theta}(\rho u) + \frac{\rho u[\cos\theta(1-\cos\theta)+(1-\cos^2\theta)]}{r\sin\theta(1-\cos\theta)}$$

$$+\frac{\partial}{r\sin\theta\partial\phi}(\rho v) = 0$$

$$\text{or} \quad \frac{\partial \rho}{\partial t} + \frac{1}{r}\frac{\partial}{\partial \theta}\frac{\rho u}{r \sin\theta}(\cos\theta + 1 + \cos\theta) + \frac{\partial}{r\sin\theta \partial\phi}(\rho v) = 0$$

$$\text{or} \quad r\sin\theta \frac{\partial \rho}{\partial t} + \sin\theta \frac{\partial}{\partial \theta}(\rho u) + \rho u(1 + 2\cos\theta) + \frac{\partial}{\partial \phi}(\rho v) = 0$$

This is the required equation of continuity.

Example 6: *If every particle moves on the surface of a sphere, prove that the equation of continuity is*

$$\frac{\partial \rho}{\partial t}\cos\theta + \frac{\partial}{\partial \theta}(\rho\omega\cos\theta) + \frac{\partial}{\partial \phi}(\rho\omega'\cos\theta) = 0$$

ρ being the density, θ, ϕ the latitude and longitude of any element and w and w′ the angular velocities of the element in latitude and longtitude respectively.

Solution: Let the point P be on the semi-circle NPS making an angle ϕ with NAS and θ and angle OP makes with OA. Take an element at P with PQ = δr, PR = $r\ \delta\theta$ and PS = $r \cos\theta\ \delta\phi$, as edges of the element, ω and ω' being angular velocities along θ and ϕ respectively, i.e. $r\omega$ and $r \cos\theta\ \omega'$ are the velocities along PR and PS respectively, there being no velocity along PQ as the particle moves on the surface of the sphere.

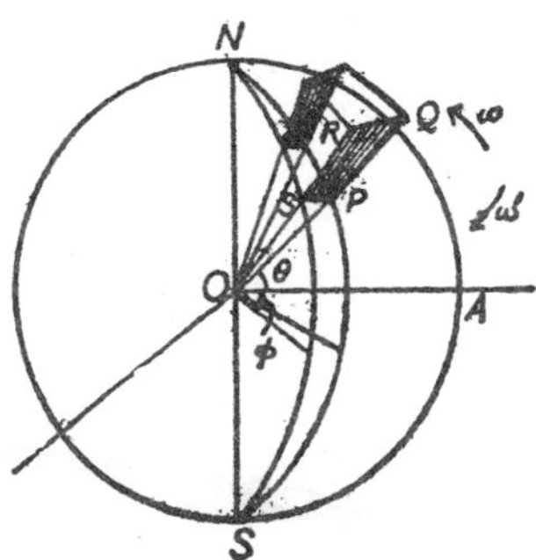

Fig. 2.8

Volume of this parallelopiped = $\rho\ \delta r.\ r\ \delta\theta.\ r \cos\theta\ \delta\phi$ and change in the mass of the element

$$= \frac{\partial}{\partial t}\ [\rho\ \delta r.\ r\ \delta\theta.\ r\cos\theta\ \delta\phi] \text{ per unit time.}$$

Now excess of flow-in over flow-out along the faces perpendicular to PR $= -r\,\delta\theta . \dfrac{\partial}{r\partial\theta}[\rho\ \delta r.\ r\cos\theta\ \delta\phi,\ r\omega]$ per unit time and excess of flow-in over flow-out along the faces perpendicular to PS $= r\cos\theta\ \delta\phi.$ $\dfrac{\partial}{r\cos\theta\partial\phi}[\ \rho.\ \delta r.\ r\ \delta\theta.\ r\cos\theta\ \omega']$ per unit time.

Excess of flow-in over flow-our along faces perp. to PQ = 0, as there is no velocity along PQ.

$\therefore$ equation of continuity is

$$\frac{\partial}{\partial t}[\rho\ \delta r.\ r\ \delta\theta.\ r\cos\theta\ \delta\phi] = -r\,\delta\theta\ \frac{\partial}{r\partial\theta}[\rho\ \delta r.\ r\cos\theta\ \delta\phi,\ r\omega]$$

$$-r\cos\theta\ \delta\phi\ \frac{\partial}{r\cos\theta\ \partial\phi}\ [\rho\ \delta r.\ r\ \delta\theta.\ r\cos\theta\ \omega']$$

$$\text{or}\ \frac{\partial}{\partial t}\ \cos\theta + \frac{\partial}{\partial\theta}(\rho\omega\cos\theta) + \frac{\partial}{\partial\phi}(\rho\omega'\cos\theta) = 0.$$

Example 7: *Each particle of a mass of liquid movesin a plane through the axis of z; find the equation of continuity.*

Solution: Let ZOAB be a plane (making an angle XOA = ω with XZ-plane) in which the particles of the liquid move.

Taking a point P (r, θ, ω) in this plane construct a parallelopiped with edges PQ, PR and PS, such that

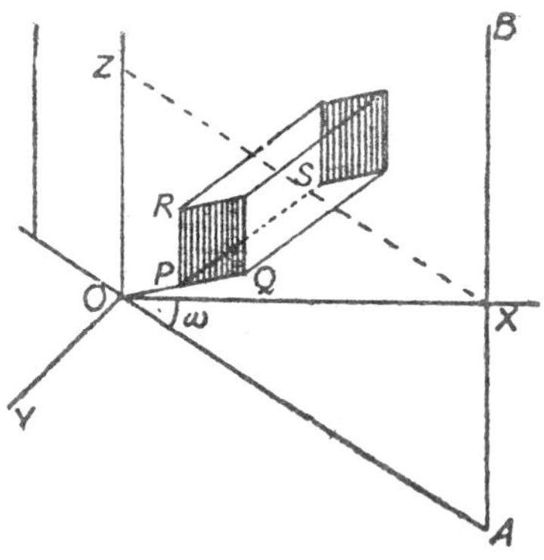

Fig. 2.9

PQ = δr, PR = $r\,\delta\theta$,

PS = r sin θ δω ;

edges PQ and PR lie in the plane while PS is perpendicular to the plane. Obviously there is no flow along PS.

Now excess of flow-in over flow-out along PQ

$$= -\,\delta r\,\frac{\partial}{\partial r}\,[\rho u r\,\delta\theta.\ r\sin\theta\,\delta\omega]\ \text{per unit time},$$

u being velocity along PQ,

and excess of flow- in over flow-out along PR

$$= -\,r\,\delta\theta\,\frac{\partial}{r\partial r}[\rho v\,\delta r.\ r\sin\theta\,\delta\omega]\ \text{per unit time},$$

v being velocity along PR.

Excess of flow-in over flow-out along PS = 0.

The mass of the element = ρ δr. r δθ r sin θ δω.

$$\therefore \text{ change in mass} = \frac{\partial}{\partial t}(\rho\,\delta r.\ r\,\delta\theta.\ r\sin\theta\,\delta\omega)\ \text{per unit time}$$

Therefore the equation of continuity is

$$\frac{\partial}{\partial t}(\rho\,\delta r.\ r\,\delta\theta.\ r\sin\theta\,\delta\omega) = -\,\delta r\,\frac{\partial}{\partial r}(\rho u r\,\delta\theta.\ r\sin\theta\,\delta\omega)$$

$$-\,r\,\delta\theta\,\frac{\partial}{r\partial\theta}(\rho v\,\delta r.\ r\sin\theta\,\delta\omega),$$

$$\text{i.e. } \frac{\partial}{\partial r} + \frac{1}{r^2}\frac{\partial}{\partial r}(\rho u r^2) + \frac{1}{r\sin\theta}\frac{\partial}{\partial\theta}(\rho v\sin\theta) = 0,$$

which is the equation of continuity.

Example 8: *Homogeneous liquid moves so that the path of any particle P lies in the plane POX, where OX is fixed axis. Prove that if OP = r and the ∠XOP = θ, the equation of continuity may be written as*

$$\frac{\partial}{\partial r}(ur^2) - \frac{\partial}{\partial\mu}(vr\sin\theta) = 0,$$

where u, v are the component velocities along and perpendicular to OP in the plane POX and μ = cos θ.

Solution: In the above example take ox instead of OZ and ρ = constant; we then have $\frac{\partial\rho}{\partial t} = 0$, so that the equation of continuity obtained in the above example now becomes

$$\frac{\partial}{\partial r}(ur^2) + r\frac{1}{\sin\theta}\frac{\partial}{\partial\theta}(v\sin\theta) = 0$$

or $\frac{\partial}{\partial r}(ur^2) - \frac{\partial}{\partial\mu}(vr\ \sin\ \theta) = 0$, where $\mu = \cos\theta$.

Example 9: *If ω is the area of cross-section of a stream filament, prove that the equation of continuity is*

$$\frac{\partial}{\partial r}(\rho\omega) + \frac{\partial}{\partial s}(\rho\omega q) = 0,$$

where ds is an element of arc of the filament in the direction of flow the q is the speed.

Solution: Let P be the point. Consider a volume bounded by the cross-section through P and at a distance ds from P.

Excess of flow-in over flow-out in this volume $= -\frac{\partial}{\partial s}[\rho\omega q]$ ds per unit time.

Also the mass of the in this volume

$= \rho.\ \omega$ ds.

$\therefore$ rate of change of mass $= \frac{\partial}{\partial t}(\rho\omega\ ds)$.

$\therefore$ equation of continuity is

$$-\frac{\partial}{\partial s}(\rho\omega q)\ ds = \frac{\partial}{\partial t}(\rho\omega\ ds),$$

i.e., $\frac{\partial}{\partial t}(\rho\omega) + \frac{\partial}{\partial s}(\rho\omega q) = 0.$

This proves the result.

Fig. 2.10

STREAM LINES OR LINES OF FLOW

A stream line or a line of flow is a curve drawn ion the fluid such that the tangent at any point of it at any time, is in the direction of flow i.e. in the direction of velocity of the fluid at that point. Thus velocity at any point is along the tangent to the stream line through that point

If we draw stream lines through every point of a closed curve C in the fluid, we obtain a stream tube as shown in the figure.

If dr be the element of arc length along a stream line, and q the fluid velocity, then q || dr. The equation of stream lines is

$\mathbf{q} \times dr = 0,$

$(v\,dz - w\,dy)\,i + (w\,dx - u\,dz)$

$j + (u\,dy - v\,dx)\,\mathbf{k} = 0,$

which gives $\dfrac{dx}{u} = \dfrac{dy}{v} = \dfrac{dz}{w},$

where $\mathbf{q} = (u, v, w)$

and $dr = (dx, dy, dz).$ ...(2)

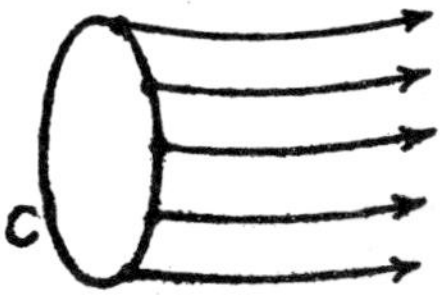

Fig. 2.11

The equations (2) have a doubly infinite set of solutions. Through each point of the flow field at which the functions u (x, y, z, t), v (x, y, z, t) do not all vanish, there passes one and only one stream line at a given instant. This result follows from the existence theorem for the system of equation (2). If the velocity vanishes at a given point, various sigularities may occur; such points are called critical points or stagnation points.

Path Lines

Path line is a curve along which a particular fluid particle travels during its motion. And therefore the differential equations of path lines are

$$\frac{dr}{dt} = q.$$

$$\text{or} \quad \frac{dx}{dt} = u, \quad \frac{dy}{dt} = v, \frac{dz}{dt} = w\,.$$

These equations have a triply infinite set of solutions.

It should be very clearly noted that stream lines are not the same as the path lines. Stream lines reveal how each particle is moving at a given instant of time while the path lines give the motion of the particles at each instant. It is only in the case of steady motion that these two coincide. Since u, v, w are always functions of time, the stream lines will alter from instant to instant.

As a matter of fact the actual path of any particle of the fluid will not in general coincide with the stream line. To show this take three consecutive points P, Q, R on a stream line at time t. A particle at P will move along PQ, but when it reaches Q at time $t + \delta t$, QR is no longer the direction of the velocity at Q and therefore now the stream line will no longer be the same.

VELOCITY POTENTIAL

Let **q** = (u, v, w) be the fluid velocity at the instant of time t. Suppose that at this instant we can find a scalar point function ϕ (x, y, z, t) such that

$$-\, d\phi = u\, dx + v\, dy + w\, dz. \qquad ...(1)$$

$$\text{i. e.} -\left(\frac{\partial\phi}{\partial x}dx + \frac{\partial\phi}{\partial y}dy + \frac{d\phi}{\partial z}dz\right) = u\, dx + v\, dy + w\, dz,$$

then we can write

$$u = \frac{\partial\phi}{\partial x}, v = -\frac{\partial\phi}{\partial y}, w = -\frac{\partial\phi}{\partial z}$$

$$\text{or } \mathbf{q} = -\nabla\phi \qquad ...(2)$$

f is called the velocity potential. For the r.h.s. of (1) to be a perfect differential a necessary and sufficient condition is

$$\nabla \times \mathbf{q} = 0. \qquad ...(3)$$

The surfaces ϕ (x, y, z, t) = const.

are called equipotential surfaces. The differential equation of the equipotential surfaces is

$$d\phi = 0,$$

$$u\, dx + v\, dy + w\, dz = 0. \qquad ...(4)$$

It is clear that these surfaces cut the stream lines

$$\frac{dx}{u} = \frac{dy}{v} = \frac{dz}{w} \qquad ...(5)$$

orthogonally.

Whenever the velocity potential ϕ exists, *i.e.* equation (2) holds, the motion is said to be irrotational.******

Example 1: *Given* $u = -\dfrac{c^2y}{r^2}, v = \dfrac{c^2y}{r^2}, v = \dfrac{c^2x}{r^2}$, *w = 0 where r denotes distance from z-axis. Find the surfaces which are orthonormal to stream lines, the liquid being homogeneous.*

Solution: $\dfrac{\partial u}{\partial x} = \dfrac{2c^2y}{r^3}.\dfrac{x}{r}$. as $x^2 + y^2 = r^2$ and $\dfrac{\partial r}{\partial x} = \dfrac{x}{r}$,

$$\frac{\partial v}{\partial y} = -\frac{2c^2x}{r^3}.\frac{y}{r}, \frac{\partial w}{\partial z} = 0$$

so that $\frac{\partial u}{\partial x}+\frac{\partial v}{\partial y}+\frac{\partial w}{\partial z}=0$, *i.e.*, the equation of continuity satisfied and therefore the motion is a possible one.

Now the differential equations of lines of flow are

$$\frac{dx}{u}=\frac{dy}{v}=\frac{dz}{w} \text{ i.e., } \frac{dx}{-y}=\frac{dy}{x}=\frac{dz}{0},$$

which on integration give $x^2 + y^2 = \text{const.} z = \text{const.}$

The surfaces cutting lines of flow orthogonally are given the differential equation

$$u\,dx + v\,dy + w\,dz = 0,$$

$$\textit{i.e.,}\ \frac{c^2 y}{x^2+y^2}dx+\frac{c^2 x}{x^2+y^2}dy=0, \text{ as } r^2 = x^2 + y^2$$

$$\textit{i.e.}\ c^2 d\left(\tan^{-1}\frac{y}{x}\right)=0.$$

This being integrable, such surfaces exist. Integrating the above equation, the surfaces are given by

$$\tan^{-1}\frac{y}{x}=\text{const., i.e. }\frac{y}{x}=\text{const.}$$

or $y = kx$ (planes through z-axis).

It is also clear here that the motion is irrotational.

Example 2: *Give $u = -\omega y$, $v = \omega x$, $w = 0$; show that the surfaces intersecting the stream lines orthogonally exist and the planes through z-axis, although the velocity potential does not exist.*

Solution: $\frac{\partial u}{\partial x}=0$, $\frac{\partial v}{\partial y}=0, \frac{\partial w}{\partial z}=0$, so the equation of continuity namely $\frac{\partial u}{\partial x}+\frac{\partial v}{\partial y}+\frac{\partial w}{\partial z}=0$ is satisfied. Therefore the motion is quite possible.

The differential equations of lines of flow are

$$\frac{dx}{du}=\frac{dy}{v}=\frac{dz}{w}, \text{ i.e. } \frac{dx}{-y}=\frac{dy}{x}=\frac{dz}{0}.$$

Integrating, $x^2 + y^2 = \text{const.}$ and $z = \text{const.}$

Now $u\,dx + v\,dy + w\,dz = -\omega y\,dx + \omega x\,dy,$

which is not a perfect differential. Therefore velocity potential does not exist.

However u dx + v dy + w dz = 0 gives

$$-\omega y\, dx + \omega x\, dy = 0, \text{ i.e. } \frac{dx}{x} - \frac{dy}{y} = 0$$

or $\frac{x}{y}$ = const. or y = kx (planes through z-axis)

which are the surfaces that cut the stream lines orthogonally.

Example 3: *Show that*

$$u = -\frac{2xyz}{(x^2+y^2)^2},\ v = \frac{(x^2-y^2)}{(x^2-y^2)^2}$$

are the velocity components of a possible liquid motion. It this motion irrotational?

Solution: The motion will be possible if u, v, w satisfy the equation of continuity which in the case (for liquid) is

$$\frac{\partial u}{\partial x} + \frac{\partial v}{\partial y} + \frac{\partial w}{\partial z} = 0.$$

Now from the given values of u, v and w,

$$\frac{\partial u}{\partial x} = 2yz\frac{3x^2-y^2}{(x^2+y^2)^3},\ \frac{\partial v}{\partial y} = \frac{y^2-3x^2}{(x^2-y^2)^3},\ \frac{\partial w}{\partial z} = 0.$$

$$\text{Now } \frac{\partial u}{\partial x} + \frac{\partial v}{\partial y} + \frac{\partial w}{\partial z} = 2yz\,\frac{3x^2-y^2}{(x^2+y^2)^3} + 2yz\frac{y^2-3x^2}{(x^2-y^2)^3} + 0 = 0.$$

Since the equation of continuity is satisfied, u, v, w are the velocity components of possible motion, we should have

Again for irrotational motion, we should have

$$\frac{\partial v}{\partial z} - \frac{\partial w}{\partial y} = 0,\ \frac{\partial w}{\partial x} - \frac{\partial u}{\partial z} = 0,\ \frac{\partial u}{\partial y} - \frac{\partial v}{\partial x} = 0.$$

$$\text{Here } \frac{\partial v}{\partial z} = \frac{x^2-y^2}{(x^2+y^2)^2},\ \frac{\partial w}{\partial y} = \frac{x^2-y^2}{(x^2+y^2)^2},$$

Showing that $\frac{\partial v}{\partial z} - \frac{\partial w}{\partial y} = 0$ satisfied.

Again $\frac{\partial w}{\partial x} = -\frac{2xy}{(x^2+y^2)^2}$ and $\frac{\partial u}{\partial z} = -\frac{2xy}{(x^2+y^2)^2}$,

showing that $\frac{\partial w}{\partial x} - \frac{\partial u}{\partial z} = 0$ is satisfied.

Similarly $\frac{\partial u}{\partial y} - \frac{\partial v}{\partial x} = 0$ is also satisfied.

Hence the motion is irrotational.

Example 4: *If the velocity of an incompressible fluid at the point (x, y, z) is given by* $\left(\frac{3xz}{r^5}, \frac{3yz}{r^5}, \frac{3z^2 - r^2}{r^5}\right)$, *prove that the liquid motion is possible and the velocity potential is* $\frac{\cos\theta}{r^2}$.

Solution: We have

$$u = \frac{3xz}{r^5}, v = \frac{3yz}{r^5}, w = \frac{3z^2 - r^2}{r^5} = \frac{3z^2}{r^5} - \frac{1}{r^3},$$

so that $\frac{\partial u}{\partial x} = \frac{3z}{r^5} - \frac{15x^2z}{r^7}$, as $r^2 = x^2 + y^2 + z^2$, $\frac{\partial r}{\partial x} = \frac{x}{r}$,

$$\frac{\partial v}{\partial y} = \frac{3z}{r^5} - \frac{15y^2y}{r^7},$$

$$\frac{\partial w}{\partial z} = \frac{6z}{r^5} - \frac{15z^3}{r^5} + \frac{3z}{r^5}.$$

Now $\frac{\partial u}{\partial x} + \frac{\partial v}{\partial y} + \frac{\partial w}{\partial z} = \frac{12z}{r^5} - \frac{15z(x^2+y^2+z^2)}{r^7} + \frac{3z}{r^5}$

$$= \frac{15z}{r^5} - \frac{15z}{r^5} = 0.$$

Since the equation of continuity is satisfied by u,v,w, the motion is a possible one.

Velocity potential is given by

$$d\phi = \frac{\partial \phi}{\partial x}dx + \frac{\partial \phi}{\partial y}dy + \frac{\partial \phi}{\partial z}dz$$

$$= -(u\,dx + v\,dy + w\,dz)$$

$$= -\frac{1}{r^5}[3z(x\,dx + y\,dy + z\,dz) - r^2\,dz]$$

$$= -\frac{3z}{2}\frac{d(r^2)}{r^5} + \frac{dz}{r^3}$$

$$= d\left(\frac{z}{r^3}\right) = d\left(\frac{r\cos\theta}{r^3}\right).$$

$$\therefore\ \phi = \frac{\cos\theta}{r^2}\quad (\because \text{ an additive constant has no significance}).$$

CONDITIONS AT A BOUNDARY SURFACE

At a point on the boundary, the normal component of the fluid velocity must be equal to the normal component of the velocity of the surface ; otherwise the fluid will break its contact with the surface.

Let P be a point on the boundary surface

F (r, t) = 0,

where the fluid velocity is **q** and velocity of surface is u.

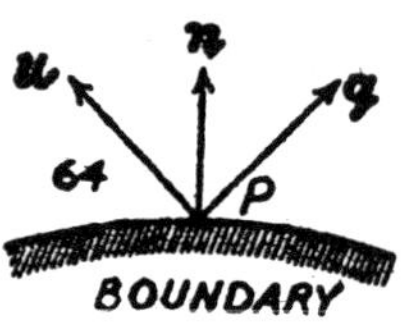

Fig. 2.12

Then **q**. **n** = **u**. **n**, ...(2)

where **n** is the unit normal at P. Now the direction ratios of n are

$$\left(\frac{\partial F}{\partial x}, \frac{\partial F}{\partial y}, \frac{\partial F}{\partial z}\right),$$

i.e. **n** $\parallel \nabla$ F,

$\therefore$ (2) gives **q**. ∇ F = **u**. ∇ F.

Let P (**r**, t) move to point P′ (**r** + δr, t + δt) in time δt. Since P′ also lies on the boundary (1), at time (t + δt), we have

F (**r** + δr, t + δt) = 0.

i.e. F (**r**, t) + δr. ∇ F + δt. $\frac{\partial F}{\partial t}$ = 0 (by Taylor's theorem)

or $\frac{\partial F}{\partial t}$ + **u**. ∇ F = 0 [using (1)]. ...(3)

Using (2), we have

$$\frac{\partial F}{\partial t} + \mathbf{q}.\nabla F = 0$$

or $\frac{\partial F}{\partial t} + u\frac{\partial F}{\partial x} + v\frac{\partial F}{\partial y} + w\frac{\partial F}{\partial z} = 0$

where q = (u, v, w).

This is the condition for F (r, t) to be a boundary surface. If the surface is at rest, then

$$\frac{\partial F}{\partial t} = 0$$

and condition for it to be boundary surface is

$$\mathbf{q}.\ \nabla F = 0$$

or $u\frac{\partial F}{\partial x} + v\frac{\partial F}{\partial y} + w\frac{\partial F}{\partial z} = 0.$

The normal velocity of the boundary is

$$\mathbf{u.\ n.} = \mathbf{u}.\ \frac{\nabla F}{|\nabla F|}$$

$$= -\frac{\partial F/\partial t}{|\nabla F|} \text{ from (3)}$$

$$= \frac{-(\partial F/\partial t)}{\sqrt{\left\{\left(\frac{\partial F}{\partial x}\right)^2 + \left(\frac{\partial F}{\partial y}\right)^2 + \left(\frac{\partial F}{\partial z}\right)^2\right\}}}.$$

Example 1: *Show that* $\frac{x^2}{a^2}\tan^2 t + \frac{y^2}{b^2}\cot^2 t = 1$ *is possible form for the bouding surface of a liquid, and find an expression for the normal velocity.*

Solution: We know that any surface F (x, y, z, t) = 0 is a possible boundary surface if it satisfies the boundary condition

$$\frac{dF}{dt} = \frac{\partial F}{\partial t} + u\frac{\partial F}{\partial x} + v\frac{\partial F}{\partial y} + w\frac{\partial F}{\partial z} = 0 \quad ...(1)$$

for values of u,v, w which satisfy the equation of continuity for the liquid, namely

$$\frac{\partial u}{\partial x} + \frac{\partial v}{\partial y} + \frac{\partial w}{\partial z} = 0. \quad ...(2)$$

Hence we have a two-dimensional equation, such that

$$F(x, y, t) = \frac{x^2}{a^2}\tan^2 t + \frac{y^2}{b^2}\cot^2 t - 1 = 0,$$

so that $\dfrac{\partial F}{\partial t} = \dfrac{x^2}{a^2}.2 \tan t \sec^2 t - \dfrac{y^2}{b^2}.\ 2 \cot t \operatorname{cosec}^2 t,$

$$\frac{\partial F}{\partial x} = \frac{2x}{a^2}\tan^2 t, \frac{\partial F}{\partial y} = \frac{2y}{b^2}\cot^2 t, \frac{\partial F}{\partial z} = 0,$$

Hence substituting these values in (1), we get

$$m\frac{x \tan t}{a^2}[x \sec^2 t + u \tan t] + \frac{y \cot t}{b^2}[- y \operatorname{cosec}^2 t + v \cot t] = 0,$$

which will hold [i. e (1) will hold] if

$x \sec^2 t + u \tan t = 0$, *i.e.* $u = \dfrac{x}{\cos t \sin t}]$

and $- y \operatorname{cosec}^2 t + v \cot t = 0$, *i.e.*, $v = + \dfrac{y}{\cos t \sin t}$. ...(3)

Now $\dfrac{\partial u}{\partial x} = \dfrac{1}{\cos t \sin t}, \dfrac{\partial v}{\partial y} = \dfrac{1}{\cos t \sin t}$.

Since there is only two-dimensional motion, (2) becomes

$$\frac{\partial u}{\partial x} + \frac{\partial v}{\partial y} = 0,$$

which is clearly satisfied by values of u and v given in (3) for which the given surface satisfies the condition of being a boundary surface.

Thus the given surface is a possible boundary surface of a liquid with velocity components u, v determined in (3).

Now normal velocity

$$u\ \frac{u\dfrac{\partial F}{\partial x} + v\dfrac{\partial F}{\partial y}}{\sqrt{\left\{\left(\dfrac{\partial F}{\partial x}\right)^2 + \left(\dfrac{\partial F}{\partial y}\right)^2\right\}}}$$

$$= \frac{\dfrac{x}{\sin t \cos t}.\dfrac{2x}{a^2}\tan^2 t + \dfrac{y}{\sin t \cos t}.\dfrac{2y}{b^2}\cot^2 t}{\sqrt{\left\{\left(\dfrac{2x}{a^2}\tan^2 t\right)^2 + \left(\dfrac{2y}{b^2}\cot^2 t\right)\right\}}}$$

Example 2: *Show that the variable ellipsoid*

$$\frac{x^2}{a^2k^2t^4} + kt^2\left[\left(\frac{y}{b}\right)^2 + \left(\frac{z}{c}\right)^2\right] = 1$$

is a possible form for the boundary surface of aliquid at any time t.

Solution: The surface

$$F\ (x,\ y,\ z\ t)\ ^{\circ}\ \frac{x^2}{a^2k^2t^4} + kt^2\left[\left(\frac{y}{b}\right)^2 + \left(\frac{z}{c}\right)^2\right] - 1 = 0 \qquad ...(1)$$

will be a possible boundary surface if it satisfies the condition

$$\frac{\partial F}{\partial t} + u\frac{\partial F}{\partial x} + v\frac{\partial F}{\partial y} + w\frac{\partial F}{\partial z} = 0. \qquad ...(2)$$

for same values of u, v, w which satisfy the equation of continuity

$$\frac{\partial u}{\partial x} + \frac{\partial v}{\partial y} + \frac{\partial w}{\partial z} = 0. \qquad ...(3)$$

$$\text{Now}\ \frac{\partial F}{\partial t} = -\frac{4x^2}{a^2k^2t^5} + 2kt\left(\frac{y^2}{b^2} + \frac{z^2}{c^2}\right),$$

$$\frac{\partial F}{\partial x} = \frac{2x}{a^2k^2t^4}\ \frac{\partial F}{\partial y} = \frac{2kt^2y}{b^2}, \frac{\partial F}{\partial z} = \frac{2kt^2z}{c^2}.$$

Putting these values in (2), we get

$$-\frac{4x^2}{a^2k^2t^5} + 2kt\left(\frac{y^2}{b^2} + \frac{z^2}{c^2}\right) + u\frac{2x}{a^2k^2t^4} + v\frac{2kt^2y}{b^2} + w\frac{2kt^2z}{c^2} = 0,$$

$$i.e.\ \frac{2x}{a^2k^2t^4}\left[u - \frac{2x}{t}\right] + \frac{2kyt}{b^2}[y + vt] + \frac{2ktz}{c^2}[z + wt] = 0$$

which will hold if

$$u = \frac{2x}{t}, v = \frac{y}{t}\ \text{and}\ w = -\frac{z}{t},$$

$$\text{Now}\ \frac{\partial u}{\partial x} = \frac{2}{t}, \frac{\partial v}{\partial y} = -\frac{1}{t}, \frac{\partial w}{\partial z} = -\frac{1}{t}.$$

Clearly (3) is satisfied for these values of u,v and w.

Therefore (1) is a possible boundary surface for a liquid with velocity components given in (4).

Example 3: *Show that the ellipsoid*

$$\frac{x^2}{a^2k^2t^{2n}} + kt^n\left(\frac{y^2}{b^2} + \frac{z^2}{c^2}\right) = 1$$

is a possible form of the boundary surface of a liquid.

Solution: *Proceed as in the above example.*

Example 4: *Determine the restriction on f_1, f_2, f_3 if*

$$\frac{x^2}{a^2}f_1(t) + \frac{y^2}{b^2}f_2(t) + \frac{z^2}{c^2}f_3(t) = 1$$

is a possible boundary surface of a liquid.

Solution: The surface

$$F\ (x, y, z, t) \equiv \frac{x^2}{a^2}f_1(t) + \frac{y^2}{b^2}f_2(t) + \frac{z^2}{c^2}f_3(t) - 1 = 0 \qquad ...(1)$$

for being boundary surface must satisfy the condition

$$\frac{\partial F}{\partial t} + u\frac{\partial F}{\partial x} + v\frac{\partial F}{\partial y} + w\frac{\partial F}{\partial z} = 0 \qquad ...(2)$$

for some values of u, v and w which must satisfy the equation of continuity

$$\frac{\partial u}{\partial x} + \frac{\partial v}{\partial y} + \frac{\partial F}{\partial z} = 0. \qquad ...(3)$$

Applying (2) for surface (1), we get

$$\frac{x^2}{a^2}f_1'(t) + \frac{y^2}{b^2}f_2'(t) + \frac{z^2}{c^2}f_3'(t) + u.\frac{2x}{a^2}f_1(t)$$

$$+ v\frac{2y}{b^2}f^2(t) + w\frac{2z}{c^2}f_3'(t) = 0,$$

$$i.e. \frac{2x}{a^2}f_1(t)\left[u + \frac{xf_1'(t)}{2f_1(t)}\right] + \frac{2y}{b^2}f_3(t)\left[v + \frac{yf_2'(t)}{2f^2(t)}\right]$$

$$+ \frac{2z}{c^2}f_3(t)\left[w + \frac{zf_3'(t)}{2f_3(t)}\right] = 0,$$

which will hold if

$$u = \frac{xf_1'}{2f_1}, v = \frac{yf_2'}{2f_2}, w = \frac{zf_3'}{2f_3}. \qquad ...(4)$$

Now $\frac{\partial u}{\partial x} = \frac{f_1'}{2f_1}, \frac{\partial v}{\partial y} = -\frac{f_2'}{2f_2}, \frac{\partial w}{\partial z} = -\frac{f_3'}{2f_3}$.

But u, v, w must satisfy (3),

$$\therefore -\frac{f_1'}{2f_1} - \frac{f_2'}{2f_2} - \frac{f_3'}{2f_3} = 0,$$

i.e. $\frac{f_1'}{f_1} + \frac{f_2'}{f_2} + \frac{f_3'}{f_3} = 0,$

i.e., $\frac{d}{dt}[\log (f_1, f_2, f_3)] = 0$

or $\log f_1 f_2 f_3 = \text{const.}$

or $f_1 f_2 f_3 = \text{const.}$,

which is the restriction of f_1, f_2 and f_3.

Example 5: *In the steady motion of homogeneous liquid if the surfaces $f_1 = a_1$, $f_2 = a_2$ define the stream lines, prove that the most general values of the velocity components u, v, w are*

$$F(f_1, f_2)\frac{\partial(f_1,f_2)}{\partial(y,z)}, F(f_1,f_2)\frac{\partial(f_1,f_2)}{\partial(z,x)}, F(f_1,f_2)\frac{\partial(f_1,f_2)}{\partial(x,y)}$$

where F is any arbitrary function and the cartesian co-ordinates are rectangular.

Solution: The motion is given to be steady, therefore the stream lines will be independent of t. Hence f_1 and f_2 are functions of x, y and z only.

Differentiating $f_1 = a1$ and f_2 a_2, we get

$$\frac{\partial f_1}{\partial x}dx + \frac{\partial f_1}{\partial y}dy + \frac{\partial f_1}{\partial z}dz = 0 \quad ...(1)$$

and $\frac{\partial f_2}{\partial x}dx + \frac{\partial f_2}{\partial y}dy + \frac{\partial f_2}{\partial z}dz = 0.$ ...(2)

solving these, we get

$$\frac{u}{\frac{\partial(f_1,f_2)}{\partial(y,z)}} = \frac{v}{\frac{\partial(f_1,f_2)}{\partial(z,x)}} = \frac{w}{\frac{\partial(f_1,f_2)}{\partial(x,y)}} = F \text{ (say)},$$

so that $u = F.\frac{\partial(f_1,f_2)}{\partial(y,z)}$,

$$v = F.\frac{\partial(f_1, f_2)}{\partial(z, x)}, w = F.\frac{\partial(f_1, f_2)}{\partial(x, y)}. \quad ...(4)$$

Here F is any arbitrary function. We shall now determine the nature of F by applying the condition

$$\frac{\partial u}{\partial x} + \frac{\partial v}{\partial y} + \frac{\partial w}{\partial z} = 0, \quad ...(3)$$

which, u, v, w must satisfy for motion.

This condition gives

$$\left\{\frac{\partial F}{\partial x}\frac{\partial(f_1, f_2)}{\partial(y, z)} + F\frac{\partial}{\partial x}\frac{\partial(f_1, f_2)}{\partial(y, z)}\right\} + \left\{\frac{\partial F}{\partial y}\frac{\partial(f_1, f_2)}{\partial(z, x)} + F\frac{\partial}{\partial y}\frac{\partial(f_1, f_2)}{\partial(z, x)}\right\}$$
$$+ \left\{\frac{\partial F}{\partial z}\frac{\partial(f_1, f_2)}{\partial(x, y)} + F\frac{\partial}{\partial z}\frac{\partial(f_1, f_2)}{\partial(x, y)}\right\} = 0.$$

But $\frac{\partial}{\partial x}\frac{\partial(f_1, f_2)}{\partial(y, z)} + \frac{\partial}{\partial y}\frac{\partial(f_1, f_2)}{\partial(z, x)} + \frac{\partial}{\partial y}\frac{\partial(f_1, f_2)}{\partial(x, y)} = 0.$

Hence the above condition gives

$$\frac{\partial F}{\partial x}\frac{\partial(f_1, f_2)}{\partial(y, z)} + \frac{\partial F}{\partial y}\frac{\partial(f_1, f_2)}{\partial(z, x)} + \frac{\partial F}{\partial y}\frac{\partial(f_1, f_2)}{\partial(x, y)} = 0,$$

i.e., $\frac{\partial(Ff_1, f_2)}{\partial(x, y, z)} = 0.$

But the vanishing of the Jacobian means that F, f_1, f_2 are not independent; therefore F is a function of f_1, f_2 only (which are independent amongst themselves).

Thus $F = F(f_1, f_2)$.

Now (4) gives the result.

Example 6: *Show that all necessary conditions can be satisfied by a velocity potential of the form*

$$\phi = \alpha x^2 + \beta y^2 + \gamma z^2,$$

and abounding surface of the form

$$F \equiv ax^4 + by^4 + cz^4 - \chi(t) = 0,$$

where $\chi(t)$ is a given function of the time and α, β, γ, a, b, c suitable functions of the time.

Solution: The necessary conditions are

(i) that ϕ satisfies the Laplace's equation

(ii) that F satisfies the condition for boundary surface, namely

$$\frac{\partial F}{\partial t}+u\frac{\partial F}{\partial x}+v\frac{\partial F}{\partial y}+w\frac{\partial F}{\partial z}=0. \qquad ...(1)$$

We have $\phi = \alpha x^2 + \beta y^2 + \gamma y^2$.

$$\therefore \quad \frac{\partial^2\phi}{\partial x^2}=2\alpha,\ \frac{\partial^2\phi}{\partial y^2}= 2\beta \text{ and } \frac{\partial^2\phi}{\partial z^2}=2\gamma.$$

The Laplace's equation $\frac{\partial^2\phi}{\partial x^2}+\frac{\partial^2\phi}{\partial y^2}+\frac{\partial^2\phi}{\partial z^2}=0$ will be satisfied if

$$2\alpha + 2\beta + 2\gamma = 0,$$

i.e., $\alpha + \beta + \gamma = 0,$ ...(2)

for which α, β, γ are to be some suitable functions of time. Again to satisfy the second condition,

$$\frac{\partial F}{\partial t} = x^4\frac{\partial a}{\partial t}+y^4\frac{\partial b}{\partial t}+z^4\frac{\partial c}{\partial t}-\chi'(t),$$

$$\frac{\partial F}{\partial x} = 4ax^3,\ \frac{\partial F}{\partial y}=4by^3,\frac{\partial F}{\partial z}=4cz^3,$$

$$u=-\frac{\partial\phi}{\partial x}=-2\alpha x,\ v=-\frac{\partial\phi}{\partial y}=-2\beta y, w=\frac{\partial\phi}{\partial z}=-2\gamma z.$$

$\therefore$ (1) gives

$$x^4\frac{\partial a}{\partial t}+y^4\frac{\partial b}{\partial t}+z^4\frac{\partial c}{\partial t}-\chi'(t)+(-2\alpha x)(4ax^3)$$

$$+ (-2\beta y)(4by^3) + (-2\gamma z)(4cz^3) = 0,$$

$$\text{i.e., } x^4\left(\frac{\partial a}{\partial t}-8a\alpha\right)+y^4\left(\frac{\partial b}{\partial t}-8b\beta\right)+z^4\left(\frac{\partial c}{\partial t}-8c\gamma\right)-\chi'(t)=0.$$

Comparing this will $F \equiv ax^4 + by^4 + cz^4 - \chi(t) = 0$, as all points on the surface satisfy these conditions simultaneously, we get

$$\frac{\frac{\partial a}{\partial t}-8a\alpha}{a}=\frac{\frac{\partial b}{\partial t}-8b\beta}{b}=\frac{\frac{\partial c}{\partial t}-8c\gamma}{c}=\frac{\chi(t)}{\chi'(t)}.$$

These conditions will also hold if a,b, c, α, β, γ are some suitable functions of time.

Hence ϕ and F = 0 satisfy the necessary conditions for velocity potential and boundary surface if a, b,c, α, β, γ are some suitable functions of time.

Example 7: *Prove that a surface of the form*

$$ax^4 + by^4 + cz^4 - \chi(t) = 0$$

is a possible form of a boundary surface of a homogeneous liquid at time t, velocity potential of the liquid motion being

$$\phi = (\beta - \gamma)\, x^2 + (\gamma - \alpha)\, y^2 + (\alpha - \beta)\, z^2,$$

where χ, α, β, γ are given functions of time and a,b, c are suitable functions of time.

Proceed as in the above examples.

Example 8: *Derive the equation of continuity in a system of orthogonal curvilinear co-ordinates. For steady motion deduce the equation in spherical polar co-ordinates. If the motion is irrotational, show that the velocity potential satisfies Laplace's equation.*

Solution: For first part see 1.11.

2nd part : Spherical Polar Co-ordinates. We know that

$$ds^2 = (h_1\, du_1)^2 + (h_2\, du_2)^2 + (h_3\, du_3)^2,$$

In spherical polar co-ordinates,

$$ds^2 = (dr)^2 + (r\, d\theta)^2 + (r \sin\theta\, d\phi)^2,$$

so that h1 = 1, h2 = r, h3 = r sin q.

Making these substitutions in (3), the equation of continuity becomes

$$\frac{\partial \rho}{\partial t} + \frac{1}{1.r.r\sin\theta}\left[\frac{\partial}{\partial r}(\rho q_1 r^2 \sin\theta) + \frac{\partial}{\partial \theta}(\rho q_2 r \sin\theta) + \frac{\partial}{\partial \phi}(\rho q_3 .r)\right] = 0,$$

i.e.,
$$\frac{\partial \rho}{\partial t} + \frac{\partial(\rho q_1)}{\partial r} + \frac{1}{r}\frac{\partial}{\partial \theta}(\rho q_3) + \frac{\operatorname{cosec}\theta}{r}\frac{\partial(\rho q_3)}{\partial \phi} + \frac{2\rho q_1}{r} + \frac{\cot\theta}{r}\rho q_2 r = 0,$$

which u, v, w are the components of the velocity along the axes.

The equation of continuity is

$$\frac{\partial u}{\partial x}+\frac{\partial v}{\partial y}+\frac{\partial w}{\partial z}=0,$$

i.e., $$\frac{\partial}{\partial x}\left(-\frac{\partial \phi}{\partial x}\right)+\frac{\partial}{\partial y}\left(-\frac{\partial \phi}{\partial y}\right)+\frac{\partial}{\partial z}\left(-\frac{\partial \phi}{\partial z}\right)=0,$$

i.e., $$\frac{\partial^2 \phi}{\partial x^2}+\frac{\partial^2 \phi}{\partial y^2}+\frac{\partial^2 \phi}{\partial z^2}=0.$$

This shows that ϕ satisfies Laplace's equation.

Example 9: *In the motion of a homogeneous liquid in two dimensions the velocity at any point is given by two components v, v′ along the directions which pass through the fixed points distant a from one another. Show that the equation of continuity is*

$$\frac{\partial v}{\partial r}+\frac{\partial v'}{\partial r'}+\frac{r^2+r'^2-a^2}{2rr'}\left(\frac{\partial v}{\partial r'}+\frac{\partial v'}{\partial r}\right)+\frac{v}{r}+\frac{v'}{r'}=0$$

where r and r′ are the distances of any point of the liquid from the fixed points.

Solution: Let A and B be the points at a distance a apart and r, r′ be the distance of any point P from A and B respectively.

With A as centre draw two circular arcs PQ and RS with r and r + δr as radii. Similarly with B as centre draw two arcs PR and QS with r′ and r′ + δr′ as radii. Clearly AR = r + δr and BQ = r′ + δr′.

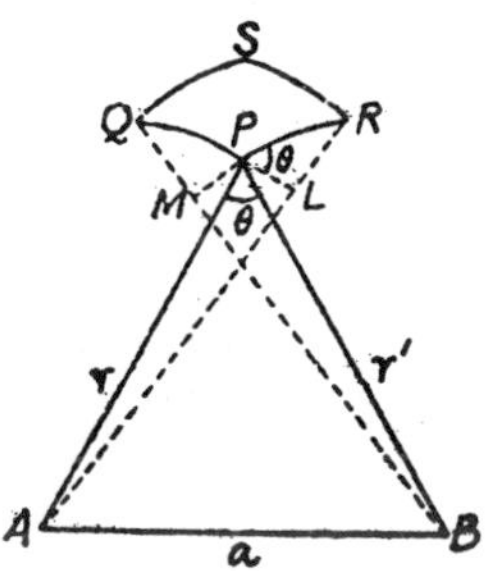

Fig. 2.13

Let ∠ APB = θ. Draw PL and PM per pendicular to AR and BQ,so that AL = r, BM = r′, BM = r′, LR = δr and MQ = δr′. Also ∠ LPR = θ.

$$\therefore \quad \frac{PR}{\sin PLR} = \frac{LR}{\sin\theta}, \text{ i.e. } \frac{PR}{\sin 90^\circ} = \frac{\delta r}{\sin\theta} \text{ or } PR = \frac{\delta r}{\sin\theta}.$$

Similarly $PQ = \dfrac{\delta r'}{\sin\theta}$.

As given v and v′ are the velocities along AP and BP.

$\therefore$ velocity along normal to PR = $v' + v\cos\theta$

and velocity along normal to PQ = $v + v'\cos\theta$.

Now mass of the liquid flowing through PQ

$= \rho PQ \times$ (velocity $\perp$ to PQ) per unit time

$$= r.\frac{\delta r'}{\sin\theta}(v + v'\cos\theta).$$

$\therefore$ excess of flow-in over flow-out across PQ and RS

$$= -\frac{\partial}{\partial r}\left[\rho\frac{\delta r'}{\sin\theta}(v + v'\cos\theta)\right]\delta r \text{ per unit time.}$$

Similarly excess of flow-in over flow-out across PR and QS

$$= -\frac{\partial}{\partial r'}\left[\rho\frac{\delta r}{\sin\theta}(v' + v\cos\theta)\right]\delta r' \text{ per unit time.}$$

Again area PQSR = PQ. PR sin QPR

$$= \frac{\delta r'}{\sin\theta}.\frac{\delta r}{\sin\theta}.\sin(\pi - \theta).$$

Hence rate of increase of mass of the fluid in PQSR

$$= \frac{\partial}{\partial t}\left[\rho\frac{\delta r'}{\sin\theta}.\frac{\delta r}{\sin\theta}\sin\theta\right].$$

Therefore the equation of continuity is

$$-\frac{\partial}{\partial r}\left[\rho\frac{\delta r'}{\sin\theta}(v + v'\cos\theta)\right]\delta r - \frac{\partial}{\partial r'}\left[\rho\frac{\delta r}{\sin\theta}(v' + v\cos\theta)\right]\delta r'$$

$$= \frac{\partial}{\partial t'}\left[\rho\frac{\delta r'}{\sin\theta}\frac{\delta r}{\sin\theta}.\sin\theta\right],$$

i.e. $$\frac{\partial}{\partial r}\left[\frac{\rho(v + v'\cos\theta)}{\sin\theta}\right] + \frac{\partial}{\partial r'}\left[\frac{\rho(v' + v\cos\theta)}{\sin\theta}\right] + \frac{1}{\sin\theta}\frac{\partial\rho}{\partial t} = 0.$$

This liquid is given to be homogeneous, i.e. ρ = const. and the equation of continuity becomes

$$\frac{\partial}{\partial r}\left[\frac{v+v'\cos\theta}{\sin\theta}\right]+\frac{\partial}{\partial r'}\left[\frac{v'+v\cos\theta}{\sin\theta}\right]=0,$$

$$\frac{1}{\sin\theta}\left[\frac{\partial v}{\partial r}+\frac{\partial v'}{\partial r}\cos\theta+v'\frac{\partial}{\partial r}(\cos\theta)\right]-\frac{v+v'\cos\theta}{\sin^2\theta}\cos\theta\frac{\partial\theta}{\partial r}$$

$$+\frac{1}{\sin\theta}\left[\frac{\partial v'}{\partial r'}+\frac{\partial v}{\partial r'}\cos\theta+v\frac{\partial}{\partial r'}(\cos\theta)\right]-\frac{(v'+v\cos\theta)}{\sin^2\theta}\cos\theta\frac{\partial\theta}{\partial r'}=0,$$

i.e. $$\left(\frac{\partial v}{\partial r}+\frac{\partial v'}{\partial r'}\right)+\left(\frac{\partial v'}{\partial r}+\frac{\partial v}{\partial r'}\right)\cos\theta+\left\{v'\frac{\partial(\cos\theta)}{\partial r}+v\frac{\partial(\cos\theta)}{\partial r}\right\}$$

$$+\frac{v+v'\cos\theta}{\sin^2\theta}\cos\theta\frac{\partial(\cos\theta)}{\partial r}+\frac{v'+v\cos\theta}{\sin^2\theta}\cos\theta\frac{\partial(\cos\theta)}{\partial r'}=0. \quad ...(1)$$

From triangle ABP, we have

$$\cos\theta=\frac{r^2+r'^2-a^2}{2rr'}=\frac{r}{2r'}+\frac{r'}{2r}-\frac{a^2}{2rr'}.$$

Differentiating w. r.t, we get

$$\frac{\partial(\cos\theta)}{\partial r}=\frac{1}{2r'}-\frac{r'}{2r^2}+\frac{a^2}{2r^2r'}$$

$$=\frac{1}{r'}-\frac{1}{r}\left(\frac{r}{2r'}+\frac{r'}{2r}-\frac{a^2}{2rr'}\right)=\frac{1}{r'}-\frac{\cos\theta}{r}.$$

Similarly $\frac{\partial}{\partial r'}(\cos\theta)=\frac{1}{r}-\frac{\cos\theta}{r'}.$

Putting these values in (1), we get

$$\frac{\partial v}{\partial r}+\frac{\partial v'}{\partial r'}+\frac{r^2+r'^2-a^2}{2rr'}\left(\frac{\partial v'}{\partial r}+\frac{\partial v}{\partial r'}\right)+\frac{v}{r}+\frac{v'}{r'}=0$$

which is the required equation of continuity.

VELOCITY POTENTIAL, IRROTATIONAL FLOW

At any instant t if **q** is fluid velocity then the equation of the streamline, at that instant, are given as

$$\frac{dx}{u}=\frac{dy}{v}=\frac{dz}{w}.$$

These curves cut the surfaces

u dx + v dy + w dz = 0.

orthogonally. Let us consider a scalar function f (x, y, z, t) at that instant, uniform throughout the entire field such that

$$u\,dx + v\,dy + w\,dz = -\,d\phi$$

$$\Rightarrow u\,dx + v\,dy + w\,dz = -\frac{\partial \phi}{\partial x}dx - \frac{\partial \phi}{\partial y}dy - \frac{\partial \phi}{\partial z}dz \quad ...(3)$$

Therefore we have $u = -\frac{\partial \phi}{\partial x}, v = -\frac{\partial \phi}{\partial y} w = -\frac{\partial \phi}{\partial z}.$ ...(4)

$\Rightarrow \mathbf{q} = -\nabla\phi,$

where ϕ is termed the velocity potential for the field **q**. The negative sign is taken as a matter of convention. This shows that ϕ decreases with an increase in the value of x, y, or z i.e., the flow is always in the direction of decreasing ϕ. The velocity potential is a scalar function of space and time.

Under the conditions of incompressibility, the velocity in terms of the potential ϕ may be substituted into the equation of continuity $\nabla . \mathbf{q} = 0$ which yield the condition that ϕ is a harmonic function satisfying Laplace equation

$$\nabla . (\nabla\phi) = 0 \Rightarrow \nabla^2\phi = 0; \quad ...(5)$$

at all points of the fluid.

The necessary and sufficient condition for (4) to hold is

$\nabla \times \mathbf{q} = 0.$

$$\text{or } \mathbf{i}\left(\frac{\partial w}{\partial y} - \frac{\partial v}{\partial z}\right) + \mathbf{j}\left(\frac{\partial u}{\partial z} - \frac{\partial w}{\partial x}\right) + \mathbf{k}\left(\frac{\partial v}{\partial x} - \frac{\partial u}{\partial y}\right) = 0. \quad ...(6)$$

If the relation (6) exists then the flow is said to be irrotational. In other words when the motion is irrotational the velocity vector is the gradient of a scalar function ϕ (x, y, z, t). The surfaces ϕ (x, y, z, t) = const. are called the equi-potentials. Potential flow is the irrotational flow of an inviscid or perfect fluid.

Rotational Flow

Let us consider a two-dimensional flow i.e., at every point P (x, y) of the fluid particle velocity will contain two components u and v only in the –X and –Y direction.

Let Q be the displaced position of the fluid particle at an interval of time t. The velocity components at the displaced position of the fluid particle Q will be given as

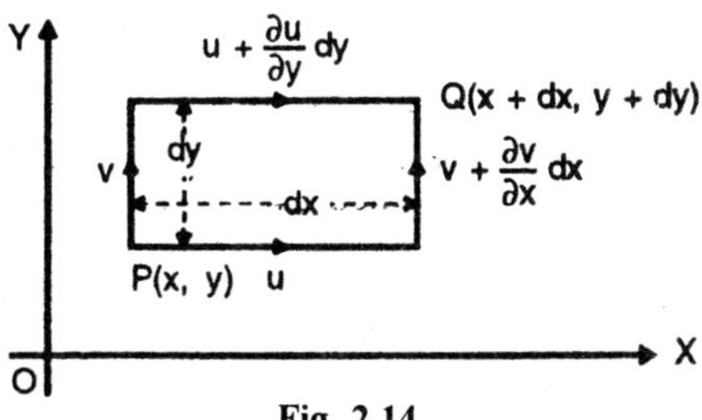

Fig. 2.14

$$u + \frac{\partial u}{\partial y}dy \text{ and } v + \frac{\partial v}{\partial x}dx.$$

The velocities will cause rotation of fluid about an axis perpendicular to the plane-XY. The rotation ω_z about Z-axis of the element is determined as fallows

$$\omega_z = \frac{\left(v + \frac{\partial v}{\partial x}dx\right) - v}{dx} = \frac{\partial v}{\partial x}, \text{ (anticlockwise)}$$

$$\text{and } \omega_z = \frac{u - \left(u + \frac{\partial u}{\partial y}dy\right)}{dy} = \frac{\partial u}{\partial y}, \text{ (clockwise)}$$

The rotation ω_z is given by the mean angular velocity then we have

$$\omega_z = \frac{1}{2}\left(\frac{\partial v}{\partial x} - \frac{\partial u}{\partial y}\right),$$

determines the condition for a flow to be rotational. The angular velocity ω_z must have some finite value other than zero.

$$\frac{\partial v}{\partial x} - \frac{\partial u}{\partial y} >< 0$$

The flow will be irrotational, when the angular velocity ωζ ωιλλ be zero. Then the condition for irrotationality is given as follows

$$\omega_z = \frac{1}{2}\left(\frac{\partial v}{\partial x} - \frac{\partial u}{\partial y}\right) = 0 \Rightarrow \frac{\partial v}{\partial x} = \frac{\partial u}{\partial y}.$$

VORTICITY

Vorticity vector : Let **q** represents the velocity of a fluid motion, then the vorticity of an element of fluid is defined as the curl of its velocity vector. The vorticity vector $\underline{\Omega}$ is represented by the relation

$$\underline{\Omega} = \text{curl } \mathbf{q}$$

Physically, vorticity may be generated in an inviscid fluid by the rotational body forces.

Vortex lines : A *vortex line* is a line whose direction coincides with the direction of the instantaneous axis of molecular rotation *i.e.*, a line to which vorticity vectors are tangent at all its points is called a vortex line. The vortex lines are determined as

$$\underline{\Omega} \times dr = 0, \ \underline{\Omega} = \mathbf{i}\xi + \mathbf{j}\eta + \mathbf{k}\zeta$$

$$\Rightarrow \frac{dx}{\xi} = \frac{dy}{\eta} = \frac{dz}{\zeta}.$$

Vortex tube : A *vortex tube* is obtained if through every point of a small closed curve, the corresponding vortex lines are drawn. The vortex lines and vortex tube cannot originate or terminate at internal points in a fluid. They can only form closed curves or terminate on boundaries. For example, in the case of smoke rings,the vortex lines form closed curves, but the vortex lines, in a whirlpoolterminates on the boundary of the fluid.

Vortex filament : The fluid contained within the vortex tube constitutes the *vortex filaments* or simply vortices. The boundary of a vortex filament is called a vortex tube.

Example 1: *Determine whether the motion specified by*

$$q = \frac{A(xj - yi)}{x^2 + y^2}, \ (A = const.)$$

is a possible motion for an incompressible fluid. If so, determine the equations of the streamlines. Also, show that the motion is of potential.

Solution: We know that

$$\nabla . \mathbf{q} = 0. \qquad ...(1)$$

$$\text{or } A\left\{-\frac{\partial}{\partial x}\left(\frac{y}{x^2 + y^2}\right) + \frac{\partial}{\partial y}\left(\frac{x}{x^2 + y^2}\right)\right\} = 0,$$

$$\text{or } A\left\{\frac{2xy}{(x^2 + y^2)^2} - \frac{2xy}{(x^2 + y^2)^2}\right\} = 0,$$

which is evident. Thus the equation of continuity for an incompressible fluid is satisfied and hence it is a possible motion for an incompressible fluid.

The equation of the streamlines are

$$\frac{dx}{u}=\frac{dy}{v}=\frac{dz}{w}$$

or $$\frac{dx}{-Ay/(x^2+y^2)}$$

$$=\frac{dy}{-Ax/(x^2+y^2)}=\frac{dz}{0}$$

or x dx + y dy = 0, dz = 0

By integrating, we have

$x^2 + y^2 =$ constant, z = 0 constant.

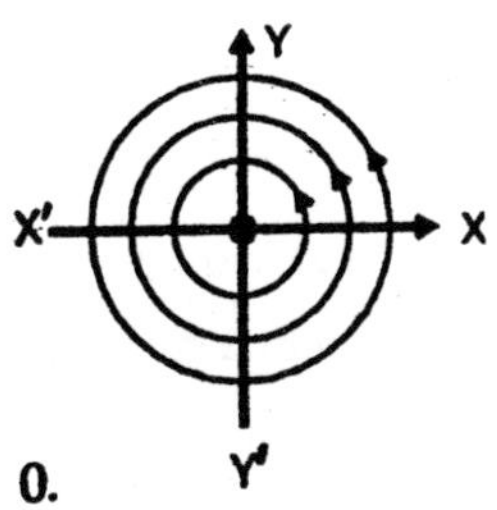

Fig. 2.15

Thus the streamlines are circle whose centres are on Z-axis, their planes being perpendicular to the axis.

$$\text{Again } \nabla \times \mathbf{q} = \begin{vmatrix} \mathbf{i} & \mathbf{j} & \mathbf{k} \\ \partial/\partial x & \partial/\partial y & \partial/\partial z \\ -\frac{Ay}{(x^2+y^2)} & \frac{Ax}{(x^2+y^2)} & 0 \end{vmatrix}$$

$$\text{or } \nabla \times \mathbf{q} = \mathbf{k}\left[\frac{\partial}{\partial x}\left\{\frac{Ax}{x^2+y^2}\right\}+\frac{\partial}{\partial y}\left\{\frac{Ay}{x^2+y^2}\right\}\right]$$

$$\text{or } \nabla \times \mathbf{q} = \mathbf{k}A\left[\frac{y^2-x^2}{(x^2+y^2)^2}+\frac{x^2-y^2}{(x^2+y^2)^2}\right]=0.$$

Thus the flow is of potential kind, so we can determine ϕ (x, y, z) such that

$$\mathbf{q} = -\nabla \phi$$

$$\text{or } \frac{\partial\phi}{\partial x}=-u=\frac{Ay}{x^2+y^2}, \frac{\partial\phi}{\partial y}=-v=-\frac{Ax}{x^2+y^2},$$

$$\frac{\partial\phi}{\partial z}=-w=0, \qquad \text{...(4, 5, 6)}$$

which shows that f is independent of z, hence

$\phi = \phi(x,y)$

Integrating the relation (4), we have

$\phi(x,y) = A \tan^{-1}(x/y) + f(y)$

or $\dfrac{\partial \phi}{\partial y} = f'(y) - Ax/(x^2 + y^2)$.

Using the relation (5), we get

$f'(y) = 0 \Rightarrow f(y) = \text{constant}$.

Therefore $\phi(x,y) = A \tan^{-1}(x/y)$. **Ans.**

Example 2: *Show that the velocity potential*

$$\phi = \frac{1}{2}a(x^2 + y^2 - 2z^2)$$

satisfies the Laplace equation. Also determine the streamlines.

Solution: Let ϕ be the velocity potential for the velocity field **q** then

$$\mathbf{q} = -\nabla\phi = -\frac{1}{2}a\nabla(x^2 + y^2 - 2z^2)$$

$$\mathbf{q} = -\frac{1}{2}a(2x\mathbf{i} + 2y\mathbf{j} - 4z\mathbf{k}).$$

Taking divergence of both the sides, we have

$\nabla^2\phi = -\nabla.\mathbf{q}$.

or $\nabla^2\phi = -1/2\, a\, \nabla.(2x\mathbf{i} + 2y\mathbf{j} - 4z\mathbf{k}) = 0$

or $\nabla^2\phi = -1/2\, a\,(2 + 2 - 4) = 0$

Hence Laplace equation is satisfied. The equation of streamlines are given by

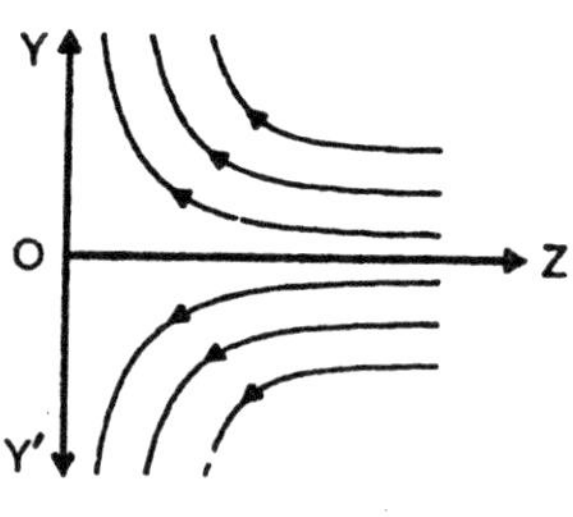

Fig. 2.16

$dx/u = dy/v = dz/w$

or $dx/(-ax) = dy/(-ay) = dz/(2az)$

(i) (ii) (iii)

From (ii) and (iii), we have

$$-\log y = \frac{1}{2}\log z - \log C,$$

where C is an integration constant.

or $y^2 z = C$,

which represents a cubic hyperbola. **Proved.**

BOUNDARY SURFACE

Physical conditions that should be satisfied on given boundaries of the fluid are called as boundary condition. At the boundary of the fluid, the equation of continuity is is replaced by a special surface condition.

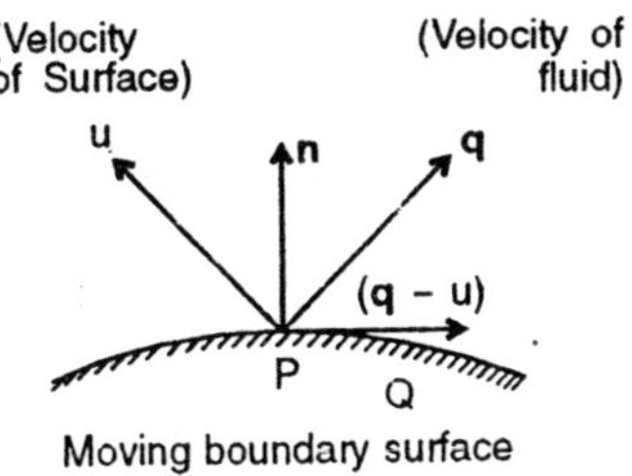

Fig. 2.17

When the fluid is in contact with an impermeable (non-porous) bounding surface the velocity of a fluid particle at any point of the boundary relative to the surface must be tangential tothe boundary. Thus at a fixed boundary, the velocity of the fluid perpendcular tothe surface must vanish and the normal component of the velocity of the fluid must be equal to the normal component of the velocity of the surface.

Let **q** be the velocity of the fluid and u be the velocity of the surface at the point P. Let **n** be unit normal vector drawn at the point P on the boundary surface F (**r**, t) = 0. Since there must be no relative normal velocity at P between boundary and fluid so we must have the two normal components equal *i.e.*,

$$\mathbf{q}.\mathbf{n} = \mathbf{u}.\mathbf{n} \Rightarrow (\mathbf{q} - \mathbf{u}).\ \mathbf{n} = 0. \quad ...(1)$$

For two fluids, in contact, a dynamical boundary is required ; viz, the pressure must be continuous across the interface

$$(\mathbf{q} - \mathbf{u}).\ \nabla F = 0,\ \mathbf{n} = \nabla F \quad ...(2)$$

The position of the point P on the moving surface at any instant of time t + δt is given by

$$F(\mathbf{r} + \delta\mathbf{r},\ t + \delta t) = 0$$

Expanding by Taylor's theorem, we have

$$F(\mathbf{r}, t) + dr.\ \nabla F + dt,\ (\partial F/\partial t) = 0$$

$$\text{or} \quad \frac{\partial F}{\partial t} + \frac{\partial r}{\partial t}.\nabla F = 0.$$

The above relation reduces to

$$\frac{\partial F}{\partial t} + u.\ \nabla\ F = 0\ ;\ \partial r \to 0,\ \partial t \to 0;\ u = dr/dt \qquad ...(3)$$

$$\frac{\partial F}{\partial t} + \mathbf{q}.\nabla F = 0 \Rightarrow \frac{\partial F}{\partial t} + u\frac{\partial F}{\partial x} + v\frac{\partial F}{\partial y} + w\frac{\partial F}{\partial z} = 0. \qquad ...(4)$$

Thus the equation of every boundary surface must satisfy the differential equation (4).

If the surface is at rest then $\partial F/\partial t = 0$, the relation (4) reduces to

$u\ (\partial F/\partial x) + v\ (\partial F/\partial y) + w\ (\partial F/\partial z) = 0$,

which represents the condition when the liquid is in contact with a rigid surface. In order that contact is maintained, thefluid and the surface must have the same velocity of the boundary is given by

$$\mathbf{u.\ n} = \frac{\mathbf{u}.\nabla F}{|\nabla F|} = -\frac{\partial F/\partial t}{\sqrt{\left\{\left(\frac{\partial F}{\partial x}\right)^2 + \left(\frac{\partial F}{\partial y}\right)^2 + \left(\frac{\partial F}{\partial z}\right)^2\right\}}}.$$

SOLVED EXAMPLES

Example 1: *Show that*

$$u = -\frac{2xyz}{(x^2+y^2)^2},\ v = \frac{(x^2-y^2)z}{(x^2+y^2)^2},\ w = \frac{y}{x^2+y^2},$$

are the velocity components of a possible liquid motion. Is this motion irrotational?

Solution: The condition for the possible liquid motion is given by

$$\frac{\partial u}{\partial x} + \frac{\partial v}{\partial y} + \frac{\partial w}{\partial z} = 0.$$

$$\Rightarrow 2yz.\ \frac{3x^2-y^2}{(x^2+y^2)^3} + 2yz.\frac{y^2-3x^2}{(x^2+y^2)^3} + 0 = 0,$$

which is an identity. Hence (u, v, w) are the velocity components of a possible liquid motion.

Again the condition for irrotational motion is

$$\frac{\partial v}{\partial z} - \frac{\partial w}{\partial y} = 0,\ \frac{\partial w}{\partial x} - \frac{\partial u}{\partial z} = 0 \text{ and } \frac{\partial u}{\partial y} - \frac{\partial u}{\partial x} = 0$$

$$\text{So } \frac{\partial v}{\partial z} - \frac{\partial w}{\partial y} = \frac{x^2 - y^2}{(x^2+y^2)^2} - \frac{x^2 - y^2}{(x^2+y^2)^2} = 0,$$

$$\frac{\partial w}{\partial x} - \frac{\partial u}{\partial z} = \frac{2xy}{(x^2+y^2)^2} - \frac{2xy}{(x^2+y^2)^2} = 0,$$

$$\text{and } \frac{\partial u}{\partial y} - \frac{\partial v}{\partial x} = \frac{2xz(3y^2 - x^2)}{(x^2+y^2)^2} - \frac{2xy(3y^2 - x^2)}{(x^2+y^2)^2} = 0.$$

Thus $\nabla \times \mathbf{q} = 0, \Rightarrow$ that the motion is irrotational. **Proved.**

Example 2: *Find the necessary and sufficient condition that vortex lines may be at right angles to the streamlines.*

Solution: The equations of the streamlines and the vortex lines are given by

$$\frac{dx}{u} = \frac{dy}{v} = \frac{dz}{w},$$

$$\text{and } \frac{dx}{\xi} = \frac{dy}{\eta} = \frac{dz}{\zeta}. \qquad \text{...(1, 2)}$$

The equation (1) and (2) are at right angles. It follows that

$$u\,\xi + v\eta + w\,\zeta = 0$$

$$\Rightarrow u\left(\frac{\partial w}{\partial y} - \frac{\partial v}{\partial y}\right) + v\left(\frac{\partial u}{\partial z} - \frac{\partial w}{\partial x}\right) + w\left(\frac{\partial v}{\partial x} - \frac{\partial u}{\partial y}\right) = 0.$$

In order that u dx + v dy + w dz may be a perfect differential, we have

$$u\,dx + v\,dy + w\,dz = \lambda\, d\phi = \lambda\left(\frac{\partial \phi}{\partial x}dx + \frac{\partial \phi}{\partial y}dy + \frac{\partial \phi}{\partial z}dz\right)$$

$$\Rightarrow u = \lambda\frac{\partial \phi}{\partial x},\ v = \lambda\frac{\partial \phi}{\partial y},\ w = \lambda\frac{\partial \phi}{\partial z},$$

which determines the necessary and sufficient condition.

Example 3: *In an incompressible fluid the vorticity at every point is constant in magnitude and direction; prove that the components of velocity u, v, w are the solutions of Laplace equation.*

Solution: Let $\underline{\Omega}$ be the vorticity any point in an incompressible fluid then

$$\underline{\Omega} = \xi\,\mathbf{i} + \eta\,\mathbf{j} + \zeta\,\mathbf{k}$$

where $\xi = \frac{\partial w}{\partial y} - \frac{\partial v}{\partial z}$, $\eta = \frac{\partial u}{\partial z} - \frac{\partial w}{\partial x}$, $\zeta = \frac{\partial v}{\partial x} - \frac{\partial u}{\partial y} = 0$.

The magnitude and direction consines of its direction are given by

$\Omega = \sqrt{\xi^2 + \eta^2 + \zeta^2}$ and $\frac{\xi}{\Omega}, \frac{\eta}{\Omega}, \frac{\zeta}{\Omega}$

Differentiating h partially with regard to z and ζ with regard to y and subtracting,we have

$$\frac{\partial}{\partial z}\left(\frac{\partial u}{\partial z} - \frac{\partial w}{\partial x}\right) - \frac{\partial}{\partial y}\left(\frac{\partial v}{\partial x} - \frac{\partial u}{\partial y}\right) = 0$$

$$\Rightarrow \frac{\partial^2 u}{\partial z^2} + \frac{\partial^2 u}{\partial y^2} - \frac{\partial}{\partial x}\left(\frac{\partial v}{\partial y} + \frac{\partial w}{\partial z}\right) = 0 \Rightarrow \frac{\partial^2 u}{\partial x^2} + \frac{\partial^2 u}{\partial y^2} + \frac{\partial^2 u}{\partial z^2} = 0.$$

Hence the velocity components satisfy Laplace Equation. **Ans.**

Example 4: *Find the vorticity components of a fluid particle when velocity distribution is*

$\boldsymbol{q} = \boldsymbol{i}\,(k_1\, x^2yt) + \boldsymbol{j}\,(k_2\, y^2zt) + \boldsymbol{k}\,(k_3zt^2)$,

where k_1, k_2, k_3 *are constants.*

Solution: The vorticity components ξ, η, ζ are given by

$$\xi = \frac{\partial w}{\partial y} - \frac{\partial v}{\partial z} = -\, k_2y^2t,$$

$$\eta = \frac{\partial u}{\partial z} - \frac{\partial w}{\partial x} = 0$$

$$\zeta = \frac{\partial v}{\partial x} - \frac{\partial u}{\partial y} = -\, k_1x^2t.$$ **Ans.**

Example 5: *Determine the equations of the vortex lines when the velocity vector of the field is given by*

$\boldsymbol{q} = \boldsymbol{i}\,(Az - By) + \boldsymbol{j}\,(Bx - Cz) + \boldsymbol{k}\,(Cy - Ax)$,

where A, B, C are constants.

Solution: The vorticity components are given by

$$\xi = \frac{\partial w}{\partial y} - \frac{\partial v}{\partial z} = C + C = 2C.$$

$$\eta = \frac{\partial u}{\partial z} - \frac{\partial w}{\partial x} = A + A = 2A,$$

$$\zeta = \frac{\partial v}{\partial x} - \frac{\partial u}{\partial y} = B + B = 2B.$$

The equations of the vortex lines are

$$\frac{dx}{\xi} = \frac{dy}{\eta} = \frac{dz}{\zeta}$$

$$\Rightarrow \frac{dx}{2C} = \frac{dy}{2A} = \frac{dz}{2B}$$

(i) (ii) (ii)

From (i) and (ii), we have

$Ax - Cy = k_1$, ...(1)

From (ii) and (iii), we have

$By - Az = k^2$, where k_1 and k_2 are integration constants ...(2)

Hence the vortex lines (1) and (2) are the straighlines. **Ans.**

Example 6: *Investigate the nature of the liquid motion given by*

$$u = \frac{ax - by}{x^2 - y^2}, v = \frac{ay - bx}{x^2 - y^2}, wm = 0$$

Also, determine the velocity potential.

Solution: Here $u = \dfrac{ax - by}{x^2 - y^2}, v = \dfrac{ay - bx}{x^2 - y^2}, w = 0$

$$\frac{\partial u}{\partial x} = \frac{a(x^2 + y^2) - 2x(ax - by)}{(x^2 + y^2)^2} = \frac{a(y^2 - x^2) + 2bxy}{(x^2 + y^2)^2},$$

$$\frac{\partial v}{\partial y} = \frac{a(x^2 + y^2) - 2y(ay - bx)}{(x^2 + y^2)^2} = \frac{a(x^2 - y^2) - 2bxy}{(x^2 + y^2)^2},$$

$$\Rightarrow \frac{\partial u}{\partial x} + \frac{\partial v}{\partial y} = 0.$$

Thus the liquid motion satisfies the continuity equation hence it is a possible motion.

Let $\underline{\Omega}$ be the vorticity then

$$\underline{\Omega} = \mathbf{i}\,\xi + \mathbf{j}\,\eta + \mathbf{k}\zeta,$$

where $\xi = \dfrac{\partial w}{\partial y} + \dfrac{\partial v}{\partial z} = 0,$

$$\eta = \frac{\partial u}{\partial z} - \frac{\partial w}{\partial x} = 0,$$

$$\zeta = \frac{\partial v}{\partial x} - \frac{\partial u}{\partial y} = 0,$$

It follows that the nature of the liquid motion is irrotational.

Let f be the velocity potential, then

$$d\phi = \frac{\partial \phi}{\partial x} dx + \frac{\partial \phi}{\partial y} dy\, x^2 + y^2) + b \tan^{-1}\left(\frac{y}{x}\right).$$ **Ans.**

Example 7: *If u dx + v dy + w dz = dθ + λ dμ where λ, θ, μ are functions of x, y, z and t, prove that the vortex lines at any time are the lines at any time are the lines of intersection of the surfaces λ = const. and μ = const.*

Solution : We know that

$$u\,dx + v\,dy + w\,dz = d\theta + \lambda\,d\mu.$$

$$\text{or } u\,dx + v\,dy + w\,dz = \left(\frac{\partial \theta}{\partial x}dx + \frac{\partial \theta}{\partial y}dy + \frac{\partial \theta}{\partial z}dz + \frac{\partial \theta}{\partial t}dt\right)$$

$$+ \lambda\left(\frac{\partial \mu}{\partial x}dx + \frac{\partial \mu}{\partial y}dy + \frac{\partial \mu}{\partial z}dz + \frac{\partial \mu}{\partial t}dt\right)$$

Equating coefficient of dx, dy dz and dt, we have

$$u = \frac{\partial \theta}{\partial x} + \lambda\frac{\partial \mu}{\partial x}, v = \frac{\partial \theta}{\partial y} + \lambda\frac{\partial \mu}{\partial y},$$

$$w = \frac{\partial \theta}{\partial z} + \lambda\frac{\partial \mu}{\partial z}, O = \frac{\partial \theta}{\partial t} + \lambda\frac{\partial \mu}{\partial t}.$$

The components of spin are

$$2\xi = \frac{\partial w}{\partial y} - \frac{\partial v}{\partial z} = \frac{\partial}{\partial y}\left(\frac{\partial \theta}{\partial z} + \lambda\frac{\partial \mu}{\partial z}\right) - \frac{\partial}{\partial z}\left(\frac{\partial \theta}{\partial y} + \lambda\frac{\partial \mu}{\partial y}\right)$$

$$\Rightarrow 2\xi = \lambda\,\frac{\partial^2 \mu}{\partial y \partial z} + \frac{\partial \lambda}{\partial y}\frac{\partial \mu}{\partial z} - \lambda\frac{\partial^2 \mu}{\partial y \partial z} - \frac{\partial \lambda}{\partial z}\frac{\partial \mu}{\partial y}$$

$$\Rightarrow 2\xi = \frac{\partial \lambda}{\partial y}\frac{\partial \mu}{\partial z} - \frac{\partial \lambda}{\partial z}\frac{\partial \mu}{\partial y}$$

$$\Rightarrow 2\xi = \begin{vmatrix} \frac{\partial\lambda}{\partial y} & \frac{\partial\lambda}{\partial z} \\ \frac{\partial\mu}{\partial y} & \frac{\partial\mu}{\partial z} \end{vmatrix}$$

Similarly $2\eta = \begin{vmatrix} \frac{\partial\lambda}{\partial z} & \frac{\partial\lambda}{\partial x} \\ \frac{\partial\mu}{\partial z} & \frac{\partial\mu}{\partial x} \end{vmatrix}$

and $\quad 2\zeta = \begin{vmatrix} \frac{\partial\lambda}{\partial x} & \frac{\partial\lambda}{\partial y} \\ \frac{\partial\mu}{\partial x} & \frac{\partial\mu}{\partial y} \end{vmatrix}$

Therefore $2\left(\xi\frac{\partial\lambda}{\partial x} + \eta\frac{\partial\lambda}{\partial y} + \zeta\frac{\partial\lambda}{\partial z}\right)$

$$= \begin{vmatrix} \lambda_x & \lambda_y & \lambda_z \\ \lambda_x & \lambda_y & \lambda_z \\ \mu_x & \mu_y & \mu_z \end{vmatrix} = 0$$

$\Rightarrow \xi\lambda_x + \eta\lambda_y + \zeta\lambda_z = 0$

Similarly $\xi\mu_x + \eta\mu_y + \xi\mu_z = 0$

It follows that the vortex lines lie on the surfaces

λ =const and μ = const. **Ans.**

Example 8: *If the velocity of an incompressible fluid at the point (x, y, z) is given by* $3xz/r^5, 3yz/r^5, (3z^2 - r^2)/r^5$*, prove that the liquid motion is possible and that the velocity potential is* $\cos\theta/r^2$*. Also, determine the stream lines.*

Solution: The condition for the possible liquid motion is

$$\frac{\partial u}{\partial x} + \frac{\partial v}{\partial y} + \frac{\partial w}{\partial z} = 0.$$

$$u = \frac{3xz}{r^5} \Rightarrow \frac{\partial u}{\partial x} = \frac{3z}{r^5} - \frac{15xz}{r^6}\cdot\frac{\partial r}{\partial x} = \frac{3z}{r^5} - \frac{15x^2z}{r^7}$$

$$\text{or} \quad \frac{3z}{r^5} - \frac{15x^2z}{r^7} + \frac{3z}{r^5} - \frac{15y^2z}{r^7} + \frac{6z}{r^5} - \frac{15z^3}{r^5} + \frac{3z}{r^5} = 0$$

or $\dfrac{15z}{r^5} - \dfrac{15z(x^2 + y^2 + z^2)}{r^7} = 0$

$\Rightarrow \dfrac{15z}{r^5} - \dfrac{15z}{r^5} = 0,$

which is an identity. Hence (u, v, w) are the velocity components of a possible liquid motion.

If ϕ be the velocity potential, then

$d\phi = (\partial\phi/\partial x)\, dx + (\partial\phi/\partial y)\, dy + (\partial\phi/\partial z) dz$

or $d\phi = -(u\, dx + v\, dy + w\, dz)$

or $d\phi = -\dfrac{1}{r^5}\{3xz\, dz + 3yz\, dy + (3z^2 - r^2)\, dz\}$

or $d\phi = -\dfrac{1}{r^5}\{3z\,(x\, dx + y\, dy + z\, dz) - r^2\, dz\}$

or $d\phi = -\dfrac{3z}{2}\dfrac{d(x^2 + y^2 + z^2)}{r^5} + \dfrac{dz}{r^3}$

or $d\phi = -\dfrac{3z}{2}\dfrac{d(r^2)}{r^5} + \dfrac{dz}{r^3} = -\dfrac{3z}{2}.\dfrac{2rdr}{r^5} + \dfrac{dz}{r^3} = d\left(\dfrac{z}{r^3}\right)$

By integrating, we have

$\phi = \dfrac{z}{r^3} = \dfrac{r\cos\theta}{r^3} = \dfrac{\cos\theta}{r^2},$

constant of integration vanishes. **Proved.**

The equations to the streamlines are given by

$$\frac{dx}{u} = \frac{dy}{v} = \frac{dz}{w}$$

or $\dfrac{dx}{3xz/r^5} = \dfrac{dy}{3yz/r^5} = \dfrac{dz}{(3z^2/r^2)/r^5}$

or $\underset{\text{(i)}}{\dfrac{dx}{3xz}} = \underset{\text{(ii)}}{\dfrac{dy}{3yz}} = \underset{\text{(iii)}}{\dfrac{dz}{3z^2 - (x^2 + y^2 + z^2)}} = \underset{\text{(iv)}}{\dfrac{xdx + ydy + zdz}{2z(x^2 + y^2 + z^2)}}$

From (i) and (ii), we have

$\dfrac{dx}{x} = \dfrac{dy}{y} \Rightarrow \log x = \log y + \log c \Rightarrow x = cy.$ ...(1)

From (i) and (iv), we have

$$\frac{dx}{3x} = \frac{xdx + ydy + zdz}{2(x^2 + y^2 + z^2)}$$

By integrating, we have

$$\frac{2}{3}\log x = \frac{1}{2}\log(x^2 + y^2 + z^2) + \log D,$$

where D is an arbitrary constant.

$$x^{2/3} = D\,(x^2 + y^2 + z^2)^{1/2}. \qquad ...(2)$$

Thus the equation (1) and (2) represents the stream lines. **Ans.**

Example 9: *For an incompressible fluid $u = -\omega y$, $v = \omega x$, $w = 0$, show that the surfaces intersecting the streamlines orthogonally exist and are the planes through Z-axis, although the velocity potential does not exist. Discuss the nature of flow.*

Solution: The motion will be possible if it satisfies the equation of continuity, that is,

$$\frac{\partial u}{\partial x} + \frac{\partial v}{\partial y} + \frac{\partial w}{\partial z} = 0,$$

which is true from the given relation. Hence the motion is a possible one.

The differential equation to the lines of flow are

$$\frac{dx}{u} = \frac{dy}{v} = \frac{dz}{w} \Rightarrow \frac{dx}{\omega y} = \frac{dy}{\omega x} = \frac{dz}{0}$$

or $x\,dx + y\,dy = 0$ and $dz = 0$

By integrating, we have

$x^2 + y^2 = \text{const.}$, and $z = \text{const.}$

The surfaces which cut the stream lines orthogonally are

$u\,dx + v\,dy + w\,dz = 0$

or $-wy\,dx + wx\,dy = 0$

By integrating, we have

$dx/x - dy/y = 0 \Rightarrow \log(x/y) = \log c$,

where c is an arbitrary constant.

Therefore $x = cy$, which represents a plane through Z-axis and cuts the stream line orthogonally.

The velocity potential will exist if u dx + v dy + w dz is a perfect differential. But u dx + v dy + w dz is not a perfect differential, therefore, the surfaces intesecting streamlines orthogonally exist and are the through Z-axis, although the velocity potential does not exist. Further

$$\nabla \times \mathbf{q} = \begin{vmatrix} \mathbf{i} & \mathbf{j} & \mathbf{k} \\ \partial/\partial x & \partial/\partial y & \partial/\partial z \\ -\omega y & \omega x & 0 \end{vmatrix} = 2w\mathbf{k}.$$

Hence the flow is not of the potential kind. It shows that a rigid body rotating about Z-axis with constant angular velocity ωk gives the same type of motion. **Ans.**

Example 10 : *Show that $\phi = x f(r)$ is a possible form for the velocity potential of an incompressible motion. Given that the liquid speed* $\mathbf{q} \to 0$ *as* $r \to \infty$, *deduce that the surface of constant speed are* $(r^2 + 3x^2) r^{-8}$ = *constant.*

Solution : Since $\phi = x\, f(r)$

$\mathbf{q} = -\nabla\phi = -\nabla\{x\, f(r)\}$

$\mathbf{q} = -f(r)\, \nabla x - x\, \nabla f(r)$

$\mathbf{q} = -f(\mathbf{r})\, i - \{x\, f'(r)/r\}\mathbf{r}$ ...(1)

It will be a possible liquid motion if $\nabla.\, \mathbf{q} = 0$

$$\nabla.\, \mathbf{q} = -\nabla\{f(r)\, \mathbf{i}\} - \left\{x\frac{f'(r)}{r}\right\}(\nabla.\, \mathbf{r}) - \mathbf{r}.\left\{\frac{xf'(r)}{r}\right\}$$

$$\nabla.\, \mathbf{q} = -\frac{x}{r}\frac{\partial f}{\partial x} - \left(\frac{\partial}{\partial x}\mathbf{j} + \frac{\partial}{\partial y}\mathbf{j} + \frac{\partial}{\partial z}\mathbf{k}\right).\left[\frac{xf'(r)}{r}(x\mathbf{i} + y\mathbf{j} + z\mathbf{k})\right]$$

$$\nabla.\, \mathbf{q} = -\frac{x}{r}\frac{\partial f}{\partial x} - \frac{\partial}{\partial x}\left\{\frac{x^2 f'(r)}{r}\right\} - \frac{\partial}{\partial y}\left\{\frac{xyf'(r)}{r}\right\} - \frac{\partial}{\partial z}\left\{\frac{xzf'(r)}{r}\right\} \quad ...(2)$$

$$\nabla.\, \mathbf{q} = -\frac{x}{r}\frac{\partial f}{\partial x} - \frac{2xf'(r)}{r} - \frac{x^3}{r^3}\{rf''(r) - f'(r)\} - \frac{xf'(r)}{r}$$

$$-\frac{xy^2}{r^3}\{rf''(r) - f'(r)\} - \frac{xf'(r)}{r} - \frac{xz^2}{r^3}\{rf''(r) - f'(r)\}$$

$$\nabla.\, \mathbf{q} = -\frac{x}{r}\frac{\partial f}{\partial x} - \frac{4x}{r}f'(r) - \frac{x}{r^2}\{(x^2 + y^2 + z^2)\, f''(r)\}$$

$$+\frac{x}{r^3}\{(x^2 + y^2 + z^2)f'(r)\}$$

$$\nabla .\, \mathbf{q} = -\frac{x}{r}\frac{\partial f}{\partial x} - \frac{4x}{r} f'(r) - x f''(r) + \frac{x}{r} f.'(r)$$

$$\nabla .\, \mathbf{q} = -\frac{x}{r} f' - \frac{4x}{r} f'(r) - x f''(r) + \frac{x}{r} f'(r)$$

$$\nabla .\, \mathbf{q} = -x\left(f'' + \frac{4f'}{r}\right).$$

For an incompressible liquid $\nabla .\, \mathbf{q} = 0$, hence

$$f'' + 4\frac{f'}{r} = 0$$

$$\Rightarrow \frac{f''}{f'} + \frac{4}{r} = 0.$$

By integrating, we have

$\log f' + 4\log r = \log C$

$\Rightarrow f' = Cr^{-4}$ ...(2)

Integrating (2), we have

$f = -\frac{1}{3}Cr^{-3} + D$, where D is a constant.

From (1) we have

$$\mathbf{q} = \left(\frac{C}{3r^3} - D\right).\mathbf{i} - \frac{Cx}{r^5}\mathbf{r} \qquad ...(3)$$

When $r \to \infty$ $\mathbf{q} \to -D\mathbf{i}$

$\Rightarrow D = 0.$

Thus $\mathbf{q} = \frac{C}{3r^3}\left(\mathbf{i} - \frac{3x\mathbf{r}}{r^2}\right).$

and $\mathbf{q}^2 = \mathbf{q}.\,\mathbf{q}.$

$$= \frac{C^2}{9r^6}\left(\mathbf{i} - \frac{3x\mathbf{r}}{r^2}\right).\left(\mathbf{i} - \frac{3x\mathbf{r}}{r^2}\right)$$

$$\mathbf{r}^2 = \frac{C^2}{9r^6}\left(1 - \frac{6x\mathbf{r}.\mathbf{i}}{r^2} + \frac{9x^2r^2}{r^4}\right).$$

Hence $\mathbf{q}^2$ = constant gives the surfaces of constant speed as

$(r^2 + 3x^2)\, r^{-8}$ = constant. **Proved.**

SYMMETRICAL FORMS OF THE EQUATION OF CONTINUITY

Spherical Symmetry : Consider two concentric spheres whose motion is symmetrical about the centre O. Let q (r, t) be the velocity in the direction OP. MASS gained by the flow through the inner surface is

$4\pi r^2 q_r \rho = F\ (r,\ t)$ (say).

Mass lost by the flow through the outer surface is

$= F\ (r + \delta r,\ t)$.

Mass between the spheres at an instant of time t is

$= 4\pi r^2\ \rho\delta r$.

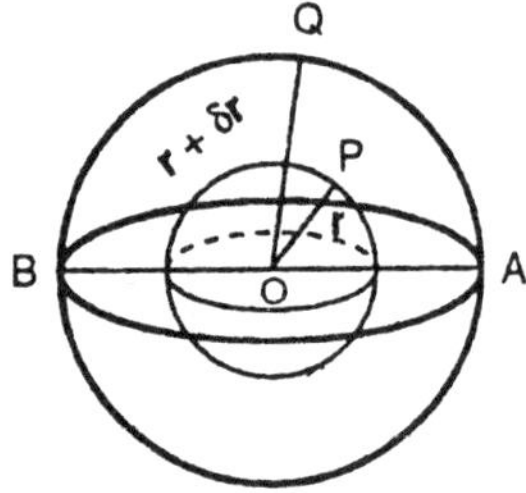

Fig. 2.18

By the principle of conservation of mass, we have

$$\frac{\partial \rho}{\partial t}(4\pi r^2 \delta r) = F\ (r,\ t) - F\ (r + dr,\ t)$$

$$\Rightarrow \frac{\partial \rho}{\partial t}(4\pi r^2 \delta r) = F\ (r,\ t) - F(r,\ t) - \frac{\partial F}{\partial r}\delta r \ .....$$

$$\Rightarrow \frac{\partial \rho}{\partial t}(4\pi r^2 \delta r) = \delta r \frac{\partial}{\partial r}(4\pi r^2 q\rho)$$

$$\Rightarrow \frac{\partial \rho}{\partial t} + \frac{1}{r^2}\frac{\partial}{\partial r}(\rho q_r r^2) = 0. \qquad ...(1)$$

For an incompressible fluid r is constant throughout the motion *i.e.*, $\partial p/\partial t = 0$, then equation (1) reduces to

$r^2 q_r = F\ (t)$.

In a steady flow F (t) shall be absolute constant.

Cylindrical Symmetry : Let q (r, t) be the velocity at any point P which is perpendicular to a fixed axis OZ, r is the perpendicular distance of P from OZ. Consider two cylinders of radii r and r + δr at a unit distance a part with OZ as axis.

Rate of flow across the inner surface = ρ**q** (2πr) = F(r, t).

Rate of flow across the outer surface = F (r + δr, t).

Mass between the two cylinders at any instant of time t

= 2π rρδr.

By the principle of continuity, we have

$$\frac{\partial}{\partial t}(2\pi r\rho\delta r) = F(r, t) - F(r + \delta r, t)$$

$$= F(r, t) - F(r, t) - \delta r\frac{\partial F}{\partial r} - \ldots$$

$$\Rightarrow \frac{\partial \rho}{\partial t}2\pi r\delta r = -\delta r\frac{\partial F}{\partial r} = -\delta r\frac{\partial}{\partial r}(\rho \mathbf{q}.\ 2\pi r)$$

$$\Rightarrow \frac{\partial \rho}{\partial t} + \frac{1}{r}\frac{\partial}{\partial r}(\rho r\mathbf{q}) = 0.$$

If the fluid be incompressible then r**q** = F (t). In a steady flow F (t) shall be an absolute constant.

Example 1: *A sphere is at rest in an infinite mass of homogeneous liquid of density ρ, the pressure at infinity being Π. Show that if, the radius R of the sphere varies in any manner, the pressure at the surface at any time t is*

$$\Pi \frac{1}{2}\rho\left\{\frac{d^2}{dt^2}(R^2) + \left(\frac{dR}{dt}\right)^2\right\} +.$$

If R = a (2 + cos nt), show that to prevent cavitation in the fluid, P must not be less that $3\rho a^2n^2$.

Solution: The only possible motion which can take place is one in which each element of liquid moves towards the centre, whence the free surface will remain spherical. In an incompressible fluid, the fluid velocity will be radial spherical outside the sphere, so it will be a function of r and the time t only. The equation of continuity reduces to

$$\frac{1}{2}\frac{d}{dr}(r^2v) = 0$$

$\Rightarrow r^2v$ = constant = F (t), (let) ...(1)

where v is the velocity of a fluid particle at a distance r from the centre at the time t.

Equation of motion is

$$\frac{\partial v}{\partial t} + v\frac{\partial v}{\partial r} = -\frac{1}{\rho}\frac{\partial p}{\partial r}, \quad \begin{cases} \text{Since } r^2 v = F(t) \\ \dfrac{dv}{dt} = \dfrac{F'(t)}{r^2} \end{cases}$$

$$\text{or } \frac{F'(t)}{r^2} + v\frac{\partial v}{\partial r} = -\frac{1}{\rho}\frac{\partial p}{\partial r}.$$

Integrating with regard to r, we have

$$-\frac{F'(t)}{r} + \frac{1}{2}v^2 = \frac{p}{\rho} + B, \qquad \text{...(2)}$$

where B is an arbitrary constant.

Initially $r \to \infty$, $v = 0$, $p = \Pi$; $B = (\Pi/\rho)$

$$\Rightarrow -\frac{F'(t)}{r} + \frac{1}{2}v^2 = \frac{\Pi}{\rho} - \frac{p}{\rho}$$

$$\Rightarrow p = \Pi + \frac{1}{2}\rho\left(\frac{2F'(t)}{r} - v^2\right)$$

Let P be the pressure on the surface of the sphere of radius R and V be the velocity, then

$$P = \Pi + \frac{1}{2}\rho\left\{\frac{2F'(t)}{R} - V^2\right\} \qquad \text{...(3)}$$

Again from (1) we have

$R^2V = F(t)$, $F(t) = R^2 (dR/dt)$.

Differentiating with regard to t, we have

$F'(t) = R^2(d^2R/dt^2) + 2R(dR/dt)^2$.

From the equation (3), we have

$$P = \Pi + \frac{1}{2}\rho\left\{2R\frac{d^2R}{dt^2} + 4\left(\frac{dR}{dt}\right)^2 - V^2\right\}$$

$$\text{or } P = \Pi + \frac{1}{2}\rho\left\{2R\frac{d^2R}{dt^2} + 4\left(\frac{dR}{dt}\right)^2 - \left(\frac{dR}{dt}\right)^2\right\}$$

$$\text{or } P = \Pi + \frac{1}{2}\rho\left\{2\left[R\frac{d^2R}{dt^2} + -\left(\frac{dR}{dt}\right)^2\right] + \left(\frac{dR}{dt}\right)^2\right\}$$

$$\text{or } P = \Pi + \frac{1}{2}\rho\left\{\frac{d^2}{dt^2}(R^2) + \left(\frac{dR}{dt}\right)^2\right\}$$ **Proved.**

Since $r'^2 v' = \text{constant}$

Now v' is maximum when r' is minimum *i.e.* $r' = R$

Again $R = a(2 + \cos nt)$, $R^2 = a^2 (2 + \cos nt)^2$ ...(4)

$$\text{or} \quad \frac{dR}{dt} = -an \sin nt$$

$$\frac{d}{dt}(R^2) = -2a^2n(2 + \cos nt) \sin nt$$

$$\text{and} \quad \frac{d^2}{dt^2}(R^2) = 2a^2n^2 \sin^2 nt - 2a^2n^2 (2 + \cos nt) \cos nt$$

Hence

$$P = \Pi + \frac{1}{2}\rho \{2a^2n^2 \sin^2 nt - a^2n^2r (2 + \cos nt) \cos nt + a^2n^2 \sin^2 nt\}$$

$$P = \text{II} + \frac{3}{2}\rho a^2n^2\sin^2 nt - a^2n^2r(2\cos nt + \cos^2 nt) \qquad ...(5)$$

From (4), we obtain that R varies form 3a to a. Thus the sphere has greatest radius 3a when $nt = 0$ or $2m\pi$. As sphere shrinks from $R = 3a$, there is a possibility of a cavitation value of pressure on the surface of the sphere *i.e.*, $t = 0$ or $nt = 2m\pi$ then from (5), we have

$$P' = \Pi - 3\rho a^2n^2$$

Hence, to prevent, cavitation in the fluid P' must be positive *i.e.*, Π must not be less than $3\rho a^2n^2$.

CAUCHY'S INTEGRAL

Consider $Q = V + \int \frac{dp}{\rho}$.

Let (x, y, z) be the co-ordinates of a particle at any time t, whose initial co-ordinates are (a, b, c). Consider the density r as a function of pressure p. Differentiating partially (1) with regard to a, we have

$$\frac{\partial Q}{\partial a} = \frac{\partial V}{\partial a} + \frac{1}{\rho}\frac{\partial p}{\partial a}. \qquad ...(2)$$

Similarly we can write other two equations.

Equations of motion are

$$\frac{\partial^2 x}{\partial t^2} = -\frac{\partial V}{\partial x} - \frac{1}{\rho}\frac{\partial p}{\partial x},$$

$$\frac{\partial^2 y}{\partial t^2} = -\frac{\partial V}{\partial y} - \frac{1}{\rho}\frac{\partial p}{\partial y},$$

$$\frac{\partial^2 z}{\partial t^2} = -\frac{\partial V}{\partial z} - \frac{1}{\rho}\frac{\partial p}{\partial z} \qquad ...(3)$$

Multiplying (3) by $\frac{\partial x}{\partial a}$, $\frac{\partial y}{\partial a}$ and $\frac{\partial z}{\partial a}$ respectively and adding, we have

$$\frac{\partial^2 x}{\partial t^2}\frac{\partial x}{\partial a} + \frac{\partial^2 y}{\partial t^2}\frac{\partial y}{\partial a} + \frac{\partial^2 z}{\partial t^2}\frac{\partial z}{\partial a} = -\left(\frac{\partial V}{\partial a} + \frac{1}{\rho}\frac{\partial p}{\partial a}\right)$$

or
$$\frac{\partial^2 x}{\partial t^2}\frac{\partial x}{\partial a} + \frac{\partial^2 y}{\partial t^2}\frac{\partial y}{\partial a} + \frac{\partial^2 z}{\partial t^2}\frac{\partial z}{\partial a} = -\frac{\partial Q}{\partial a},$$

Similarly
$$\frac{\partial^2 x}{\partial t^2}\frac{\partial x}{\partial b} + \frac{\partial^2 y}{\partial t^2}\frac{\partial y}{\partial b} + \frac{\partial^2 z}{\partial t^2}\frac{\partial z}{\partial b} = -\frac{\partial Q}{\partial b},$$

and
$$\frac{\partial^2 x}{\partial t^2}\frac{\partial x}{\partial c} + \frac{\partial^2 y}{\partial t^2}\frac{\partial y}{\partial c} + \frac{\partial^2 z}{\partial t^2}\frac{\partial z}{\partial c} = -\frac{\partial Q}{\partial c}. \qquad ...(4, 5, 6)$$

Since $\frac{\partial}{\partial b}\left(\frac{\partial Q}{\partial c}\right) = \frac{\partial}{\partial c}\left(\frac{\partial Q}{\partial b}\right)$.

Q will be eliminated by differentiating (5) and (6) partially with regard to c and b respectively and subtracting, we have

$$\left(\frac{\partial^2 u}{\partial b \partial t}\frac{\partial x}{\partial c} - \frac{\partial^2 u}{\partial c \partial t}\frac{\partial x}{\partial b}\right) + \left(\frac{\partial^2 v}{\partial b \partial t}\frac{\partial x}{\partial c} - \frac{\partial^2 v}{\partial c \partial t}\frac{\partial y}{\partial b}\right)$$

$$+ \left(\frac{\partial^2 w}{\partial b \partial t}\frac{\partial z}{\partial c} - \frac{\partial^2 w}{\partial c \partial t}\frac{\partial z}{\partial b}\right) = 0$$

or
$$\left(\frac{\partial}{\partial t}\left(\frac{\partial u}{\partial b}\frac{\partial x}{\partial c} - \frac{\partial u}{\partial c}\frac{\partial x}{\partial b}\right) - \frac{\partial u}{\partial b}\frac{\partial^2 x}{\partial t \partial c} + \frac{\partial u}{\partial c}\frac{\partial^2 x}{\partial t \partial b}\right)$$

$$+ \text{two similarexpression} = 0$$

or $\frac{\partial}{\partial t}\left(\frac{\partial u}{\partial b}\frac{\partial x}{\partial c}-\frac{\partial u}{\partial c}\frac{\partial x}{\partial b}\right)+\frac{\partial}{\partial t}\left(\frac{\partial v}{\partial b}\frac{\partial y}{\partial c}-\frac{\partial v}{\partial c}\frac{\partial y}{\partial b}\right)$

$$+\frac{\partial}{\partial t}\left(\frac{\partial w}{\partial b}\frac{\partial z}{\partial c}-\frac{\partial w}{\partial c}\frac{\partial z}{\partial b}\right)=0.$$

$$\left(\text{Since}\frac{\partial^2 x}{\partial t\partial c}=\frac{\partial u}{\partial c},\frac{\partial^2 x}{\partial t\partial b}=\frac{\partial u}{\partial b}\right)$$

By integrating with regard to t, we have

$$\frac{\partial u}{\partial b}\frac{\partial x}{\partial c}-\frac{\partial u}{\partial c}\frac{\partial x}{\partial b}+\frac{\partial v}{\partial b}\frac{\partial y}{\partial c}-\frac{\partial v}{\partial c}\frac{\partial y}{\partial b}+\frac{\partial w}{\partial b}\frac{\partial z}{\partial c}-\frac{\partial w}{\partial c}\frac{\partial z}{\partial b}$$

$$=\frac{\partial w_0}{\partial b}-\frac{\partial v_0}{\partial c}, \quad ...(4)$$

where u_0, v_0, w_0 are initial values.

Initially $\frac{\partial x}{\partial a}=1,\ \frac{\partial x}{\partial b}=0,\ \frac{\partial x}{\partial c}=0$ etc. as $x = a$, $y = b$, $z = c$

Since $\frac{\partial u}{\partial a}=\frac{\partial u}{\partial x}\frac{\partial x}{\partial a}+\frac{\partial u}{\partial y}\frac{\partial y}{\partial a}+\frac{\partial u}{\partial z}\frac{\partial z}{\partial a}$.

Then relation (2) becomes

$$\left(\frac{\partial w}{\partial y}-\frac{\partial v}{\partial z}\right)\frac{\partial(yz)}{\partial(bc)}+\left(\frac{\partial u}{\partial z}-\frac{\partial w}{\partial x}\right)\frac{\partial(zx)}{\partial(bc)}$$

$$+\left(\frac{\partial v}{\partial x}-\frac{\partial u}{\partial y}\right)\frac{\partial(xy)}{\partial(bc)}=\frac{\partial w_0}{\partial b}-\frac{\partial v_0}{\partial c}$$

or $\xi\frac{\partial(yz)}{\partial(bc)}+\eta\frac{\partial(zx)}{\partial(bc)}+\zeta\frac{\partial(xy)}{\partial(bc)}=\xi_0,$

where ξ, η, ζ are the vorticity components. ...(5)

Similarly other two expressions are

$$\xi\frac{\partial(yz)}{\partial(ca)}+\eta\frac{\partial(zx)}{\partial(ca)}+\zeta\frac{\partial(xy)}{\partial(ca)}=\eta_0, \quad ...(6)$$

and $\xi\frac{\partial(yz)}{\partial(ab)}+\eta\frac{\partial(zx)}{\partial(ab)}+\zeta\frac{\partial(xy)}{\partial(ab)}=\zeta_0,$ ...(7)

where $\xi=\frac{\partial w}{\partial y}-\frac{\partial v}{\partial z},\ \eta=\frac{\partial u}{\partial z}-\frac{\partial w}{\partial x}$

and $\quad \zeta = \dfrac{\partial v}{\partial x} - \dfrac{\partial u}{\partial y}$

Also, the equation of continuity in Lagrangian form is

$$\rho \; \frac{\partial(xyz)}{\partial(abc)} = \rho_0.$$

Multiplying (5), (6) and (7) by $\dfrac{\partial x}{\partial a}$, $\dfrac{\partial x}{\partial b}$, $\dfrac{\partial x}{\partial c}$, respectively and adding, we have

$$\frac{\xi}{\rho} = \frac{\xi_0}{\rho_0}\frac{\partial x}{\partial a} + \frac{\eta_0}{\rho_0}\frac{\partial x}{\partial b} + \frac{\zeta_0}{\rho_0}\frac{\partial x}{\partial c},$$

$$\frac{\eta}{\rho} = \frac{\xi_0}{\rho_0}\frac{\partial y}{\partial a} + \frac{\eta_0}{\rho_0}\frac{\partial y}{\partial b} + \frac{\zeta_0}{\rho_0}\frac{\partial y}{\partial c}$$

$$\frac{\zeta}{\rho} = \frac{\xi_0}{\rho_0}\frac{\partial z}{\partial a} + \frac{\eta_0}{\rho_0}\frac{\partial z}{\partial b} + \frac{\zeta_0}{\rho_0}\frac{\partial z}{\partial c}. \qquad \text{...(9)}$$

Thus the components ξ, η, ζ of a particle can be obtained at any instant if its path is known. These are known as Cauchy's integral When velocity potential exists then $\xi = \eta = \zeta = 0$. Thus from (9) we conclude that these quantities are always zero if their initial values are zero.

Example 2: *Determine the pressure if the velocity field*

$q_r = 0, q_\theta = Ar + B,\ q_z = 0,$

satisfies the equation of motion $\rho \dfrac{q_\theta^2}{r} = \dfrac{dp}{dr}$, *where A and B are arbitrary constants.*

Solution: $\dfrac{dp}{dr} = \rho \dfrac{1}{r}\left(Ar + \dfrac{B}{r}\right)^2$

or $\dfrac{dp}{dr} = \rho\left(A^2 r + \dfrac{B^2}{r^2} + 2AB\dfrac{1}{r}\right)$

By integrating, we have

$$p = \rho\left(\frac{1}{2}A^2 r^2 + \frac{B^2}{2r^2} + 2AB\log r\right) + C,$$

where C is an integration constant.

Example 3: *Prove that the equation of motion is satisfied for an inviscid, incompressible, steady flow with negligible body force whose velocity components are given by*

$$q_r = U\left(1-\frac{A^3}{r^3}\right)\cos\theta,\ q_\theta = -U\left(1+\frac{A^3}{2r^3}\right)\sin\theta,\ q_\phi = 0,$$

where A is constant. Find the resultant velocity when $r \to \infty$.

Solution: The equations of motion for an inviscid, incompressible and steady flow with negligible external force, in spherical polar coordinates, are given as

$$q_r\frac{\partial p_r}{\partial r}+\frac{q\theta}{r}\frac{\partial q_r}{\partial\theta}+\frac{q\phi}{r\sin\theta}-\frac{\partial q_r}{\partial\phi}-\frac{q_\theta^2+q_\phi^2}{r} = -\frac{1}{\rho}\frac{\partial p}{\partial r},$$

$$q_r\frac{\partial q_\theta}{\partial r}+\frac{q_\theta}{r}\frac{\partial q_\theta}{\partial\theta}+\frac{q\phi}{r\sin\theta}\frac{\partial q\theta}{\partial\phi}+\frac{q_r q_\theta}{r}-\frac{q_\phi^2\cot\theta}{r} = -\frac{1}{\rho}\frac{\partial p}{\partial\theta},$$

$$q_r\frac{\partial q_\phi}{\partial r}+\frac{q_\theta}{r}\frac{\partial q\phi}{\partial\theta}+\frac{q\phi}{r\sin\theta}\frac{\partial q\phi}{\partial\phi}+\frac{q_\phi q_r}{r}-\frac{q_\theta q_\phi\cot\theta}{r}$$

$$= -\frac{1}{\rho}\frac{1}{r\sin\theta}\frac{\partial p}{\partial\phi} \qquad \text{...(1, 2, 3)}$$

Here $q_r = U\left(1-\frac{A^3}{r^3}\right)\cos\theta,\ q_\theta = -U\left(1+\frac{A^3}{2r^3}\right)$

$\sin\theta,\ q_\phi = 0,$ ...(4)

From the relation (4), equations (1,2,3) reduce to

$$U\left(1-\frac{A^3}{r^3}\right)\cos\theta\left(\frac{3UA^3}{r^4}\right)\cos\theta+\frac{U}{r}\left(1+\frac{A^3}{2r^3}\right)\sin\theta$$

$$\times U\left(1-\frac{A^3}{r^3}\right)\sin\theta-\frac{U^2}{r}\left(1+\frac{A^3}{2r^3}\right)\sin^2\theta = -\frac{1}{\rho}\frac{\partial p}{\partial r},$$

$$U\left(1-\frac{A^3}{r^3}\right)\cos\theta\left(\frac{3UA^3}{r^4}\right)\sin\theta+\frac{U}{r}\left(1+\frac{A^3}{2r^3}\right)\sin\theta$$

$$\times U\left(1+\frac{A^3}{2r^3}\right)\cos\theta-\frac{U^2}{r}\left(1-\frac{A^3}{r^3}\right)\left(1+\frac{A^3}{2r^3}\right)\sin\theta\cos\theta$$

$$= -\frac{1}{\rho}\frac{\partial p}{r\partial\theta},$$

$$0 = \frac{1}{\rho}\frac{1}{r\sin\theta}\frac{\partial p}{r\phi}. \qquad ...(5, 6, 7)$$

Equation (7) shows that the pressure p is independent of ϕ, therefore $p = p\ (r, \theta)$. On simplifying the equation (5) and (6), we have

$$\frac{3U^2A^3}{r^4}\left(1-\frac{A^3}{r^3}\right)\cos^2\theta + \frac{U^2}{r}\left(1-\frac{A^3}{2r^3}-\frac{A^6}{2r^6}\right)\sin^2\theta$$

$$-\frac{U^2}{r}\left(1+\frac{A^3}{r^3}+\frac{A^6}{4r^6}\right)\sin^2\theta = -\frac{1}{\rho}\frac{\partial p}{\partial r},$$

or $$\frac{3U^2A^3}{r^4}\left(1-\frac{A^3}{r^3}\right)\cos^2\theta + \frac{U^2}{r}\left(-\frac{3A^3}{2r^3}-\frac{3A^6}{4r^6}\right)\sin^2\theta = \frac{1}{\rho}\frac{\partial p}{\partial r},$$

or $$\frac{3U^2A^3}{r^4}\left(1-\frac{A^3}{r^3}\right)\cos^2\theta - \frac{3U^2A^3}{2r^4}\left(1+\frac{A^3}{2r^3}\right)\sin^2\theta$$

$$= -\frac{1}{\rho}\frac{\partial p}{\partial r}. \qquad ...(8)$$

and $$\frac{3U^2A^3}{2r^4}\left(1-\frac{A^3}{r^3}\right)\sin\theta\cos\theta + \frac{U^2}{r}\left(1+\frac{A^3}{r^3}+\frac{A^6}{4r^6}\right)\sin\theta\cos\theta$$

$$-\frac{U^2}{r}\left(1-\frac{A^3}{2r^3}-\frac{A^6}{2r^6}\right)\sin\theta\cos\theta \quad -\frac{1}{\rho}\frac{\partial p}{\partial\theta}.$$

or $$\frac{3U^2A^3}{2r4}\left(1-\frac{A^3}{r^3}\right)\sin\theta\cos\theta + \frac{3U^2A^3}{2r^4}\left(1+\frac{A^3}{2r^3}\right)\sin\theta\cos\theta =$$

$$-\frac{1}{\rho}\frac{\partial p}{r\partial\theta}. \qquad ...(9)$$

Differentiating equation (8) with regard to θ, we have

$$-\frac{6U^2A^3}{r^4}\left(1-\frac{A^3}{r^3}\right)\cos\theta\sin\theta - \frac{3U^2A^3}{r^4}\left(1+\frac{A^3}{2r^3}\right)\sin\theta\cos\theta$$

$$= -\frac{1}{\rho}\frac{\partial^2 p}{\partial r\partial\theta}$$

or $$\left(-\frac{9U^2A^3}{r^4}+\frac{9U^2A^6}{2r^7}\right)\sin\theta\cos\theta = -\frac{1}{\rho}\frac{\partial^2 p}{\partial r\partial\theta} \quad - \qquad ...(10)$$

Differentiating equation (9) with regard to r, we have

$$\frac{3U^2A^3}{2}\left(-\frac{3}{r^4}+\frac{6A^3}{r^7}\right)\cos\theta\sin\theta + \frac{3U^2A^3}{2}\left(-\frac{3}{r^4}+\frac{3A^3}{r^7}\right)\sin\theta\cos\theta = -\frac{1}{\rho}\frac{\partial^2 p}{\partial r\partial\theta}$$

or $$\frac{9}{2}\frac{U^2A^3}{2}\left(1-\frac{2A^3}{r^3}\right)\cos\theta\sin\theta - \frac{9}{2}\frac{U^2A^3}{2}\left(1+\frac{A^3}{r^3}\right)\sin\theta\cos\theta = -\frac{1}{\rho}\frac{\partial^2 p}{\partial r\partial\theta}$$

or $$\left(-\frac{9U^2A^3}{r^4}+\frac{9U^2A^6}{2r^7}\right)\sin\theta\cos\theta = -\frac{1}{\rho}\frac{\partial^2 p}{\partial r\partial\theta} \quad ...(11)$$

Equation (10) and (11) are identical. Hence, the equation of motion is satisfied.

When $r \to \infty$, the resultant velocity is equal to U. **Proved.**

Example 4: *Prove that the velocity components*

$$q_r\ (r,\theta) = -U\left(1-\frac{a^2}{r^2}\right)\cos\theta, q_\theta(r,\theta) = U\left(1+\frac{a^2}{r^2}\right)\sin\theta,$$

satisfy the equation of motion for a two-dimensional inviscid incompressible flow. Find the pressure associated with this velocity field. U and a are constants.

Solution: The equations of motion for a two-dimensional steady, inviscid, incompressible flow in the absence of external force, in spherical polar coordinates, are given by

$$q_r\frac{\partial q_r}{\partial r}+\frac{q_\theta}{r}\frac{\partial q_r}{\partial\theta}-\frac{q_\theta^2}{r} = -\frac{1}{\rho}\frac{\partial p}{\partial r},$$

$$q_r\frac{\partial q_\theta}{\partial r}+\frac{q_\theta}{r}\frac{\partial q_\theta}{\partial\theta}+\frac{q_r q_\theta}{r} = -\frac{1}{\rho}\frac{\partial p}{r\partial\theta},$$

$$0 = -\frac{1}{\rho}\frac{1}{r\sin\theta}\frac{\partial p}{\partial\phi}. \quad ...(1, 2, 3)$$

Here $q_r\ (r, \theta) = -U\left(1-\frac{a^2}{r^2}\right)\cos\theta,$

$$q_\theta\ (r, .\theta) = U\left(1+\frac{a^2}{r^2}\right)\sin\theta. \qquad ...(4)$$

From (3), it follows that the pressure p is independent of ϕ *i.e.*, p = p (r, θ).

Using the relation (4) into (1) and (2), we have

$$U\left(1-\frac{a^2}{r^2}\right)\cos\theta+\left(\frac{2Ua^2}{r^3}\right)\cos\theta+\frac{U}{r}\left(1+\frac{a^2}{r^2}\right)\sin\theta$$

$$\times\, U\left(1-\frac{a^2}{r^2}\right)\sin\theta-\frac{U^2}{r}\left(1+\frac{a^4}{r^2}\right)^2\sin^2\theta=-\frac{1}{\rho}\frac{\partial p}{\partial r},$$

or
$$\frac{2U^2a^2}{r^3}\left(1-\frac{a^2}{r^2}\right)\cos^2\theta-\frac{2U^2a^2}{r^3}\left(1+\frac{a^2}{r^2}\right)\sin^2\theta$$

$$=-\frac{1}{\rho}\frac{\partial p}{\partial r}. \qquad ...(5)$$

and
$$\frac{2U^2a^2}{r^3}\left(1-\frac{a^2}{r^2}\right)\sin\theta\cos\theta+\frac{U^2}{r}\left(1+\frac{a^2}{r^2}\right)\sin\theta\cos\theta$$

$$-\frac{U^2}{r}\left(1-\frac{a^2}{r^4}\right)\sin\theta\cos\theta=-\frac{1}{\rho}\frac{\partial p}{r\partial\theta}$$

or
$$\frac{2U^2a^2}{r^3}\left(1-\frac{a^2}{r^2}\right)\sin\theta\cos\theta+\frac{2U^2a^2}{r}\left(1+\frac{a^2}{r^2}\right)\sin\theta\cos\theta$$

$$=-\frac{1}{\rho}\frac{\partial p}{r\partial\theta}$$

$$\frac{4U^2a^2}{r^3}\sin\theta\cos\theta=-\frac{1}{\rho}\frac{\partial p}{r\partial\theta}. \qquad ...(6)$$

Differentiating (5) with regard to θ, we have

$$-\frac{4U^2a^2}{r^3}\left(1-\frac{a^2}{r^2}\right)\cos\theta\sin\theta\frac{4U^2a^2}{r^2}\left(1+\frac{a^2}{r^2}\right)\sin\theta\cos\theta$$

$$=-\frac{1}{\rho}\frac{\partial^2 p}{\partial r\partial\theta}$$

$$\frac{8U^2a^2}{r^3}\cos\theta\sin\theta = \frac{1}{\rho}\frac{\partial^2 p}{\partial r\partial\theta}. \qquad ...(7)$$

Differentiating (6) with regard to r, we have

$$\frac{8U^2a^2}{r^3}\cos\theta\sin\theta = \frac{1}{\rho}\frac{\partial^2 p}{\partial r\partial\theta}. \qquad ...(8)$$

Equation (7) and (8) are identical. Hence, the equation of motion are satisfied.

Again, p is a function or r and θ, we have

$$dp = \left(\frac{\partial p}{\partial r}\right) dr + \left(\frac{\partial p}{\partial \theta}\right) d\theta$$

$$\text{or } dp = -2r\ U2a2 \left[\left(\frac{1}{r^3} - \frac{a^2}{r^5}\right)\cos^2\theta - \left(\frac{1}{r^3} + \frac{a^2}{r^5}\right)\sin^2\theta\right]dr$$

$$- \frac{4\rho U^2a^2}{r^2}(\sin\theta\cos\theta)\ d\theta$$

By integrating, we have

$$p = -2\rho U^2a^2\left[\left(-\frac{1}{2r^2} + \frac{a^2}{4r^4}\right)\cos^2\theta - \left(-\frac{1}{2r^2} - \frac{a^2}{4r^4}\right)\sin^2\theta\right] + C$$

$$\text{or } p = \frac{4\rho U^2a^2}{r}(\cos^2\theta - \sin^2\theta) + C,$$

where C is an integration constant.

$$\text{or } p = 2\rho U^2a^2$$

$$\left[\left(\frac{1}{2r^r} - \frac{a^2}{4r^4} - \frac{2}{r^2}\right)\cos^2\theta - \left(\frac{1}{2r^2} - \frac{a^2}{4r^4} - \frac{2}{r^2}\right)\sin^2\theta\right] + C$$

$$\text{or } p = 2\rho U^2a^2\left[-\frac{3}{2r^2}\cos 2\theta - \frac{a^2}{4r^4}\right] + C$$

$$\text{or } p = -\rho U^2a^2\left[-\frac{3}{r^2}\cos 2\theta + \frac{a^2}{2r^4}\right] + C,$$

Which gives the required pressure distribution.

Example 5(a): *Show that the velocity field*

$$u\ (x,y) = \frac{A(x^2-y^2)}{(x^2+y^2)^2},\ v(x,y) = \frac{2Axy}{(x^2+y^2)^2}, w = 0,$$

satisfies the equation of motion for inviscid incompressible flow. Determine the pressure associated with this velocity field.

Solution: Here $u\ (x,y) = \frac{A(x^2-y^2)}{(x^2+y^2)^2}$, v (x, y)

$$= \frac{2Axy}{(x^2+y^2)^2},\ w = 0. \qquad ...(1)$$

$$\frac{\partial u}{\partial x} = A\frac{-2x(x^2+y^2)^2 - 4x(x^2-y^2)(x^2+y^2)}{(x^2-y^2)^4} = \frac{2Ax(3y^2-x^2)}{(x^2+y^2)^3},$$

$$\frac{\partial u}{\partial y} = A\frac{-2y(x^2+y^2)^2 - 4y(x^2-y^2)(x^2+y^2)}{(x^2-y^2)^4} = -\frac{2Ay(3x^2-x^2)}{(x^2+y^2)^3},$$

$$\frac{\partial v}{\partial x} = 2A\frac{y(x^2+y^2) - 4x^2y(x^2+y^2)}{(x^2+y^2)^4} = \frac{2Ay(y^2-3x^2)}{(x^2-y^2)^3},$$

$$\frac{\partial v}{\partial y} = 2A\frac{x(x^2+y^2)^2 - 4xy^2(x^2+y^2)}{(x^2+y^2)^4} = \frac{2Ax(x^2-3y^2)}{(x^2+y^2)^3}$$

The equations of motion for steady flow are given by

$$u\frac{\partial u}{\partial x} + v\frac{\partial u}{\partial y} + w\frac{\partial u}{\partial z} = -\frac{1}{\rho}\frac{\partial p}{\partial x},$$

$$u\frac{\partial v}{\partial x} + v\frac{\partial v}{\partial y} + w\frac{\partial v}{\partial z} = -\frac{1}{\rho}\frac{\partial p}{\partial y},$$

$$u\frac{\partial w}{\partial x} + v\frac{\partial w}{\partial y} + w\frac{\partial w}{\partial z} = -\frac{1}{\rho}\frac{\partial p}{\partial z}, \qquad ...(2,3,4)$$

From (1) and (2, 3, 4), we have

$$\frac{2A^2x}{(x^2+y^2)^3} = \frac{1}{\rho}\frac{\partial p}{\partial x},$$

$$\frac{2A^2y}{(x^2+y^2)^3} = \frac{1}{\rho}\frac{\partial p}{\partial y},$$

$$0 = -\frac{1}{\rho}\frac{\partial p}{\partial z}. \qquad ...(5,\ 6,7)$$

Equation (7) shows that pressure p is independent of z, *i.e.*, p = p (x,y). Therefore

$$dp = (\partial p/\partial x)\,dx + (\partial p/\partial y)\,dy$$

$$\text{or } dp = \frac{2A^2\rho x}{(x^2+y^2)^3}dx + \frac{2A^2\rho y}{(x^2+y^2)^3}dy$$

$$\text{or } dp = 2A^2\rho\frac{xdx + ydy}{(x^2+y^2)^3}$$

By integrating, we have

$$p = \frac{A^2\rho}{2(x^2+y^2)^2}.$$ **Proved**

Example 5(b): *The particle velocity for a fluid motion referred to rectangular axes is given by the components.*

$$u = A\cos\frac{\pi x}{2a}\cos\frac{\pi z}{2a},\quad v = 0,$$

$$w = A\sin\frac{\pi x}{2a}\sin\frac{\pi z}{2a},$$

where A is a constant. Show that this is a possible motion of an incompressible fluid under no body forces in an infinite fixed rigid tube, $-a \le x \le a$, $0 \le z \le 2a$. *Also, find the pressure associated with this velocity field.*

Solution: The equations of motion for a two-dimensional steady, inviscid, incompressible flow under no body force, in cartesian coordinates, are given by

$$u\frac{\partial u}{\partial x} + v + \frac{\partial u}{dy} + w\frac{\partial u}{\partial z} = -\frac{1}{\rho}\frac{\partial p}{\partial x},\quad 0 = -\frac{1}{\rho}\frac{\partial p}{\partial y},$$

$$u\frac{\partial w}{\partial x} + v + \frac{\partial w}{dy} + w\frac{\partial w}{\partial z} = -\frac{1}{\rho}\frac{\partial p}{\partial z}, \qquad \text{...(1, 2, 3)}$$

Here $$u = A\cos\frac{\pi x}{2a}\cos\frac{\pi z}{2a},$$

$$v = 0,\ w = A\sin\frac{\pi x}{2a}\sin\frac{\pi z}{2a}, \qquad \text{...(4)}$$

From the equation (2), it follows that the pressure p is independent of y *i.e.*, p = p (x, z).

Using (4) into (1) and (3), we have

$$\left(A\cos\frac{\pi x}{2a}\cos\frac{\pi z}{2a}\right)\left(-\frac{\pi A}{2a}\sin\frac{\pi x}{2a}\cos\frac{\pi z}{2a}\right)+\left(A\sin\frac{\pi x}{2a}\sin\frac{\pi z}{2a}\right)$$
$$\times\left(-\frac{\pi A}{2a}\cos\frac{\pi x}{2a}\sin\frac{\pi z}{2a}\right)=-\frac{1}{\rho}\frac{\partial p}{\partial x},$$

or $$\frac{\pi A^2}{2a}\left[\cos\frac{\pi x}{2a}\sin\frac{\pi x}{2a}\cos^2\frac{\pi z}{2a}+\cos\frac{\pi x}{2a}\sin\frac{\pi x}{2a}\sin^2\frac{\pi z}{2a}\right]=\frac{1}{\rho}\frac{\partial p}{\partial x}$$

or $$\frac{\pi A^2}{2a}\cos\frac{\pi x}{2a}\sin\frac{\pi x}{2a}=\frac{1}{\rho}\frac{\partial p}{\partial x}, \quad ...(5)$$

and $$\left(A\cos\frac{\pi x}{2a}\cos\frac{\pi z}{2a}\right)\left(\frac{\pi A}{2a}\cos\frac{\pi x}{2a}\sin\frac{\pi z}{2a}\right)+\left(A\sin\frac{\pi x}{2a}\sin\frac{\pi z}{2a}\right)$$
$$\times\left(\frac{\pi A}{2a}\sin\frac{\pi x}{2a}\cos\frac{\pi z}{2a}\right)=-\frac{1}{\rho}\frac{\partial p}{\partial z},$$

or $$\frac{\pi A^2}{2a}\left[\cos\frac{\pi z}{2a}\sin\frac{\pi z}{2a}\cos^2\frac{\pi x}{2a}+\cos\frac{\pi z}{2a}\sin\frac{\pi z}{2a}\sin^2\frac{\pi x}{2a}\right]=-\frac{1}{\rho}\frac{\partial p}{\partial z},$$

or $$\frac{\pi A^2}{2a}\cos\frac{\pi z}{2a}\sin\frac{\pi z}{2a}=\frac{1}{\rho}\frac{\partial p}{\partial z}, \quad ...(6)$$

The equations (5) and (6) show that the velocity components satisfy the equations of motion.

Again $dp = (\partial p/\partial x)dx + (\partial p/\partial z)\, dz$

or $$dp=\frac{\pi\rho A^2}{2a}\left[\cos\frac{\pi x}{2a}\sin\frac{\pi x}{2a}dx-\cos\frac{\pi z}{2a}\sin\frac{\pi z}{2a}dz\right]$$

By integrating, we have

$$p=\frac{1}{2}\rho A^2\left[\cos^2\frac{\pi z}{2a}-\cos^2\frac{\pi x}{2a}\right]+C,$$

where C is an integration constant. This gives the required pressure distribution.

Proved

INTEGRATION OF EULER'S EQUATION

The Euler's equation of motion is given as follows :

$$\frac{\partial \mathbf{q}}{\partial t}+\nabla\left(\frac{1}{2}\mathbf{q}^2\right)-\mathbf{q}\times\text{curl}\,\mathbf{q}=F-\frac{1}{\rho}\nabla p. \quad ...(1)$$

Suppose that the body forces are conservative and that the flow is of the potential kind then there exist scalar functions W and f, such that

$$F = \nabla\, \Omega, \qquad \mathbf{q} = -\,\nabla\phi.$$

From (1) and (2), we have ...(2)

$$\frac{\partial \mathbf{q}}{\partial t} + \nabla\left(\frac{1}{2}\mathbf{q}^2\right) - \mathbf{q} \times \text{curl}\,\mathbf{q} = -\,\nabla\Omega - \frac{1}{\rho}\nabla p$$

$$\text{or } \frac{\partial \mathbf{q}}{\partial t} - \mathbf{q} \times \text{curl}\,\mathbf{q} = -\,\nabla\left[\Omega + \int \frac{dp}{p} + \frac{1}{2}\mathbf{q}^2\right]$$

$$\text{or } \frac{\partial}{\partial t}(-\nabla\phi) - \mathbf{q} \times \text{curl}\,\mathbf{q} = -\nabla\left[\Omega + \int \frac{dp}{p} + \frac{1}{2}\mathbf{q}^2\right]$$

$$\text{or } \nabla\left[-\frac{\partial \phi}{\partial t} + \Omega + \int \frac{dp}{\rho} + \frac{1}{2}\mathbf{q}^2\right] = \mathbf{q} \times \text{curl}\,\mathbf{q}. \qquad ...(1)$$

STEADY AND IRROTATIONAL FLOW

If the motion be steady and irrotational then $\partial\phi/\partial t = 0$, curl $\mathbf{q} = 0$. Since $\mathbf{q} \times$ curl $\mathbf{q} = 0$, it follows that $\mathbf{q}$ and curl $\mathbf{q}$ are parallel *i.e.*, stream lines and vortex lines coincide. For such a motion $\mathbf{q}$ is known a Beltrami vector and the flow is *Beltrami flow.*

Thus the equation (5) reduces to

$$\int \frac{dp}{\rho} + \frac{1}{2}\mathbf{q}^2 + \Omega = \text{const.}, \qquad ...(6)$$

where the constant is an absolute constant *i.e.*, independent of the time t.

If the fluid be incompressible and homogeneous then the equation (6) reduces to

$$\frac{p}{\rho} + \frac{1}{2}\mathbf{q}^2 + \Omega = \text{const.} = C, \qquad ...(7)$$

where C is a constant along a streamline but varies from one streamline to another. The equation (7) is termed as Bernoulli*'s equation for steady and irrotational flows* and holds irrespective of whether the motion is rotational or irrotational.

HELMHOLTZ EQUATIONS

Euler's equations of motion are given by

$$\frac{\partial u}{\partial t} + u\frac{\partial u}{\partial x} + v\frac{\partial u}{\partial y} + w\frac{\partial u}{\partial z} = X - \frac{1}{\rho}\frac{\partial p}{\partial x},$$

$$\frac{\partial v}{\partial t} + u\frac{\partial v}{\partial x} + v\frac{\partial v}{\partial y} + w\frac{\partial v}{\partial z} = y - \frac{1}{\rho}\frac{\partial p}{\partial y},$$

$$\frac{\partial w}{\partial t} + u\frac{\partial w}{\partial x} + v\frac{\partial w}{\partial y} + w\frac{\partial w}{\partial z} = Z - \frac{1}{\rho}\frac{\partial p}{\partial z}, \qquad ...(1, 2, 3)$$

Let V be the potential function of the external forces and the density ρ be the function of the pressure p. Equation (1) may be written as

$$\frac{\partial u}{\partial t} + \left(u\frac{\partial u}{\partial x} + v\frac{\partial u}{\partial x} + w\frac{\partial w}{\partial x}\right) + v\left(\frac{\partial u}{\partial y} - \frac{\partial v}{\partial x}\right) + w\left(\frac{\partial u}{\partial z} - \frac{\partial w}{\partial x}\right)$$

$$= -\frac{\partial V}{\partial x} - \frac{1}{\rho}\frac{\partial p}{\partial x}$$

$$\Rightarrow \frac{\partial u}{\partial t} - \frac{1}{\rho}\frac{\partial}{\partial x}(q^2) - 2v\zeta + 2w\eta = -\frac{\partial V}{\partial x} - \frac{1}{\rho}\frac{\partial}{\partial x},$$

where $\underline{\Omega}\,(\xi, \eta, \zeta)$ are the spin components and $q^2 = u^2 + v^2 + w^2$.

$$\Rightarrow \frac{\partial u}{\partial t} - 2v\zeta + 2w\eta = -\frac{\partial}{\partial x}\left(V + \frac{1}{2}q^2 + \int\frac{dp}{\rho}\right) = -\frac{\partial Q}{\partial x}\text{ (let)},$$

where $Q = V + \frac{1}{2}\mathbf{q}^2 + \int\frac{dp}{\rho}$.

Thus $\frac{\partial u}{dt} - 2v\zeta + 2w\eta = -\frac{\partial Q}{dx}$,

Similarly $\frac{\partial v}{dt} - 2w\xi + 2u\zeta = -\frac{\partial Q}{dy}$,

and $\frac{\partial w}{dt} - 2u\eta + 2v\,x = -\frac{\partial Q}{\partial z}$. ...(4, 5,6)

Differentiating (5) and (6) partially with to z and y, we have

$$\frac{\partial^2 v}{\partial z\partial t} - 2w\frac{\partial \xi}{\partial z} - 2\xi\frac{\partial w}{\partial z} + 2u\frac{\partial \zeta}{\partial z} + 2\zeta\frac{\partial u}{\partial z}$$

$$= \frac{\partial^2 w}{\partial y\partial t} - 2u\frac{\partial \eta}{\partial y} - 2\eta\frac{\partial u}{\partial y} + 2v\frac{\partial \xi}{\partial y} + 2\xi\frac{\partial v}{\partial y}$$

$$\Rightarrow \frac{\partial}{\partial t}\left(\frac{\partial w}{\partial y} - \frac{\partial v}{\partial z}\right) - 2u\left(\frac{\partial \eta}{\partial y} + \frac{\partial \zeta}{\partial z}\right) + 2v\left(\frac{\partial \xi}{\partial y}\right) + 2w\left(\frac{\partial \xi}{\partial z}\right)$$

$$+ 2\xi\left(\frac{\partial v}{\partial y} + \frac{\partial w}{\partial z}\right) - 2\eta\left(\frac{\partial u}{\partial y}\right) - 2\zeta\frac{\partial u}{\partial z} = 0. \qquad ...(7)$$

But $\frac{\partial \xi}{\partial x}+\frac{\partial \eta}{\partial y}+\frac{\partial \zeta}{\partial z} = 0$ ...(8)

From (7) and (8), we have

$$2\frac{\partial \xi}{\partial t}+2u\frac{\partial \xi}{\partial x}+2v\frac{\partial \xi}{\partial y}+2w\frac{\partial \xi}{\partial z}+2\xi\left(\frac{\partial u}{\partial x}+\frac{\partial v}{\partial y}+\frac{\partial w}{\partial z}\right)$$

$$-2\xi\frac{\partial u}{\partial x}-2\eta\frac{\partial y}{\partial y}-2\zeta\frac{\partial u}{\partial z}=0$$

$$\Rightarrow \frac{D\xi}{Dt}+\xi\left(\frac{\partial u}{\partial x}+\frac{\partial v}{\partial y}+\frac{\partial w}{\partial z}\right) = \xi\frac{\partial u}{\partial x}+\eta\frac{\partial u}{\partial y}+\zeta\frac{\partial u}{\partial z} \quad ...(9)$$

The equation of continuity is

$$\frac{D\rho}{Dt}+\rho\left(\frac{\partial u}{\partial x}+\frac{\partial v}{\partial y}+\frac{\partial w}{\partial z}\right)= 0. \quad ...(10)$$

From (9) and (10), we have

$$\frac{D\xi}{Dt}+\frac{\xi}{\rho}\frac{D\rho}{Dt}=\xi\frac{\partial u}{\partial x}\eta\frac{\partial u}{\partial y}+\zeta\frac{\partial u}{\partial z}$$

$$\Rightarrow \frac{1}{\rho}\frac{D\xi}{Dt}-\frac{\xi}{\rho^2}\frac{D\rho}{Dt} = \frac{\xi}{\rho}\frac{\partial u}{\partial x}+\frac{\eta}{\rho}\frac{\partial u}{\partial y}+\frac{\zeta}{\rho}\frac{\partial u}{\partial z} \quad ...(11)$$

$$\Rightarrow \frac{D}{Dt}\left(\frac{\xi}{\rho}\right)= \frac{\xi}{\rho}\frac{\partial u}{\partial x}+\frac{\eta}{\rho}\frac{\partial u}{\partial y}+\frac{\zeta}{\rho}\frac{\partial u}{\partial z},$$

Similarly $\frac{D}{Dt}\left(\frac{\eta}{\rho}\right)=\frac{\xi}{\rho}\frac{\partial v}{\partial x}+\frac{\eta}{\rho}\frac{\partial v}{\partial y}+\frac{\zeta}{\rho}\frac{\partial v}{\partial z},$

and $\frac{D}{Dt}\left(\frac{\zeta}{\rho}\right)= \frac{\xi}{\rho}\frac{\partial w}{\partial x}+\frac{\eta}{\rho}\frac{\partial w}{\partial y}+\frac{\zeta}{\rho}\frac{\partial w}{\partial z}.$ ...(12, 13)

But $\frac{\eta}{\rho}\frac{\partial u}{\partial y}+\frac{\zeta}{\rho}\frac{\partial u}{\partial z} =\frac{\eta}{\rho}\left\{\left(\frac{\partial u}{\partial y}-\frac{\partial v}{\partial x}\right)+\frac{\partial v}{\partial x}\right\}+\frac{\zeta}{\rho}\left\{\left(\frac{\partial u}{\partial z}-\frac{\partial w}{\partial x}\right)+\frac{\partial w}{\partial x}\right\}$

$$=\frac{\eta}{\rho}(-2\zeta)+\frac{\eta}{\rho}\frac{\partial v}{\partial x}+\frac{\zeta}{\rho}(2\eta)+\frac{\zeta}{\rho}\frac{\partial w}{\partial x},$$

$$= \frac{\eta}{\rho}\frac{\partial v}{\partial x}+\frac{\zeta}{\rho}\frac{\zeta}{\rho}\frac{\partial w}{\partial x} \quad ...(14)$$

Using the relation (14), the equations (11, 12,13) reduce to

$$\frac{D}{Dt}\left(\frac{\xi}{\rho}\right) = \frac{\xi}{\rho}\frac{\partial u}{\partial y} + \frac{\eta}{\rho}\frac{\partial v}{\partial y} + \frac{\zeta}{\rho}\frac{\partial w}{\partial y}$$

$$\frac{D}{Dt}\left(\frac{\eta}{\rho}\right) = \frac{\xi}{\rho}\frac{\partial u}{\partial y} + \frac{\eta}{\rho}\frac{\partial v}{\partial y} + \frac{\zeta}{\rho}\frac{\partial w}{\partial y},$$

$$\frac{D}{Dt}\left(\frac{\eta}{\rho}\right) = \frac{\xi}{\rho}\frac{\partial u}{\partial z} + \frac{\eta}{\rho}\frac{\partial v}{\partial z} + \frac{\zeta}{\rho}\frac{\partial w}{\partial z}, \qquad \text{...(15, 16, 17)}$$

Equations (15), (16), (17) are known as Helmholtz's equation. Let $\xi = \eta = \zeta = 0$ at an instant of time t then

$$\frac{D}{Dt}\left(\frac{\xi}{\rho}\right) = \frac{D}{Dt}\left(\frac{\eta}{\rho}\right) = \frac{D}{Dt}\left(\frac{\zeta}{\rho}\right) = 0$$

$$\frac{D\zeta}{Dt} = \frac{D\eta}{Dt} = \frac{D\zeta}{Dt} = 0, \ \rho = \text{const.}$$

ξ, η, ζ must be constant. Since they are all zero at an instant of time t and have to remain constant.

In general, let $\frac{\partial u}{\partial x}, \frac{\partial v}{\partial x}$,...are all finite and less than a quantity P then $\frac{\xi}{\rho}, \frac{\eta}{\rho}, \frac{\zeta}{\rho}$ can not increase faster than if they satisfy the equations.

$$\frac{D}{Dt}\left(\frac{\xi}{\rho}\right) =]\frac{D}{Dt}\left(\frac{\eta}{\rho}\right) = \frac{D}{Dt}\left(\frac{\zeta}{\rho}\right) = \frac{P}{\rho}(\xi + \eta + \zeta)$$

Let $\xi + \eta + \zeta = PW$, then

$$\frac{D}{Dt}\left(\frac{\xi}{\rho} + \frac{\eta}{\rho} + \frac{\zeta}{\rho}\right) = \frac{D}{Dt}(W) = 3PW$$

$\Rightarrow W = ke^{3Pt}, \ W \neq 0$

When $t = 0$,

$W = 0 \Rightarrow k = 0$ *i.e.*,

W be zero at time $t = 0$,

it may be so for all time.

Since W is the sum of three quantities ξ, γ, ζ which cannot be negative. Hence $W = 0$, it follows that each of these three quantities must be zero $\xi = 0 = \eta = \zeta$.

Hence if the motion is irrotational at any instant, it must be so for all time *i.e.*, if once, the velocity potential exists it exists for all time. This is known as *the principle of Permanance of irrotational motion.*

Example 1: *An elastic fluid, the weight of which is neglected, obeying Boyle's law is in motion in a uniform straight tube; show that on the hypothesis of parallel sections the velocity at any time t at a distance r from a fixed point in the tube is defined by the equation*

$$\frac{\partial^2 v}{\partial t^2} + \frac{\partial}{\partial r}\left(2v\frac{\partial v}{\partial t} + v^2\frac{\partial v}{\partial r}\right) = k\,\frac{\partial^2 v}{\partial r^2}.$$

Solution: Since the fluid obeys Boyle's law then

$$p = k\rho. \qquad ...(1)$$

The equation of continuity and the equation of motion is given by

$$\frac{\partial \rho}{\partial t} + \frac{\partial}{\partial r}(\rho v) = 0, \qquad ...(2)$$

and
$$\frac{\partial v}{\partial t} + v\frac{\partial v}{\partial r} = -\frac{1}{\rho}\frac{\partial p}{\partial r} \qquad ...(3)$$

Using the relation (1), we have

$$\frac{\partial v}{\partial t} + v\frac{\partial v}{\partial r} = -\frac{k}{\rho}\frac{\partial \rho}{\partial r}. \qquad ...(4)$$

Differentiating (4) partially with regard to t, we have

$$\frac{\partial^2 v}{\partial t^2} + \frac{\partial}{\partial t}\left(v\frac{\partial v}{\partial r} + \frac{k}{\rho}\frac{\partial \rho}{\partial r}\right) = 0,$$

$$\Rightarrow \quad \frac{\partial^2 v}{\partial t^2} + \frac{\partial}{\partial r}\left(v\frac{\partial v}{\partial t} + \frac{k}{\rho}\frac{\partial \rho}{\partial t}\right) = 0$$

$$\Rightarrow \quad \frac{\partial^2 v}{\partial t^2} + \frac{\partial}{\partial r}\left\{v\frac{\partial v^*}{\partial t} + \frac{k}{\rho}\left(-\frac{\partial}{\partial r}(\rho v)\right)^t\right\} = 0$$

$$\Rightarrow \quad \frac{\partial^2 v}{\partial t^2} + \frac{\partial}{\partial r}\left\{v\frac{\partial v}{\partial t} + \frac{k}{\rho}\left(\rho\frac{\partial}{\partial r} + v\frac{\partial \rho}{\partial r}\right)^t\right\} = 0$$

$$\Rightarrow \quad \frac{\partial^2 v}{\partial t^2} + \frac{\partial}{\partial r}\left\{v\frac{\partial v}{\partial t} - k\frac{\partial v}{\partial r} - \frac{k}{\rho}\frac{\partial \rho}{\partial r}v\right\} = 0$$

$$\Rightarrow \frac{\partial^2 v}{\partial t^2} + \frac{\partial}{\partial r}\left\{ v\frac{\partial v}{\partial t} - k\frac{\partial v}{\partial r} + \left(\frac{\partial v}{\partial t} + v\frac{\partial v}{\partial r}\right) v \right\} = 0$$

$$\Rightarrow \qquad \frac{\partial^2 v}{\partial t^2} + \frac{\partial}{\partial r}\left(2v\frac{\partial v}{\partial t} + v^2\frac{\partial v}{\partial r} - k\frac{\partial v}{\partial r}\right) = 0$$

$$\Rightarrow \qquad \frac{\partial^2 v}{\partial t^2} + \frac{\partial}{\partial r}\left(2v\frac{\partial v}{\partial t} + v^2\frac{\partial v}{\partial r}\right) = k\,\frac{\partial^2 v}{\partial r^2}.$$

Proved.

IRROTATIONAL FLOW

For an irrotational flows the vorticity vector $\underline{\omega}$ = curl $\mathbf{q}$ = 0 then the equation (3) becomes

$$\nabla\left[-\frac{\partial\phi}{\partial t} + \Omega + \int\frac{dp}{\rho} + \frac{1}{2}\mathbf{q}^2\right] = 0.$$

By integrating, we have

$$-\frac{\partial\phi}{\partial t} + \Omega + \int\frac{dp}{\rho} + \frac{1}{2}\mathbf{q}^2 = \chi\,(t), \qquad ...(4)$$

where the constant χ (t) is an arbitrary function of time only, c (t) can be absorbed in $(\partial\phi/\partial t)$ then the equation (4) reduces to the form

$$-\frac{\partial\phi}{\partial t} + \Omega + \int\frac{dp}{\rho} + \frac{1}{2}\mathbf{q}^2 = \text{constant}, \qquad ...(5)$$

which is known as Bernoulli's equation for unsteady, irrotational flows.

Example 1: *Air obeying Boyles's law is in motion in a uniform tube of small section, Prove that if ρ be the density and v the velocity at a distance x from a fixed point at time t.*

$$\frac{\partial^2\rho}{\partial t^2} = \frac{\partial^2}{\partial x^2}\{\rho(v^2 + k)\}.$$

Solution: Let p be the pressure and v be the velocity at a distance x from the end of the tube at any time t. The equation of motion and the equation of continuity is given by

$$\frac{\partial v}{\partial t} + v\frac{\partial v}{\partial x} = -\frac{1}{\rho}\frac{\partial p}{\partial x}, \qquad ...(1)$$

and

$$\frac{\partial\rho}{\partial t} + \frac{\partial}{\partial x}(\rho v) = 0. \qquad ...(2)$$

Since the air obeys Boyle's law, then

$$p = k\rho \Rightarrow dp = kd\rho \qquad ...(3)$$

From (1) and (3), we have

$$\frac{\partial v}{\partial t} + v\frac{\partial v}{\partial t} = -\frac{k}{\rho}\frac{\partial p}{\partial x}. \qquad ...(4)$$

Differentiating (2) partially with regard to t, we have

$$\frac{\partial^2 \rho}{\partial t^2} + \frac{\partial}{\partial t}\left\{\frac{\partial}{\partial x}(\rho v)\right\} = 0$$

$$\Rightarrow \frac{\partial^2 \rho}{\partial t^2} + \frac{\partial}{\partial x}\left\{\frac{\partial}{\partial t}(\rho v)\right\} = 0$$

$$\Rightarrow \frac{\partial^2 \rho}{\partial t^2} + \frac{\partial}{\partial x}\left\{\rho\frac{\partial}{\partial t}v\frac{\partial \rho}{\partial t}\right\} = 0$$

From (2) and (4), we have

$$\frac{\partial^2 \rho}{\partial t^2} + \frac{\partial}{\partial x}\left\{\rho\left(-v\frac{\partial v}{\partial x} - \frac{k}{\rho}\frac{\partial \rho}{\partial x}\right) - v\frac{a}{\partial x}(pv)\right\} = 0$$

$$\Rightarrow \frac{\partial^2 \rho}{\partial t^2} + \frac{\partial}{\partial x}\left\{\rho v\frac{\partial v}{\partial x} + v\frac{\partial}{\partial x}(pv) + k\frac{\partial \rho}{\partial x}\right\}$$

$$\Rightarrow \frac{\partial^2 \rho}{\partial t^2} + \frac{\partial}{\partial x}\left\{\frac{\partial}{\partial x}(\rho v.v + k\frac{\partial \rho}{\partial x}\right\} = \frac{\partial^2}{\partial x^2}\{\rho\,(v^2 + k)\}.$$ **Proved.**

BERNOULLI'S EQUATION

Consider a stream tube with A1 and A_2 as its two cross-sections. Let $\rho_1 p_1$, q_1 and ρ_2,p_2, q_2 be the density, the pressure, and the velocities at its two ends. Since the velocity normal to the curved surface of the stream tube is always zero therefore the flow will take place only through A_1 and A_2. From the law of conservation, we have

$$\rho_1 A_1 = \rho_2 A_2 = k \text{ (const.)} \qquad ...(1)$$

Fig. 2.19

The net force on the tube due to pressure is

$$p_1A_1 - p_2A_2 = \frac{p_1}{\rho_1}(\rho_1 A_1) - \frac{p_2}{\rho_2}(\rho_2 A_2)$$

$$\Rightarrow p_1A_1 - p_2A_2 = k\left(\frac{p_1}{\rho_1} - \frac{p_2}{\rho_2}\right).$$

The difference of the forces acting on the two cross-sections must be equal to the gain in the energy of the fluid passing through the tube. Thus, we have

$$\frac{p_1}{\rho_1} - \frac{p_2}{\rho_2} = k_2 - k_1$$

$$\text{or } \frac{p_1}{\rho_1} + k_1 = \frac{p_2}{\rho_2} + k_2 = \text{constant}, \qquad \ldots(1)$$

where k_1 and k_2 are the sum of kinetic $\left(\frac{1}{2}q^2\right)$ and potential (Ω) energies per unit mass of the fluid entering and leaving the tube. Thus, we have

$$\frac{p}{\rho} + \frac{1}{2}q^2 + \Omega = \text{const.}$$

Again, let the cross-section A_1 and A_2 of the stream tube tends to zero such that it reduces to a stream line, even then equation (3) holds. Hence the equation holds along a stream line.

IMPULSIVE MOTION OF A FLUID

Consider ρ be the density of the fluid at the point P (x, y,z). Let u_1, v_1, w_1 and u_2, v_2, w_2 be the velocity components at the point P just before and just after the impulsive action. Let I_x, I_y, I_z be the components of the external impulsive forces per unit mass of the fluid. Construct a small parallelopiped with edges of lengths δx, δy, δz parallel to the respective

Force on the face PQRS $= \bar{\omega}\,\delta y \delta z$

$= f\,(x, y, z)$ (let)

Force on the opposite face P′Q′R′S′,

$= f\,(x, \delta, y, z)$ (let)

$$= f(x, y, z) - f(x, y, z) + \delta x.\frac{\partial}{\partial x} f(x, y, z) + \ldots$$

$$= f(x, y, z) - f(x, y, z) - \delta x.\ \frac{\partial}{\partial x} f(x,y,z)\ldots$$

$$= -\ \delta x\ \frac{\partial}{\partial x} f(x,y,z)$$

$$= -\ \delta x\ \frac{\partial}{\partial x}(\overline{\omega}\ \delta y\ \delta z)$$

$$= -\ \frac{\partial \overline{\omega}}{\partial x} \delta x \delta y \delta z \text{ along X-axis}$$

The impulse on the parallelopiped along X-axis due to external impulsive body force I_x

$$= \rho I_x\ \delta x\ \delta y\ \delta z$$

Change in momentum along X-axis

$$= \rho\ \delta x\ \delta y\ \delta z\ (u_2 - u_1)$$

By Newton's second law of motion, we have

Total impulse applied along X-axis

Change of momentum along X-axis

$$\Rightarrow \rho I_x \delta x \delta y \delta z - \frac{\partial \overline{\omega}}{\partial x} \delta x \delta y \delta z = \rho\ (u_2 - u_1)\ \delta x \delta y \delta z$$

$$\Rightarrow \rho\ (u_2 - u_1) = \rho I_x - \frac{\partial \overline{\omega}}{\partial x}$$

$$\Rightarrow u_2 - u_1 = I_x - \frac{1}{\rho} \frac{\partial \overline{\omega}}{\partial x}, \text{ along X-axis}$$

$$\text{Similarly } v_2 - v_1 = I_y - \frac{1}{\rho} \frac{\partial \overline{\omega}}{\partial y}, \text{ along Y-axis}$$

$$\text{and } w_2 - w_1 = I_z - \frac{1}{\rho} \frac{\partial \overline{\omega}}{\partial z}, \text{ along Z-axis}$$

known as the equations of motion of an inviscid fluid under the action of impulsive forces.

CONSERVATIVE FIELD OF FORCE

If the work done by the force F of the field in taking a unit mass from one point A to another point B independent of the path then it is termed as conservative field of force

$$\int_{ACB} F.dr$$

$$= \int_{ADB} F.dr = -\Omega(\text{say}),$$

where W is a scalar point function whose value depends on the initial and final position A and B. Thus

$$F = -\nabla\Omega.$$

where W is known as force potential which measures the potential energy of the field.

ENERGY EQUATION

The principle of energy enunciates that the change in total energy is equal to work done by the extraneous forces. The potential due to external forces is supposed to be independent of time.

The Euler's equation of motion is

$$\frac{\partial \mathbf{q}}{\partial t} = -\nabla\,\Omega - \frac{1}{\rho}\nabla p. \qquad \text{...(1)}$$

where the extraneous forces are conservative.

$$\rho\,\frac{\partial \mathbf{q}}{\partial t} = \rho\,(\nabla\,\Omega) - \nabla p,$$

Multiplying the equation (1) scalarly by $\mathbf{q}$, we get

$$\rho\mathbf{q}\,\frac{\partial \mathbf{q}}{\partial t} = \rho\mathbf{q}\,(\nabla\,\Omega) - \mathbf{q}.\,\nabla p$$

$$\text{or}\quad \frac{1}{2}\rho\frac{d}{dt}(\mathbf{q}^2) + \rho\mathbf{q}.(\nabla\Omega) = -\,\mathbf{q}.\nabla p$$

$$\frac{1}{2}\rho\frac{d}{dt}(\mathbf{q}^2) + \rho\frac{d\Omega}{dt} = -\,\mathbf{q}.\nabla p$$

$$\text{or}\quad \rho\frac{d}{dt}\left(\frac{1}{2}\mathbf{q}^2 + \Omega\right) = -\,\mathbf{q}.\nabla p \qquad \text{...(2)}$$

Integrating the relation (2) over V, we have

$$\int_V \rho\frac{d}{dt}\left(\frac{1}{2}\mathbf{q}^2 + \Omega\right)dv = \int_V (\mathbf{q}.\nabla p)\,dv$$

$$\text{or}\quad \frac{d}{dt}\left[\int_V \frac{1}{2}\rho\mathbf{q}^2 dv + \int_V \rho\Omega dv\right] = -\int_V (\mathbf{q}.\nabla p)\,dv$$

$$\text{or } \frac{d}{dt}(T + W) = -\int_V \nabla.(p\mathbf{q})dv - \int_V \frac{p}{\rho}\frac{d\rho}{dt}dv.$$

By virtue of divergence theorem and the equation of continuity the R. H. S. may be expressed as

$$\frac{d}{dt}(T + W) = -\int_S p.\mathbf{q}.\mathbf{n}dS - \frac{dI}{dt}$$

$$\text{or } \frac{d}{dt}(T + W + I)^{++} = \int_S p.\mathbf{q}.\mathbf{n}dS^{\dagger}$$

which is known as energy equation.

Thus, the rate of change of total energy (kinetic, potential and intrinsic) of a portion of perfect fluid is equal to the rate at which the work is being done by the pressure on the boundary.

SOLVED EXAMPLES

Example 1: *Stream is rushing from a boiler through a conical pipe, the diameters of the ends of which are D and d. If V and v be the corresponding velocities of the stream and if the motion be supposed to be that of divergence from the vertex of the cone, prove that*

$$\frac{v}{V} = \frac{D^2}{d^2}\exp.\left(\frac{v^2 - V^2}{2k}\right),$$

where k is the pressure divided by the density, and supposed constant.

Solution: Let ρ_1 and ρ_2 be the densities of steam at the ends of the conical pipe AB and CD. By the principle of conservation of mass the mass of the steam that enters and leaves at the ends AB and CD are of the same. Thus we have

$$\pi\left(\frac{1}{2}d\right)^2 v\rho_1 = \pi\left(\frac{1}{2}D\right)^2 v\rho_2$$

$$\text{or } \qquad \frac{v}{V} = \frac{D^2}{d^2}\frac{\rho^2}{\rho_1}. \qquad ...(1)$$

Let p be the pressure ρ the density and u the velocity at distance r from AB, then the equation of motion is given by

$$u\frac{\partial u}{\partial r} = -\frac{1}{\rho}\frac{\partial p}{\partial r},$$

Since $k = p/\rho \Rightarrow p = k\rho$

$$\Rightarrow u\frac{\partial u}{\partial r} = -\frac{k}{\rho}\frac{\partial \rho}{\partial r}.$$

By integrating, we have

$$\frac{1}{2}u^2 = -k \log \rho + k \log A$$

where A is an arbitrary constant.

$\Rightarrow \log(\rho/A) = -(u^2/2k)$

$\Rightarrow \rho = A \exp.(-u^2/2k).$

Again $\rho = \rho 1$ ωηεν $u = v$ then $\rho_1 = A \exp(-v^2/2k)$,

and $\rho = \rho^2$ when $u = V$ then $\rho_2 = A \exp(-V^2/2k)$,

$$\Rightarrow \frac{\rho_1}{\rho_2} = \frac{\exp(-v2/2k)}{\exp(-V^2/2k)}.$$

$$\Rightarrow \frac{\rho_2}{\rho_1} = \exp\{(v2 - V^2)/2k\}. \quad ...(2)$$

From (1) and (2), we have

$$\frac{v}{V} = \frac{D^2}{d^2}\exp\left(\frac{v^2 - V^2}{2k}\right).$$ **Proved.**

Example 2: *A stream in a horizontal pipe, after passing a contraction in the pipe at which its sectional area is A, is delivered at atmospheric pressure at a place where the sectional area is B. Show that if a side is connected with the pipe at the former place, water will be sucked up through it into the pipe from a reservoir at a depth* $\frac{S^2}{2g}\left(\frac{1}{A^2} - \frac{1}{B^2}\right)$ *below the pipe; S being the delivery per second.*

Solution: Let v and p be the velocity and pressure in the tube of sectional area A and V and Π be the velocity and the atmospheric pressure at the sectional area B. The equation of motion is

$$v\frac{\partial v}{\partial r} = \frac{1}{\rho}\frac{\partial p}{\partial r}. \quad ...(1)$$

By integrating, we have

$$\frac{1}{2}v^2 = -\frac{p}{\rho} + C, \quad ...(2)$$

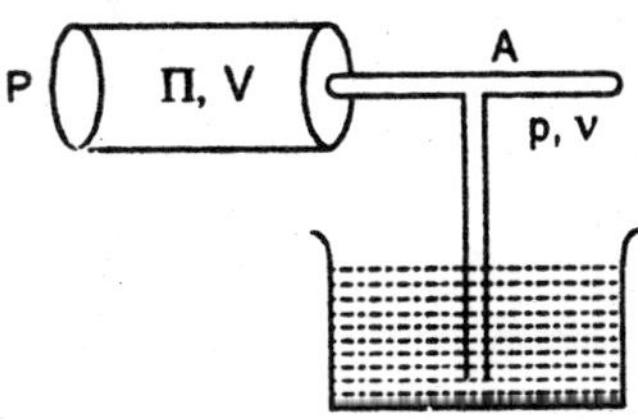

Fig. 2.20

where C is an arbitrary constant.

Since v = V, p = P at the sectional area B then

$$\frac{1}{2}v^2 + \frac{p}{\rho} = \frac{1}{\rho}V^2 + \frac{\Pi}{\rho}; C = \frac{1}{2}V^2 + \frac{\Pi}{\rho} \qquad ...(3)$$

$$\frac{1}{2}(v^2 - V2) = \frac{1}{\rho}(\Pi - p). \qquad ...(4)$$

The delivery of stream per second in the tube of sectional area A and B is given as S. From the equation of continuity, we have

Av = BV = S.

$$\Rightarrow \frac{1}{\rho} = \frac{1}{2}\left(\frac{S^2}{A^2} - \frac{S^2}{B^2}\right). \qquad ...(6)$$

Let h be the height through which the water is sucked up, then gph = dff. of pressure in the tube of section A and section B.

From (6) and (7), we have

$$\frac{1}{\rho}gph = \frac{S^2}{2}\left(\frac{1}{A^2} - \frac{1}{B^2}\right) \Rightarrow h = \frac{S^2}{2g}\left(\frac{1}{A^2} - \frac{1}{B^2}\right) \qquad \textbf{Proved.}$$

Example 3: *A pulse travelling along a fine straight uniform, tube filled with gas causes the at time t and distance x from the origin where the velocity is u_0 to become $\rho_0\ \phi\ (vt - x)$. Prove that the velocity u (at time t and distance x from the origin) is given by*

$$v + \frac{(u_0 - v)\phi(vt)}{\phi(vt - x)}.$$

Solution: Let u be the velocity and r be the density of the gas at a distance x then

$$\rho = p_0\phi\ (vt - x). \qquad ...(1)$$

The equation lf continuity is given by

$$\frac{\partial \rho}{\partial t} = \frac{\partial}{\partial x}(\rho u) = 0$$

$$\Rightarrow \frac{\partial \rho}{\partial t} + \rho\frac{\partial u}{\partial x} + u\frac{\partial \rho}{\partial x} = 0. \qquad ...(2)$$

Differentiating (1) partially with regard to t and x respectively, we have

$$\frac{\partial \rho}{\partial t} = \rho_0 v\, \phi,\ (vt - x);\ \frac{\partial \rho}{\partial x} = -\ \rho_0\ \phi'(vt - x). \qquad ...(3)$$

From (2) and (3), we have

$$\frac{du}{v-u} + \frac{\phi'(vt-x)}{\phi(vt-x)}\ dx = 0$$

$$\Rightarrow (v- u)\ \phi\ (vt - x) = A, \qquad ...(4)$$

where A is integration constant to be determined by the given initial condition; $x = 0$, $u = u_0$; $A = (v - u_0)\ \phi\ (vt)$

$$\text{or } u = v + \frac{(u_0 - v)\phi(vt)}{\phi(vt-x)}.$$ **Proved.**

Example 4: *An infinite mass of fluid is acted on by a force $\mu r^{-3/2}$ per unit mass directed to the origin. If initially the fluid is at rest and there is a cavity to the form of the sphere $r = c$ in it. Show that the cavity will be filled up after an interval of time*

$(2/5\mu)^{1/2}\ c^{5/4}$.

Solution: Let v' be the velocity at a distance r' from the origin at any time t and p be the pressure, then the equation of continuity is

$$r'^2 v' = f\ (t) = r^2 v. \qquad ...(1)$$

Equation of motion is

$$\frac{\partial v'}{\partial t} + v\frac{\partial v'}{\partial r'} = \mu r'^{-3/2} - \frac{1}{\rho}\frac{\partial p}{\partial r'}$$

$$\Rightarrow -\frac{f'(t)}{r'2} + v'\frac{\partial v'}{\partial r'} = -\ \mu r t'^{3/2} - \frac{1}{\rho}\frac{\partial p}{\partial r'}$$

By integrating with regard to r¢, we have

$$-\frac{f'(t)}{r'} + \frac{1}{2}v'^2 = \frac{2\mu}{r'^{1/2}} - \frac{p}{\rho} + A. \qquad ...(2)$$

Initially $r' = \infty$, $v' = 0$,

$p = 0;\ A = 0.$

Equation (2) reduces to

$$-\frac{f'(t)}{r'}+\frac{1}{2}v'^2 = \frac{2\mu}{r'^{1/2}}-\frac{p}{\rho}. \qquad \text{...(3)}$$

Let r be the radius of the cavity and v be the velocity at any time t. The motion of the cavity is

$$-\frac{f'(t)}{r}+\frac{1}{2}v^2=\frac{2\mu}{r^{1/2}}$$

$$\Rightarrow -\left(rv\frac{dv}{dr}+2v^2\right)+\frac{1}{2}v^2=\frac{2\mu}{r^{1/2}}.$$

$$\Rightarrow rv\frac{dv}{dr}+\frac{3}{2}v^2=\frac{2\mu}{r^{1/2}}.$$

Multiplying both the sides with $2r^2$ dr and integrating, we get

$$2r^3vdv + 3r^2v^2\, dr = 4\mu r^{3/2}dr \qquad \text{...(4)}$$

$$r^3v^2 = -(8\mu/5)\, r^{5/2} + B.$$

Now $r = c$, $v = 0$; $B = (8\mu/5)c^{5/2}$.

$$\Rightarrow r^3v^2 = \frac{8\mu}{5}(c^{5/2}-r^{5/2})$$

$$\Rightarrow v = \frac{dr}{dt} = \pm\sqrt{\left(\frac{8\mu}{5}\right)}.\sqrt{\left(\frac{c^{5/2}-r^{5/2}}{r^3}\right)},$$

– ve sign is taken since velocity increases as r decreases.

If t be the time of filling up the cavity then

$$t = -\sqrt{\left(\frac{5}{8\mu}\right)}\int_c^0\frac{r^{3/2}dr}{\sqrt{(c^{5/2}-r^{5/2})}}$$

Let $r^{5/2} = c^{5/2}\sin^2\theta$, $\frac{5}{2}r^{3/2}\,dr = c^{5/2}\sin\theta\cos\theta\, d\theta$

$$\text{or } t = \frac{4}{5}\sqrt{\left(\frac{5}{8\mu}\right)}\int_0^{\pi/2}\frac{c^{5/2}\sin\theta\cos\theta}{c^{5/4}\cos\theta}d\theta$$

$$\Rightarrow t = \frac{4}{5}c^{5/4}\sqrt{\left(\frac{5}{8\mu}\right)}\int_0^{\pi/2}\sin\theta d\theta = \sqrt{\left(\frac{2}{5\mu}\right)}c^{564}.$$

Example 5(a): *A venturi meter measures the flow of oil (sp. gr. 0.85) in a 8cm. diameter pipe. The difference of head between inlet and throat of the meter is measured by a U-tube mercury manometer. What should be the diameter of throat of the meter in order that the difference in level of the mercury in U tube shallbe 25 cm. When the quantity of oil flowing in the pipe is 10 litres/sec. Assume the coefficient of venturi meter as 0.97.*

Solution: Discharge through a venturimeter is given by

$$Q = C_d \frac{S_1}{\sqrt{(r^2 - 1)}} \sqrt{(2gH)} \text{ , } r = s_1/s_2.$$

Here $H = 3.75$ m, $S_1 = p/4\ (0.08)^2\ m^2$

$Q = 10$ litres/sec $= 0.01\ m^3/sec.$, $C_d = 0.97$

$$\text{or } r^2 = \frac{2gHC_d^2S_1^2}{Q^2} + 1 = 18.6.$$

Let D_1 be the diameter of the pipe and D_2 be the diameter of the throat, then

$$\frac{S_1^2}{S_2^2} = \frac{D_1^4}{D_2^4} = 18.6 \text{ .}$$

or $D_1/D_2 = 2.08$

$\Rightarrow D_2 = 8/2.08 = 3.85$ cm. **Ans.**

Example 5(b): *Water flows over a re ctangular wire 1m. wide at a depth of 15 cm. and after wards passes thought a triangular right angled weir. Find the depth of water through the triangular weir. The C_d for the rectangular and triangular weirs are 0.62 and 0.59 respectively.*

Solution: The same discharge occurs over the rectangular weir as that over the triangular weir

$$Q = Q_1 = \frac{2}{3} C_{d_1} \sqrt{(2g)} LH_1^{3}/2$$

$$Q = Q_2 = \frac{2}{15} C_{d_2} \sqrt{(2g)} H_2^{5}/^{2} \tan \theta.$$

Here $L = 1m$, $H_1 = 0.15$ m., $\theta = \pi/2$,

$C_{d1} = 0.62$ and $C_{d2} = 0.59$.

$$H_2^{5/2} = \frac{2}{3} \times \frac{15}{8} \times \frac{0.62}{0.59} \times 1 \times (0.15)^{3/2} = 0.0763$$

or $H_2 = 0.3573$ m = 35.73 cm. **Ans.**

Example 6: *A quantity of liquid occupies a length 2l of a straight tube of uniform small bore under the action of a force to a point in the tube varying as the distance from that point. Determine the motion and the pressure.*

Solution: Consider p be the pressure and u be the velocity at a distance x from the fixed point O. Let x_0 be the distance of the nearer free surface from O. The equation of continuity is $\partial u/\partial x = 0$. The equation of motion becomes

$$\frac{\partial u}{\partial t} = -\mu x - \frac{1}{\rho}\frac{\partial p}{\partial x}. \qquad ...(1)$$

Integrating (1) with regard to x, we have

$$x\frac{\partial u}{\partial t} = -\frac{1}{2}\mu x^2 - \frac{p}{\rho} + A.$$

Again $p = 0,\ x = x_0$

and $p = 0,\ x = x_0 + 2l$. ...(3)

From (2) and (3), we have

$$x_0\frac{\partial u}{\partial t} = -\frac{1}{2}\mu x_0^2 + A,$$

and $$(x_0 + 2l)\frac{\partial u}{\partial t} = \frac{1}{2}\mu(x_0 + 2l)^2 + A$$

Eliminating the arbitrary constant A, we have

$$\frac{\partial^2 x_0}{\partial t^2} + \mu(x_0 + l) = 0;\ u = x_0 \qquad ...(4)$$

whose solution is given by

$$x_0 + l = B\cos(\sqrt{\mu}t + \alpha), \qquad ...(5)$$

where the constants are determined by the initial position and velocity. The pressure at any point is determined as

$$p/\rho = -\frac{1}{2}\mu(x^2 - x_0^2) - (x - x_0)(\partial u/\partial t), \qquad \textbf{Ans.}$$

$$\Rightarrow p/\rho = -\frac{1}{2}\mu(x^2 - x_0^2) + \mu(x - x_0)(x_0 + l).$$

Example 7: *A centre of force attracting inversely as the square of the distance is at the centre of a spherical cavity within an infinite mass* *of incompressible fluid, the pressure on which at an infinite distance is Π, and is such that the work done by this pressure on a unit of area through a unit of length in one-half the work done by the attractive force on a unit volume of the fluid from infinity to the initial boundary of the cavity, prove that the time of filling up the cavity will be*

$$\pi a \sqrt{\left(\frac{\rho}{\Pi}\right)\left\{2-\left(\frac{3}{2}\right)^{3/2}\right\}};$$

a being the initial radius of the cavity, and ρ the density of the fluid.

Solution: The only possible motion which can take place is one in which each element of liquid moves towards the centre, whence the free surfaces will remain spherical. Let v′ be the velocity at a distance r′ at any time t and p be the pressure there. The equation of continuity reduce to

$$\frac{1}{r^2}\frac{d}{dr'}(r'^2 v') = 0,$$

$$\Rightarrow r'^2 v' = f(t) = \text{Const.} \qquad \text{....(1)}$$

$$\Rightarrow r'^2 v' = f(t) = r^2 v.$$

Equation of motion is

$$\frac{\partial v'}{\partial t} + v'\frac{\partial v'}{\partial r'} = -\frac{\mu}{r'^2} - \frac{1}{\rho}\frac{\partial v'}{\partial r'} \qquad \text{...(2)}$$

$$\Rightarrow \frac{f'(t)}{r'^2} + v'\frac{\partial v'}{\partial r'} = -\frac{\mu}{r'^2} - \frac{1}{\rho}\frac{\partial v'}{\partial r'}.$$

Integrating with regard to r′, we have

$$-\frac{f'(t)}{r'} + \frac{1}{2}v'^2 = -\frac{\mu}{r'} - \frac{p}{\rho} + A,$$

where A is an arbitrary constant.

Initially r′ = ∞, v′ = 0 and p = Π; A = Π/ρ.

$$\Rightarrow -\frac{f'(t)}{r'} + \frac{1}{2}v'^2 = \frac{\mu}{r'} + \frac{\Pi - p}{\rho}. \qquad \text{...(3)}$$

Let r be radius of the spherical cavity and v be the velocity at any time t. The pressure p vanishes on the surface of the cavity. Hence equation (3) reduces to

$$-\frac{f'(t)}{r} + \frac{1}{2}v^2 = \frac{\mu}{r} + \frac{\Pi}{\rho}$$

$$\Rightarrow -\frac{1}{r}\left(r^2 v\frac{dv}{dr}+2rv^3\right)+\frac{1}{2}v^2 = \frac{\mu}{r}+\frac{\Pi}{\rho}\quad \left\{\begin{aligned} &\text{Since} f(t)=r^2 v\\ &f'(t)=r^2\frac{dv}{dt}+2rv\frac{dr}{dt}\\ &f'(t)=r^2\frac{dv}{dr}.\frac{dr}{dt}+2rv^2\\ &f'(t)=r^2 v\frac{dv}{dr}+2rv^2.\end{aligned}\right.$$

$$\Rightarrow 2rv\ dv + 3v^2\ dr = -2\left(\frac{\mu}{r}+\frac{\Pi}{\rho}\right)dr$$

Multiplying both the sides with r^2, we get

$$2r^3 v\ dv + 3v^2 r^2\ dr = -2\left(\mu r+\frac{\Pi}{\rho}r^2\right)dr.$$

By integrating, we have

$$r^3 v^2 = -\{\mu r^2 + (2\Pi/3\rho)\, r^3\} + B, \quad \text{...(4)}$$

where B is an arbitrary constant.

Initially $r = a$, $v = 0$; $B = \mu a^2 + (2\Pi/3\rho)\, a^3$.

Equation (4) becomes

$$r^3 v^2 = \mu\,(a^2 - r^2) + (2\Pi/3\rho)\,(a^3 - r^3). \quad \text{...(5)}$$

The work done by the pressure Π on a unit of area through a unit of length $= \frac{1}{2}\times$ work done by the attractive force on a unit volume of the fluid from infinity to the initial boundary of the cavity *i.e.*,

$$\Pi \times 1 \times 1 = \frac{1}{2}\int_{\infty}^{a}\left(\frac{\mu}{r^2}\right)\rho dr = \frac{\mu\rho}{2a} \Rightarrow \mu = \frac{2a\Pi}{\rho}.$$

Substituting the value of m in (5), we get

$$r^3 v^2 = \frac{2a\Pi}{\rho}(a^2 - r^2) + \frac{2\Pi}{3\rho}(a^3 - r^3)$$

$$\Rightarrow r^3 v^2 = \frac{2\Pi}{3\rho}\{3a\,(a^2 - r^2) + (a^3 - r^3)\}$$

$$\Rightarrow v^2 = \frac{2\Pi}{3\rho}\frac{(3a(a^2 - r^2) + (a^3 - r^3)\}}{r^3}$$

$$\Rightarrow \frac{dr}{dt} = \sqrt{\left(\frac{2\Pi}{3\rho}\right)}.\frac{\{3a(a^2 - r^2) + (a^3 - r^2)^{1/2}}{r^{3/2}}$$

$$\Rightarrow dt = -\sqrt{\left(\frac{3\rho}{2\Pi}\right)}\int_a^0 \frac{r^{3/2}dr}{\{3a(a^2-r^2)+(a^3-r^3)^{1/2}}$$

Negative sign taken as r decreases when t increases.

Let t be the time of filling up the cavity then

$$t = \sqrt{\left(\frac{3\rho}{2\Pi}\right)}\int_0^a \frac{r^{3/2}dr}{\sqrt{\{3a(a^2-r^2)+(a^3-r^3)\}}}$$

$$\Rightarrow t = \sqrt{\left(\frac{3\rho}{2\Pi}\right)}\int_0^a \frac{r^{3/2}dr}{(r+2a)\sqrt{(a-r)}}$$

Let $r = a\sin^2\theta$, $dr = 2a\sin\theta\cos\theta\, d\theta$

$$\Rightarrow t = 2a\sqrt{\left(\frac{3\rho}{2\Pi}\right)}\int_0^{\pi/2} \frac{\sin^4\theta}{\sin^2\theta+2}d\theta$$

$$\Rightarrow t = 2a\sqrt{\left(\frac{3\rho}{2\Pi}\right)}\int_0^{\pi/2}\left(\sin^2\theta - 2 + \frac{4}{2+\sin^2\theta}\right)d\theta$$

$$\Rightarrow t = 2a\sqrt{\left(\frac{3\rho}{2\Pi}\right)}\left[\frac{\pi}{4} - \pi + \int_0^{\pi/2}\frac{4\sec^2\theta}{2\sec^2\theta+\tan^2\theta}d\theta\right]$$

$$\Rightarrow t = 2a\sqrt{\left(\frac{3\rho}{2\Pi}\right)}\left[-\frac{3\pi}{4} + \int_0^{\pi/2}\frac{4\sec^2\theta}{2+3\tan^2\theta}d\theta\right]$$

$$\Rightarrow t = 2a\sqrt{\left(\frac{3\rho}{2\Pi}\right)}\left[-\frac{3\pi}{4} + \frac{4}{3}\sqrt{\left(\frac{3}{2}\right)}\cdot\frac{\pi}{2}\right]$$

$$\Rightarrow t = \pi a\sqrt{\left(\frac{\rho}{\Pi}\right)}\left\{2-\left(\frac{3}{2}\right)^{3/2}\right\}.$$

Proved.

Example 8(a): *A volume* $\frac{4}{3}\pi c^3$ *of gravitating liquid, of density r is initially in the form of a spherical shell of infinitely great radius. If the liquid shell contract under the influence of its own attraction, there being no external or internal pressure, show that when the radius of the inner spherical surface is x, its velocity will be given by*

$$V^2 = \frac{4\pi\gamma\rho z}{15x^3}(2z^4 + 2z^2x^2 - 3zx^3 - 3x^4),$$

where γ is the constant of gravitation, and $z^3 = x^3 + c^3$.

Solution: Let v′ be the velocity at a distance r′ from the centre of the spherical shell at any time t and p be the pressure there. The equation of continuity is

$$\frac{1}{r'^2}\cdot\frac{d}{dr}(r'^2 v') = 0,\ r'^2 v' = f(t).$$

Let r be the radius of the inner surface. The attraction at the point distance r′ from the centre due to liquid is given by

$$= \frac{\frac{4}{3}\gamma\pi\rho(r'^3 - r^3)}{r'^2} = \frac{4}{3}\gamma\pi\rho\left(r' - \frac{r^3}{r'^2}\right)$$

The equation of motion is

$$\frac{\partial v'}{\partial t} + v'\frac{\partial v'}{\partial r'} = -\frac{4}{3}\gamma\pi\rho\left(r' - \frac{r^3}{r'^2}\right) - \frac{1}{\rho}\frac{\partial p}{\partial r'}.$$

$$\text{or}\quad \frac{f'(t)}{r'^2} + v'\frac{\partial v'}{\partial r'} = -\frac{4}{3}\gamma\pi\rho\left(r' - \frac{r^3}{r'^2}\right) - \frac{1}{\rho}\frac{\partial p}{\partial r'}.$$

Integrating with regard to r′, we get

$$-\frac{f'(t)}{r} + \frac{1}{2}v^2 = \frac{4}{3}\gamma\pi\rho\left(\frac{1}{2}r'^2 + \frac{r^3}{r'}\right) - \frac{p}{\rho} + A. \qquad ...(1)$$

The initial conditions are given as

(i) $r' = r,\ v' = v,\ p = 0,$

(ii) $r' = z,\ v' = v_1,\ p = 0.$

Using these conditions into (1), we have

$$-\frac{f'(t)}{r} + \frac{1}{2}v^2 = \frac{4}{3}\gamma\pi\rho\left(\frac{1}{2}r^2 + r^2\right) + A$$

$$-\frac{f'(t)}{z} + \frac{1}{2}v_1^2 = \frac{4}{3}\gamma\pi\rho\left(\frac{1}{2}z^2 + \frac{r^3}{z}\right) + A.$$

Eliminating the constant, we get

$$-f'(t)\left\{\frac{1}{r} - \frac{1}{z}\right\} + \frac{1}{2}\left\{v^2 - v_1^2\right\} = -\frac{4}{3}\gamma\pi\rho\left\{\frac{3r^2}{2} - \frac{z^2}{2} - \frac{r^3}{z}\right\} \qquad ...(2)$$

$$\Rightarrow -f'(t)\left\{\frac{1}{r} - \frac{1}{z}\right\} + \frac{1}{2}\left\{\frac{f^2(t)}{r^4} - \frac{f^2(t)}{z^4}\right\} = -\frac{4}{3}\gamma\pi\rho\left\{\frac{3r^2}{2} - \frac{z^2}{2} - \frac{r^3}{z}\right\}$$

Since $\quad r^2v = z^2v_1 = f\ (t)$

$\Rightarrow \quad 2r^2\ dr = 2z^2dz = 2f\ (t)\ dt.$ ...(3)

Multiplying both the the sides of (2) by the relation (3) and integrating, we have

$$f^2\ (t)\left\{\frac{1}{r}-\frac{1}{z}\right\} = \frac{4}{3}\gamma\pi\rho\int(3r^4dr - dz\frac{r^3}{z} 2z^2dz), z^3 - r^3 = c^3\ .$$

$$\Rightarrow f^2\ (t)\left\{\frac{1}{r}-\frac{1}{z}\right\} = \frac{4}{3}\gamma\pi\rho\left\{\frac{3}{5}r^5 - \frac{1}{5}z^5 - 2\int z(z^3-c^3)dz\right\}$$

$$\Rightarrow f^2\ (t)\left\{\frac{1}{r}-\frac{1}{z}\right\} = \frac{4}{3}\gamma\pi\rho\left\{\frac{3}{5}r^5 - \frac{1}{5}z^5 - \frac{2}{5}z^5 + c^3z^2\right\}$$

$$\Rightarrow f^2\ (t)\left\{\frac{1}{r}-\frac{1}{z}\right\} = \frac{4}{15}\gamma\pi\rho\left\{3(r^5 - z^5\right\} + 5_c^3z^2\}. \quad ...(4)$$

If the radius of the inner spherical surface be x (*i.e.*, r = 1) then the equation of continuity becomes $x^2V = f\ (t)$ and equation (4) reduces

$$\Rightarrow x^4V^2\left\{\frac{1}{x}-\frac{1}{z}\right\} = \frac{4}{15}\gamma\pi\rho\{3(x^5 - z^5) + 5c^3z^2\}$$

$$\Rightarrow V^2 = \frac{4}{15}\frac{\pi\gamma\rho z}{x^3}\left\{\frac{3(x^5 - z^5) + 5c^3z^2}{z - x}\right\}$$

$$\Rightarrow V^2 = \frac{4}{15}\frac{\pi\gamma\rho z}{x^3}\ \{-3(x^4 + x^3z + x^2\ z^2 + xz^3 + z^4) + 5z^2(z^2 + zx + x^2)\}$$

$$\Rightarrow V^2 = \frac{4\pi\gamma\rho z}{15x^3}\{2z^2 + 2z^3x + 2z^2x^2 - 3zx^3 - 3x^4\}.$$ **Proved.**

Example 8(b): *An infinite mass of homogeneous incompressible fluid is at rest subject to a uniform pressure Π and contains a spherical cavity of radius a, filled with gas at a pressure m Π : prove that, if the inertia of the gas be neglected, and Boyle's law be supposed to hold throughout the ensuing motion, the radius of the sphere will oscillate between the values a and na, where n is determined by the equation*

$1 + 3m \log n - n^3 = 0.$

If m be nearly to 1, the time of an oscillation will be

$2\pi\ \sqrt{(a^2\rho/3\Pi}$

π being the density of the fluid.

Solution: In an incompressible fluid the velocity v will be radial outside the spherical cavity, so v will be a function of r radial distance from the centre of cavity (*i. e.*, origin) and the time t only. The continuity equation reduces to

$$\frac{1}{r}\frac{d}{dr}(r^2 v) = 0 \Rightarrow r^2 v = f(t).$$

Let v′ be the velocity at a distance r′ from the centre of the cavity at any time t and p be the pressure there, then we have

$$r'^2 v' = f(t) = r^2 v.$$

Equation of motion is

$$\frac{\partial v'}{\partial t} + v'\frac{\partial v'}{\partial r'} = -\frac{1}{\rho}\frac{\partial p}{\partial r'},$$

or $$\frac{f'(t)}{r'^2} + v'\frac{\partial v}{\partial r'} = -\frac{1}{\rho}\frac{\partial p}{\partial r'},$$

Integrating with regard to r′, we have

$$-\frac{f'(t)}{r'} + \frac{1}{2}v'^2 = -\frac{p}{\rho} + A,$$

where A is an arbitrary constant,

Initially when $r' = \infty$, $v' = 0$,

$$p = \Pi \;;\; A = \Pi/\rho.$$

Equation (3) becomes

$$-\frac{f'(t)}{r'} + \frac{1}{2}v'^2 = \frac{\Pi - p}{\rho}.$$

Hence motion on the surface of the cavity is

$$-\frac{f'(t)}{r} + \frac{1}{2}v'^2 = \frac{\Pi - p}{\rho}.$$

From Boyle's law, we have

$$\frac{4}{3}\pi r^3 p = \frac{4}{3}\pi a^3 m\Pi \Rightarrow p = (a^3/r^3)m\Pi.$$

Substituting the value of f ′ (t) and p in (5), we get

$$-\left(2rv^2 + r^2 v\frac{dv}{dr}\right) + \frac{1}{2}v^2 = \frac{\Pi}{r} - \frac{a^3}{\rho r^3}m\Pi$$

$$\Rightarrow \ rv\ \frac{dv}{dr}+\frac{3}{2}v^2 = -\frac{\Pi}{\rho}-\frac{a^3}{\rho r^3}m\Pi$$

Multiplying with $2r^2$ dr both the sides, we have

$$2r^3v\ dv + 3r^2v^2\ dr = \left(-\frac{2\Pi r^2}{\rho}+\frac{2a^2m\Pi}{\rho r}\right)dr.$$

By integrating, we have

$$r^3v^2 = -\frac{2\Pi}{3\rho}r^3+\frac{2a^3m\Pi}{\rho}\log r + B.$$

Initially r = a, v = 0;

$$B = \frac{2\Pi}{3\rho.}a^3-\frac{2a^2m\Pi}{\rho}\log a$$

$$r^3v^2 = \frac{2\Pi}{3\rho.}(a^3-r^3)\frac{2a^2m\Pi}{\rho}\log\left(\frac{r}{a}\right). \quad ...(6)$$

Since the radius of the sphere oscillates between a andna it follows that at r = a, v = 0 and r = na, v = 0.

Substituting r = na, v = 0 in relation (6), we get

$$\frac{2\Pi}{3\rho.}\left\{a^3-n^3a^3+3ma^3\log\left(\frac{na}{a}\right)\right\} = 0$$

$$1 + 3m \log n - n^3 = 0,\ a \neq 0$$

Which proves the required condition.

Again if m be nearly equal to 1, let r = a + x where x is small then from (5), we have

$$(a+x)^3x^2 = \frac{2\Pi}{3\rho}\{a^3-(a+x)^3\}+\frac{2a^3\Pi}{\rho}\log\left(\frac{a+x}{a}\right).$$

Since r = a + x, v = dr/dt = x

$$\Rightarrow a^3\left(1+\frac{x}{a}\right)^3x^2 = \frac{2\Pi}{3\rho}a^3\left\{1-\left(1+\frac{x}{a}\right)^3\right\}+\frac{2a^3\Pi}{\rho}\log\left(1+\frac{x}{a}\right)$$

$$\Rightarrow \left(1+\frac{x}{a}\right)^3x^2 = \frac{2\Pi}{2\rho}\left\{1-\left(1+\frac{3x}{a}+\frac{3x^2}{a^2}+...\right)\right\}$$

$$+\frac{2a^3\Pi}{\rho}\left\{\frac{x}{a}-\frac{1}{2}\frac{x^2}{a^2}+...\right\}$$

$$\Rightarrow x^2 = \frac{2\Pi}{3\rho}\left[\left(-\frac{9}{2}\frac{x^2}{a^2}+...\right)\left(1+\frac{x}{a}\right)^{-3}\right]$$

$$\Rightarrow x^2 = \frac{2\Pi}{3\rho}\left[\left(-\frac{9}{2^2}\frac{x^2}{a^2}+...\right)\left(1+\frac{3x}{a}+\frac{6x^2}{a^2}+...\right)\right] = \frac{3\Pi}{\rho a^2}x^2,$$

neglecting higher orders of x, being very small.

Differentiating with regard to t, we have

$$2x\,x = -(3\Pi/\rho a^2).2xx$$

$$\Rightarrow \quad x = -(3\Pi/\rho a^2)\,x,\ x \neq 0,$$

which represents the standard equation of S. H. M. whose time period is

$2p\sqrt{(\rho a^2/3\Pi)}$. **Proved.**

Example 9: *An infinite fluid in which as a spherical hollow of radius a is initially at rest under the action of no forces. If a constant pressure Π is applied at infinity, show that the time of filling up the cavity is*

$$\pi a^2 (\rho/\Pi)^{1/2}\, 2^{5/6}\left\{\Gamma\left(\frac{1}{3}\right)\right\}^{-3}$$

Solution: In an incompressible fluid, the fluid velocity v will be radial outside the hollow sphere and will be a function or a (radial distance from the centre of the sphere *i.e.*, origin) and the time t and p be the pressure there. From the continuity equation, we have

$$r'^2v' = f(t) = r^2v. \qquad ...(1)$$

The equation of motion is

$$\frac{\partial v'}{\partial r} + v'\frac{\partial v'}{\partial r'} = -\frac{1}{\rho}\frac{\partial p}{\partial r'}$$

or
$$\frac{f'(t)}{r'^2} + v'\frac{\partial v'}{\partial r'} = -\frac{1}{\rho}\frac{\partial p}{\partial r'}$$

Integrating with regard to r′, we get

$$-\frac{f'(t)}{r'} + \frac{1}{2}v'^2 = -\frac{p}{\rho} + A,$$

where A is an arbitrary constant.

initially $r' = \infty$, $v' = 0$ $p = \Pi$; $A = \dfrac{\Pi}{\rho}$.

$$\text{So } -\frac{f'(t)}{r'}+\frac{1}{2}v'^2 = \frac{\Pi-p}{\rho} \qquad ...(2)$$

Let v be the velocity and r be the radius of spherical cavity at any time t $v' = v$, $r' = r$, $p = 0$ (being hollow).

From (2), we get

$$-\frac{f'(t)}{r}+\frac{1}{2}v^2 = \frac{\Pi}{\rho}. \qquad ...(3)$$

$$rv\,\frac{dv}{dr}+\frac{3}{2}v^2 = -\frac{\Pi}{\rho}.$$

⇒ Multiplying both the sides by $2r^2dr$ and integrating, we have

$$2r^3vdv + 3r^2v^2\,dr = (2\Pi/\rho)r^2\,dr \qquad ...(4)$$

$$\Rightarrow r^3v^2 = -(2\Pi/3\rho)\,r^3 + B,$$

Initially $r = a$, $v = 0$;

$$B = (2\pi/3\rho)\,a^3.$$

Equation (4) reduces to

$$r^3v^2 = \frac{2\Pi}{3\rho}(a^3-r^3)$$

$$\Rightarrow v = \frac{dr}{dt} = -\sqrt{\left(\frac{2\Pi}{3\rho}\right)}\sqrt{\left(\frac{a^3-r^3}{r^3}\right)}$$

(–) ve sign is taken since v increases as r decreases.

Let t be the time of filling up the cavity

$$t = -\sqrt{\left(\frac{3\rho}{2\Pi}\right)}\int_a^0\sqrt{\left(\frac{r^3}{a^3-r^3}\right)}dr$$

Let $r = a\sin^{2/3}\theta$, $dr = -\frac{2a}{3}\sin^{-1/3}\theta\cos\theta d\theta$

$$t = \frac{2a}{3}\sqrt{\left(\frac{3\rho}{2\Pi}\right)}\int_0^{\pi/2}\sin^{2/3}\theta d\theta$$

By applying Gamma function, we have

$$= \frac{2a}{3}\sqrt{\left(\frac{3\rho}{2\Pi}\right)}.\frac{\Gamma\frac{5}{6}\Gamma\frac{1}{2}}{2\Gamma\left(\frac{5}{6}+\frac{1}{2}\right)}$$

$$= \frac{2a}{3}\sqrt{\left(\frac{3\rho}{2\Pi}\right)}.\frac{\Gamma\frac{5}{6}\Gamma\frac{1}{2}}{2\Gamma\frac{4}{3}}$$

$$= \frac{2a}{3}\sqrt{\left(\frac{3\rho}{2\Pi}\right)}.\frac{\Gamma\frac{5}{6}\Gamma\frac{1}{2}}{2\Gamma\frac{1}{3}\Gamma\frac{1}{3}}$$

$$= \frac{2a}{3}\frac{3}{2}\sqrt{\left(\frac{3\rho}{2\Pi}\right)}.\frac{\sqrt{\pi}\Gamma\left(\frac{1}{3}+\frac{1}{2}\right)}{\Gamma\frac{1}{3}}$$

Multiplying N^r and D^r with $\Gamma\frac{1}{3}$, we have

$$= a\sqrt{\left(\frac{3\rho}{2\Pi}\right)}.\sqrt{\pi}\frac{\Gamma\left(\frac{1}{3}+\frac{1}{2}\right)\Gamma\frac{1}{3}}{\left(\Gamma\frac{1}{3}\right)^2} \quad \left[\text{as}\Gamma n\Gamma\left(n+\frac{1}{2}\right)=\sqrt{\pi}\frac{\Gamma n}{2^{2n-1}}\right]$$

$$= a\sqrt{\left(\frac{3\rho}{2\Pi}\right)}.\sqrt{\pi}.\frac{\sqrt{\pi}.\Gamma\frac{2}{3}}{\left(\Gamma\frac{1}{3}\right)^2 2^{2/3-1}}$$

$$= \pi a\sqrt{\left(\frac{3\rho}{2\Pi}\right)}.\frac{\Gamma\frac{2}{3}}{\left(\Gamma\frac{1}{3}\right)^2 2^{-1/3}}$$

Multiplying Nr and Dr with $\Gamma\frac{1}{3}$, we have

$$= \pi a\sqrt{\left(\frac{3\rho}{2\Pi}\right)}.\frac{\Gamma\left(1-\frac{1}{3}\right)\Gamma\frac{1}{3}}{2^{-1/3}\left(\Gamma\frac{1}{3}\right)_3^3} \qquad \left[\text{as}\Gamma n\Gamma(1-n)=\frac{\pi}{\sin n\pi}\right]$$

$$= \pi a\sqrt{\left(\frac{3\rho}{2\Pi}\right)}.2^{1/3}\Gamma\left(\frac{1}{3}\right)^{-3}.\frac{\pi}{\sin\pi/3}$$

$$= \pi^2 a \sqrt{\left(\frac{3\rho}{2\Pi}\right)} \cdot \left(\Gamma \frac{1}{3}\right)^{-3} .2^{1/3} . \frac{2}{\sqrt{3}}$$

$$= \pi^2 a(\rho/\Pi)^{1/2} 2^{5/6} \left(\Gamma \frac{1}{3}\right)^{-3}$$ **Proved.**

Example 10(a): *A venturi flume 1 m, wide at entrance and 1 m. wide at the throat-section has its base horizontal. If the depth of water at the entrance and the throat is 1 m. and 0.9 , respectively. Calculate the discharge.*

Solution: The discharge is given by

$$Q = \frac{S_1 S_2}{\sqrt{(S_1^2 - S_2^2)}} \sqrt{(2gh)}$$

Here h = 1.0 = 0.9 = 0.1 m.

$S_1 = 2 \times 1 = 2m^2$, $S_2 = 1 \times 0.9 = 0.9\ m^2$.

$$\text{or } Q = \frac{2 \times 09}{\sqrt{(2^3 - 0.9^2)}} \sqrt{(2 \times 9.81 \times 0.1)} = 1.41\ m^3/s.$$ **Ans.**

Example 10(b): *A venturi meter of throat diameter 4 cm. filled into a pipe line of 10 cm. diameter. The coefficient of discharge is 0.96. Calculate the flow through the meter when the reading on a mercury-water mano-meter connected across the upstream and throat tapping is 25 cm.*

Solution : We know that

$$Q = 0.96 \times \frac{\pi}{4} \times (0.10)^2 \sqrt{\left\{\frac{19.62 \times 0.25 \times 12.6}{(10/4)^4 - 1}\right\}}$$

$Q = 9.61 \times 10^{-3} m^3/s$. **Ans.**

Example 10(c): *The venturi meter has an entrance of 6 in. diameter and a throat of 3 in. diameter whose centre is 18 in. above the centre of the entrance. Find the velocity of water at the throat when $p_1 - p_2$ = 5lbf/in². The coefficient of discharge C_{dv} is 0.97.*

Solution: Here $C_{dv} = 0.97$, $S_2/S_1 = (3/6)^2 = 1/4$,.

$$\delta h \left(\frac{\gamma_m}{\gamma} - 1\right) = \frac{5 \times 144}{62.4} - \frac{18}{12} = 10.04\text{ft}.$$

$$\text{Therefore } q_2 = \frac{0.97}{\sqrt{\left\{1-\left(\frac{1}{4}\right)^2\right\}}}\sqrt{\{2g(10.0.3)\}} = 25.4 \text{ ft/sec.} \qquad \textbf{Ans.}$$

Example 11: *A solid sphere of radius a is surrounded by a mass of liquid whose volume is* $\frac{4}{5}\pi c^3$ *and its centre is a centre of attractive force varying directly as the square of the distance. If the solid sphere be suddenly annihilated, show that the velocity of the inner surface, when its radius is x, is given by*

$$x^2x^3\ [x^3 + c^3)^{1/3} - x] = \left[\frac{2\Pi}{3\rho} + \frac{2\mu c^3}{9}\right](a^3 - x^3)(c^3 - x^3)^{1/3},$$

where ρ is the density, P The external pressure and m the absolute force.

Solution: Let v′ be the velocity at a distance r′ at any time t and p be the pressure there, then the equation of continuity is given by

$$r'^2 v' = f(t). \qquad ...(1)$$

The equation of motion is given by

$$\frac{\partial v'}{\partial t} + v'\frac{\partial v'}{\partial r'} = -\ \mu r'^2 - \frac{1}{\rho}\frac{\partial p}{\partial r'},$$

where $\mu r'^2$ is the attractive force.

$$\frac{f'(t)}{r'^2} + v'\frac{\partial v'}{\partial r'} = -\mu r'^2 - \frac{1}{\rho}\frac{\partial p}{\partial r'}$$

Integrating with regard to r′, we have

$$-\frac{f'(t)}{r'^2} + \frac{1}{2}v'^2 = -\ \frac{1}{3}\mu r'3 - \frac{p}{\rho} + A \qquad ...(1)$$

Let r and R be the internal and external radii of the shell and v and V be the velocities there at any subsequent time t.

Initially I. $r' = r$, $v' = v$ and $p = 0$

II. $r' = R$ $v' = v$ and $p = \text{II}$

Using the initial conditions I and II, the equation (1) reduces to

$$-\frac{f'(t)}{r} + \frac{1}{2}v^2 = \frac{1}{3}\mu r^3 + A,$$

$$-\frac{f'(t)}{R} + \frac{1}{2}V^2 = \frac{1}{3}\mu R^3 - \frac{\text{II}}{\rho} + A. \qquad ...(2, 3)$$

Eliminating the constant A, by subtracting (2) and (3), we have

$$f'(t)\left[\frac{1}{r}-\frac{1}{R}\right]-\frac{1}{2}(v^2-V^2)=\frac{\mu}{3}(r^3-R^3)-\frac{\Pi}{\rho} \quad ...(4)$$

Since the volume of the liquid is constant, so

$$\frac{4}{3}\pi R^3-\frac{4}{3}\pi r^3=\frac{4}{3}\pi c^3 \Rightarrow R^3-r^3=c^3. \quad ...(5)$$

We know that

$$r^2 v = f(t) = R^2 V$$

$$\Rightarrow r^2\frac{dr}{dt} = f(t) = R^2\frac{dR}{dt}$$

$$\Rightarrow r^2\, dr = f(t)\, dt = R^2 dR$$

$$f'(t) = \frac{d}{dt}(r^2 v) = \frac{d}{dt}(r^2 v)\frac{dr}{dt} = v\frac{d}{dr}(r^2 v);\ V=\frac{r^2 v}{R^2}. \quad ...(6)$$

From (4) and (5), we have

$$v\frac{d}{dr}(r^2 v).\left(\frac{1}{r}-\frac{1}{R}\right)-\frac{1}{2}\left(v^2-\frac{r^4 v^2}{R^4}\right) = -\frac{1}{3}\mu c^3-\frac{\Pi}{\rho}$$

Multiplying both the sides by r^2, we have

$$2r^2 v\ \frac{d}{dr}(r^2 v).\left(\frac{1}{r}-\frac{1}{R}\right)-v^2 r^4\left(\frac{1}{r^2}-\frac{r^2}{R^4}\right) = \frac{2}{3}\mu c^3 r^2-\frac{2\Pi}{\rho}r^2$$

$$\Rightarrow \frac{d}{dr}(r^2 v)^2\left(\frac{1}{r}-\frac{1}{R}\right)-(r^2 v)^2\left(\frac{1}{r^2}\frac{r^2}{R^4}\right) = \frac{2}{3\rho}(\mu\rho c^3+3\Pi)r^2$$

By integrating, we have

$$r^4 v^2\left(\frac{1}{r}-\frac{1}{R}\right) = -\frac{2}{3\rho}(\mu\rho c^3+3\Pi).\frac{r^3}{3}+B$$

when r = a, v = 0

$$\Rightarrow B = \frac{2}{9\rho}(\mu\rho c^3+3\Pi).a^3$$

$$\Rightarrow r^4 v^2\left(\frac{1}{2}-\frac{1}{R}\right) = \frac{2}{9\rho}(\mu\rho c^3+3\Pi)(a^3-r^3)$$

$$\Rightarrow r^4 v^2\left[\frac{1}{r}-\frac{1}{(r^3+c^3)^{1/3}}\right] = \left(\frac{2\mu c^3}{9}+\frac{2\Pi}{3\rho}\right)(a^3-r^3)$$

For the inner surface r = x, v = x, we have

$$x^2x^3\ [(x^3 + c^3) - x] = \left[\frac{2\mu c^3}{9} + \frac{2\Pi}{3\rho}\right](a^3 - x^3)(x^3 + c^3)^{1/3}.$$

Proved.

Example 12: *A sphere is at rest in an infinite mass of homogeneous liquid of density ρ, the pressure at infinity being Π. If the radius R of the sphere varies in such a way that R = a + b cos nt where b < a, show that pressure at the surface of the sphere at any time is*

$$P + \frac{1}{4}bn^2\rho(b - 4a\cos nt - 5b\cos 2nt).$$

Solution: Let v¢ be the velocity at a distance r¢ at any time t and p¢ be the pressure there. Let v be the velocity on the surface of the sphere of radius R where R = a + b cos nt.

The equation of continuity becomes

$$r'\ v' = f\ (t) = R^2v.$$

The equation of motion is given by ...(1)

$$\frac{\partial v'}{\partial t} + v'\frac{\partial v'}{\partial r'} = \frac{1}{\rho}\frac{\partial p'}{\partial r'}$$

$$\frac{f'(t)}{r'^2} + v'\frac{\partial v'}{\partial r'} = -\frac{1}{\rho}\frac{\partial p'}{\partial r'}$$

Integrating with regard to r′, we have

$$-\frac{f'(t)}{r} + \frac{1}{2}v'^2 = \frac{p'}{\rho} + A$$

Initially r′ = ∞, v = 0, p′ = P then $A = \frac{P}{\rho}$

$$\frac{-f'(t)}{r'} + \frac{1}{2}v'^2 = \frac{P - p'}{\rho}$$

Again r′ = R, p′ = p and v′ = v

then $$-\frac{f'(t)}{R} + \frac{1}{2}v^2 = \frac{1}{\rho}(P - p)$$

$$p = P + \rho\left[\frac{f'(t)}{R} - \frac{1}{2}v^2\right] \qquad ...(2)$$

$$\text{R. H. S.} \frac{f'(t)}{R} + \frac{1}{2}v^2 = 2\left(\frac{dR}{dt}\right)^2 + R\frac{d^2R}{dt^2} - \frac{1}{2}\left(\frac{dR}{dt}\right)^2$$

$$= \frac{3}{2}\left(\frac{dR}{dt}\right)^2 + R\frac{d^2R}{dt^2}$$

$$= \frac{3}{2}(-\,bn \sin nt)^2 + (a + b\cos nt)(-\,bn^2 \cos nt)$$

$$= \frac{1}{2}\,bn^2\,(3b\sin^2 nt - 2b\cos^2 nt - 2a\cos nt)$$

$$= \frac{1}{4}\,bn^2\,[3b\,(1 - \cos 2nt) - 2b\,(1 + \cos 2nt) - 4a\cos nt]$$

$$= \frac{1}{4}bn^2\,[b - 4a\cos nt - 5\,b\cos 2nt]$$

Hence $p = P + \frac{1}{4}\,bn^2\rho\,[b - 4a\cos nt - 5b\cos 2nt]$. **Proved.**

Example 13: *A mass of fluid of density r and volume 4/3pc³ is in the form of a spherical shell. A constant pressure P is exerted on the external surface of the shell. There is no pressure on the internal surface and no other forces act on liquid. Initially the liquid is at rest and the internal radius of the shell is 2c. Prove the velocity of the internal surface when its radius is c is .*

$$\sqrt{\left(\frac{14\Pi}{3\rho}.\frac{2^{1/3}}{2^{1/3}-1}\right)}.$$

Solution: Let v′ be the velocity at a distance r′ at any time t and p be the pressure there then the equation of continuity is

$$r'^2v' = f(t) \qquad ...(1)$$

The equation of motion is

$$\frac{\partial v'}{\partial t} + v'\frac{\partial v'}{\partial r'} = -\frac{1}{\rho}\frac{\partial p}{\partial r'} \qquad ...(2)$$

or
$$\frac{f'(t)}{r'^2} + v'\frac{\partial v'}{\partial r'} = -\frac{1}{\rho}\frac{\partial p}{\partial r'}$$

or
$$-\frac{f'(t)}{r'} + \frac{1}{2}v'^2 = -\frac{p}{\rho} + A \qquad ...(3)$$

Let r and R be the internal and external radii of the shell and v and V be the velocities there at any subsequent time t.

initially (i) $r' = R$, $v' = V$, $p = \Pi$ and (ii) $r' = r$, $v' = v$, $p = 0$. (Since there is no pressure on the internal surface)

The value of the constant A may be determined by using the above conditions with the relation (3). Thus, we get

$$\frac{f'(t)}{R} + \frac{1}{2}V^2 = -\frac{\Pi}{\rho} + A,$$

and
$$\frac{f'(t)}{r} + \frac{1}{2}v^2 = A.$$

Eliminating the constant, we have

$$-f'(t)\left\{\frac{1}{r} - \frac{1}{R}\right\} + \frac{1}{2}\{v^2 - V^2\} = \frac{\Pi}{\rho} \qquad ...(4)$$

We know that

$$r^2 v = R^2 V = f(t) \qquad ...(5)$$

$$\Rightarrow \quad 2r^2 dr = 2R^2 dR = 2f(t)\, dt \qquad ...(6)$$

From (4) and (5), we get

$$-f'(t)\left\{\frac{1}{r} - \frac{1}{R}\right\} + \frac{1}{2}\left\{\frac{f^2(t)}{r^4} - \frac{f^2(t)}{R^4}\right\} = \frac{\Pi}{\rho}.$$

Multiplying both sides by the relation (6) and integrating, we have

$$-2f(t)\, f'(t)\left\{\frac{1}{r} - \frac{1}{R}\right\} + \frac{1}{2}f^2(t)\left\{\frac{2r^2 dr}{r^4} - \frac{2R^2 dR}{R^4}\right\} = \frac{\Pi}{\rho}.\, 2r^2\, dr$$

$$\Rightarrow -2f(t)\, f'(t)\left\{\frac{1}{r} - \frac{1}{R}\right\} + \frac{1}{2}f^2(t)\left\{\frac{2dr}{r^2} - \frac{2dR}{R^2}\right\} = \frac{2\Pi}{\rho}r^2 dr$$

$$\Rightarrow f^2(t)\left\{\frac{1}{r} - \frac{1}{R}\right\} = -\frac{2\Pi}{3\rho}r^3 + B.$$

Initially $r = 2c$, $v = 0$ *i.e.*, $f(t) = 0 \Rightarrow B = \frac{2\Pi}{3\rho}8c^3$.

$$\Rightarrow f^2(t)\left\{\frac{1}{r} - \frac{1}{R}\right\} = \frac{2\Pi}{3\rho}(8c^3 - r^3)$$

$$\Rightarrow r^4 v^2\left\{\frac{1}{r} - \frac{1}{R}\right\} = \frac{2\Pi}{3\rho}(8c^3 - r^3).$$

Since the volume of the liquid is constant,

$$\frac{4}{3}\pi R^3 - \frac{4}{3}\pi r^3 = \frac{4}{3}\pi c^3,\ R^3 - r^3 = c^3.$$

$$\Rightarrow r^4v^2\left\{\frac{1}{r}-\frac{1}{(r^3+c^3)^{1/3}}\right\}=\frac{2\Pi}{3\rho}(8c^3-r^3)$$

$$\Rightarrow v^2=\frac{2\Pi}{3\rho}\frac{8c^3-r^3}{r^3\left\{\frac{1}{r}-\frac{1}{(r^3+c^3)^{1/3}}\right\}} \qquad ...(7)$$

which gives the velocity in terms of radius r at the inner surface of the cavity.

Hence the velocity of the internal surface (when r = c) is given by

$$v^2=\frac{2\Pi}{3\rho}\frac{7c^3}{c^4\left\{\frac{1}{c}-\frac{1}{c.2^{1/3}}\right\}} \Rightarrow v=\sqrt{\left(\frac{14\Pi}{3\rho}\cdot\frac{2^{1/3}}{2^{1/3}-1}\right)}$$ **Proved.**

Example 14(a): *A spherical mass of liquid of radius b has a concentric spherical cavity of radius a; which contains gas at pressure p whose mass may be neglected; at every point of the external boundary of the liquid an impulsive pressure* $\underline{\omega}$ *per unit area is applied. Assuming that the gas obeys Boyle's law, shew that when the liquid first comes to rest, the radius of the internal surface will be*

$$a \exp.\left\{-\frac{\underline{\omega}^2 b}{2p\rho a^2(b-a)}\right\},$$

where ρ is the density of the liquid.

Solution: Let v′ be the velocity at a distance r′ from the centre of the spherical cavity at any time t. The equation of continuity is

$$r'^2v' = f(t) = b^2V. \qquad ...(1)$$

Let $\underline{\omega}'$ the impulsive pressure at a distance r′ then

$$d\,\underline{\omega}' = -\rho v'\,dr' = -r\,(b^2V/r'^2)\cdot dr'.$$

Integrating with regard to r′, we have

$$\underline{\omega}' = (\rho b^2V/r') + C, \qquad ...(2)$$

where C is an arbitrary constant.

Again $r' = a$, $\underline{\omega}' = 0$ and $r' = b$, $\underline{\omega}' = \underline{\omega}$. ...(3)

Using (2) and (3), we can determine the arbitrary constant as

$$0 = (\rho b^2V/a) + C \text{ and } \underline{\omega} = (\rho b^2V/b) + C,$$

$$\text{or } \underline{\omega} = \rho b^2V\left(\frac{1}{b}-\frac{1}{a}\right) = \frac{\rho bV}{a}(a-b). \quad ...(4)$$

The initial kinetic energy is

$$= \frac{1}{2}\int_a^b (4\pi r'^2 \rho dr').v'^2 = 2\pi\rho \int_a^b r'^2.\frac{b^4V^2}{r'^4}dr'$$

$$= 2\pi\rho b^4V^2 \int_a^b \frac{dr'}{r'^2} = \frac{2\pi\rho b^2V^2}{a}(b-a). \quad ...(5)$$

Let r be the radius of the internal spherical cavity and p_1 be the pressure of the gas there. Since the gas obeys Boyle's law, we have

$$\frac{4}{3}\pi r^3.p_1 = \frac{4}{3}\pi \dot{a}^3.p$$

$$\Rightarrow \quad p_1 = \frac{a^3p}{r^3}$$

$$\text{Total work done} = \int_a^r 4\pi r^2.p_1dr = 4\pi a^3p \log (r/a)$$

Again change in K.E. = total work done

$$\frac{2\pi b^3V^2}{a}(b-a) = \pi a^3\rho \log (r/a)$$

$$\text{or } \log (r/a) = \frac{2\pi\rho b^3V^2(b-a)}{4\pi a^4p}$$

$$= \frac{2\pi\rho b^3(b-a)}{4\pi p a^4}.\frac{a^2\underline{\omega}^2}{\rho^2b^2(a-b)^2}$$

$$= -\frac{\underline{\omega}^2b}{2p\rho a^2(b-a)}$$

$$\text{or } r = a \exp.\left\{\frac{\underline{\omega}^2b}{2p\rho a^2(b-a)}\right\}.$$ **Proved.**

Example 14(b): *Show that the rate per unit of time at which work is done by the internal pressures between the parts of a compressible fluid obeying Boyle's law is*

$$\iiint p\left(\frac{\partial u}{\partial x}+\frac{\partial v}{\partial y}+\frac{\partial w}{\partial z}\right)dx\,dy\,dz,$$

where p is the pressure and (u,v, w), the velocity at any point and the where p is the pressure through the volume of the fluid.

Solution: Let W is the work done, p is the pressure and dv an elementary volume. Work done in compressing th fluid is

$$W = \int p(-dv)$$

Rate per unit time of work done is

$$\frac{DW}{Dt} = \iiint \frac{Dp}{Dt} dv. \qquad ...(1)$$

Equation of continuity is

$$\frac{D\rho}{Dt} + \rho\left(\frac{\partial u}{\partial x} + \frac{\partial v}{\partial y} + \frac{\partial w}{\partial z}\right) = 0$$

$$\frac{Dp}{Dt} + p\left(\frac{\partial u}{\partial x} + \frac{\partial v}{\partial y} + \frac{\partial w}{\partial z}\right) = 0,\ p = kr \qquad ...(2)$$

From (1) and (2), we have

$$\frac{DW}{Dt} = \iiint p\left(\frac{\partial u}{\partial x} + \frac{\partial v}{\partial y} + \frac{\partial w}{\partial z}\right) dv$$

Hence the rate per unit time of work done is given by

$$\iiint p\left(\frac{\partial u}{\partial x} + \frac{\partial v}{\partial y} + \frac{\partial w}{\partial z}\right) dx\, dy\, dz.$$ **Proved.**

Example 14(c): *A mass of gravitating fluid is at rest under its own attraction only, the free surface being a sphere of radius b and the inner surface a rigid concentric shell of radius a. Show that if the shell suddenly disappears, the initial pressure at any point of the fluid at distance from the centre is*

$$\frac{2}{3}\pi\gamma\rho^2(b-a)(r-a)\left\{\frac{a+b}{r}+1\right\}.$$

Solution: Let v′ be the velocity at a distance r′ from the centre of the spherical shell at any time t and p be the pressure there. The equation of continuity is

$$\frac{1}{r'^2}\frac{d}{dr}(r'^2 v') = 0$$

$\Rightarrow r'^2 v' = f(t)$.

Let r be the radius of the inner surface. The attraction at the point distance r′ from the centre due to liquid is given by

$$\frac{\frac{4}{3}\pi\gamma\rho(r'^3 - r^3)}{r'^2} = \frac{4}{3}\pi\gamma\rho\left(r' - \frac{r^3}{r'^2}\right)$$

The equation of motion is given by

$$\frac{\partial v'}{\partial t} + v'\frac{\partial v'}{\partial r'} = -\frac{4}{3}\pi\gamma\rho\left(r' - \frac{r^3}{r'^2}\right) - \frac{1}{\rho}\frac{\partial p}{\partial r'}$$

$$\Rightarrow \frac{f'(t)}{r'^2} + v'\frac{\partial v'}{\partial r'} = -\frac{4}{3}\pi\gamma\rho\left(r' - \frac{r^3}{r'^2}\right) - \frac{1}{\rho}\frac{\partial p}{\partial r'}$$

Integrating with regard to r′, we have

$$-\frac{f'(t)}{r'} + \frac{1}{2}v'^2 = \frac{4}{3}\pi\gamma\rho\left(\frac{1}{2}r'^2 + \frac{r^3}{r'}\right) - \frac{p}{\rho} + A \qquad ...(1)$$

Initially t = 0, v′ = 0, r = a

and p = P (let)

$$-\frac{f'(0)}{a} = -\frac{4}{3}\pi\gamma\rho\left(\frac{1}{2}r'^2 + \frac{a^3}{r'}\right) - \frac{P}{\rho} + A \qquad ...(2)$$

Again P = 0, r′ = a and r′ = a

and r′ = b, v′ = 0

From (2), we have

$$-\frac{f'(0)}{a} = -\frac{4}{3}\pi\gamma\rho\left(\frac{1}{2}a^2 + a^2\right) + A$$

$$-\frac{f'(0)}{b} = -\frac{4}{3}\pi\gamma\rho\left(\frac{1}{2}b^2 + \frac{a^3}{b}\right) + A. \qquad ...(3, 4)$$

Subtracting (3) and (4), we have

$$f'(0)\left[\frac{1}{b} - \frac{1}{a}\right] = \frac{4}{3}\pi\gamma\rho\left\{\frac{1}{2}(b^2 - a^2) + a^2\left(\frac{a}{b} - 1\right)\right\}$$

$$\Rightarrow f'(0) = -\frac{2}{3}\pi\gamma\rho ab(a + b) + \frac{4}{3}\pi\gamma\rho a^3 \qquad ...(5)$$

Multiplying (3) by a and (4) by b and subtracting, we have

$$0 = \frac{4}{3}\pi\gamma\rho.\frac{1}{2}(b^3 - a^3) + A(a-b)$$

$$\Rightarrow A = \frac{2}{3}\pi\gamma\rho(a^2 + b^2 + ab) \qquad ...(6)$$

Substituting the values of f ′ (0) and the constant A in (2), we have

$$-\frac{1}{r'}\left\{-\frac{2}{3}\pi\gamma\rho ab(a+b) + \frac{4}{3}\pi\gamma\rho a^3\right\} = -\frac{4}{3}\pi\gamma\rho\left(\frac{1}{2}r'^2 + \frac{a^3}{r'}\right) - \frac{P}{\rho}$$

$$+ \frac{2}{3}\pi\gamma\rho(a^2 + b^2 + ab)$$

$$\Rightarrow P = \frac{2}{3}\pi\gamma\rho\left\{a^2 + b^2 + ab - \frac{ab(a+b)}{r'} - r'^2\right\}$$

Replacing r′ by r the initial pressure at any point of the fluid at a distance r from the centre is obtained

$$P = \frac{2}{3}\pi\gamma\rho^2(r-a)(b-r)\left\{\frac{a+b}{r} + 1\right\}$$ **Proved.**

Example 15: *Liquid is contained between two parallel planes; the free surfaceis a circular cylinder of radius a whose axis is perpendicular to the planes. All the liquid within a concentric circular cylinder of radius b is suddenly annihilated. Prove that if Π be the pressure at the outer surface, the initial pressure at any point of the liquid distance r from the centre is*

$$\Pi\frac{\log r - \log b}{\log a - \log b}.$$

Solution: In an incompressible fluid, the fluid velocity will be radial outside the cylinder | z | = b. It will be a function of r and the time t only. Let v′ be the velocity at a distance r′ and p be the pressure there at any time t, then equation of continuity is

$$r'\,v' = f(t). \qquad ...(1)$$

Equation of motion is

$$\frac{\partial v'}{\partial t} + v'\frac{\partial v'}{\partial r} = -\frac{1}{\rho}\frac{\partial p}{\partial r'},$$

$$\text{or} \quad \frac{f'(t)}{r'} + v'\frac{\partial v'}{\partial r} = -\frac{1}{\rho}\frac{\partial p}{\partial r'},$$

Integrating with regard to r′, we have

$$f'(t)\log r' + \frac{1}{2}v'^2 = -\frac{p}{\rho} + A. \qquad \text{...(2)}$$

Initially t = 0, v′ = 0,

r′ = a, we have

$$f'(0)\log r = \frac{p}{\rho} + A. \qquad \text{...(3)}$$

Again p = Π, when r = a,

and p = 0, when r = b.

By using the above condition, equation (3) becomes

$$f'(0)\log a = \frac{\Pi}{\rho} + A \qquad \text{...(4)}$$

and $$f'(0)\log b = -A. \qquad \text{...(5)}$$

From (3) and (5), we have

$$f'(0)\{\log r - \log b\}. = \frac{p}{\rho}.$$

Also from (4) and (5), we get

$$f'(0)\{\log a - \log b\} = -\frac{\Pi}{\rho}.$$

Eliminating f ′ (0), we have

$$\frac{\log r - \log b}{\log a - \log b} = \frac{p}{\Pi}.$$

Thus the initial pressure at any point of the liquid at a distance r from the centre is

$$p = \Pi\,\frac{\log r - \log b}{\log a - \log b}. \qquad \textbf{Proved.}$$

Example 16: *A mass of liquid of density ρ whose external surface is a ling circular cylinder of radius a, which is subject to a constant pressure cylinder is suddenly destroyed. Show that if V is the velocity at the internal surface when the radius is r, then*

$$V^2 = \frac{2\Pi(b^2 - r^2)}{\rho r^2 \log\{(r^2 + a^2 - b^2)/r^2\}}.$$

Solution: When the internal cylinder is destroyed, suddenly the motion of the liquid will be along the radii of the normal sections of the cylinder. The velocity and p the pressure at any point distance r′ from the axis of the cylinder. The equation of continuity is

$$r' v' = f(t). \quad \text{...(1)}$$

Equation of motion is

$$\frac{\partial v'}{\partial t} + v' \frac{\partial v'}{\partial r'} = \frac{1}{\rho} \frac{\partial p}{\partial r'},$$

or $$\frac{f'(t)}{r'} + v' \frac{\partial v'}{\partial r'} = -\frac{1}{\rho} \frac{\partial p}{\partial r'}.$$

Integrating with regard to r′, we have

$$f'(t) \log r' + \frac{1}{2} v'^2 = \frac{p}{\rho} + A. \quad \text{...(2)}$$

Let r and R be the radii of the internal and external surface of the cylinder and v and v_1, the velocities there at any time t.

The conditions are defined as

$$r' = r,\ v' = v,\ p = 0$$

and $$r' = R,\ v' = V,\ p = \Pi.$$

$$f'(t) \log r + \frac{1}{2} v^2 = A$$

$$f'(t) \log R + \frac{1}{2} V^2 = -\frac{\Pi}{\rho} + A,$$

Eliminating the constant A, we have

$$f'(t) (\log r - \log R) + \frac{1}{2}(v^2 - V^2) = \frac{\Pi}{\rho} \quad \text{...(3)}$$

From the equation of continuity, we have

$$rv = RV = f(t)$$
$$2r\, dr = 2\, R\, dR = 2f(t)\, dt \quad \text{...(4)}$$

and $R^2 - r^2 = a^2 - b^2$.

Multiplying both the side of (3) by the relation (4), we have

$$2f(t) f'(t) (\log r - \log R) + \frac{1}{2} f^2(t) \left\{ \frac{2r\, dr}{r^2} - \frac{2R dR}{R^2} \right\} = \frac{2\Pi}{\rho} r dr$$

$$\Rightarrow 2f(t) f'(t) (\log r - \log R) + f^2(t) \left\{ \frac{dr}{r} - \frac{dR}{R} \right\} = \frac{2\Pi}{\rho} r\, dr.$$

By integrating, we have

$$f^2(t)(\log r - \log R) = \frac{\Pi}{\rho} r^2 B. \qquad ...(5)$$

At $r = b$, $v = 0$; $B = -\frac{\Pi}{\rho} b^2$.

Substituting the value of B into (5), we have

$$f^2(t)(\log r - \log R) = \frac{\Pi}{\rho}(r^2 - b^2)$$

$$\Rightarrow f^2(t) = \frac{\Pi}{\rho} \frac{r^2 - b^2}{\log r - \log R},$$

$$\Rightarrow v^2 r^2 = \frac{\Pi}{\rho} \frac{2(r^2 - b^2)}{\log(r^2/R^2)},$$

$$\Rightarrow v^2 = \frac{2\Pi(r^2 - b^2)}{\rho r^2 \log\{r^2/(r^2 + a^2 - b^2)\}},$$

$$\Rightarrow v^2 = \frac{2\Pi(p^2 - r^2 - b^2)}{\rho r^2 \log\{(r^2 + a^2 - b^2)/r^2\}}. \qquad \textbf{Proved.}$$

Example 17: *A portion of homogeneous fluid is confined between two concentric spheres of radii A and a, and is attracted towards their centre by a force varying inversely as square of the distance. The inner spherical surface is suddenly annihilated, and when the radii of the inner and outer surface of the fluid are r and R, the fluid impinges on a solid ball concentric with their surfaces. Prove that the impulsive pressure at any point of the ball for different values of R and r varies as*

$$\sqrt{\left\{(a^2 - r^2 - A^2 + R^2)\left[\frac{1}{r} - \frac{1}{R}\right]\right\}}.$$

Solution: The equation of continuity is

$$\frac{1}{r'} \frac{d}{dr'}(r'^2 v') = 0$$

$$\Rightarrow \quad r'^2 v' = f(t). \qquad ...(1)$$

and the equation of motion is

$$\frac{\partial v'}{\partial t} + v' \frac{\partial v'}{\partial r'} = -\frac{\mu}{r'^2} - \frac{1}{\rho} \frac{\partial p}{\partial r'},$$

or

$$\frac{f'(t)}{r'^2} + v' \frac{\partial v'}{\partial r'} = -\frac{\mu}{r'^2} - \frac{1}{\rho} \frac{\partial p}{\partial r'}. \qquad ...(2)$$

Integrating (2) with regard to r′, we have

$$-\frac{f'(t)}{r'}+\frac{1}{2}v'^2=\frac{\mu}{r'}-\frac{p}{\rho}+A \qquad ...(3)$$

Let r and R be the internal and outer radii and v and V be the velocities there at any time t. The conditions are defined as

(i) $r' = r, v' = v, p = 0$

(ii) $r' = R, v' = V, p = 0$.

The integration constant can be determined by applying the conditions. Therefore

$$-\frac{f'(t)}{r}+\frac{1}{2}v^2 = A + \frac{\mu}{r}.$$

and $-\frac{f'(t)}{R}+\frac{1}{2}V^2 = A + \frac{\mu}{r}.$

By eliminating the constant, we get

$$-f¢\ (t)\left(\frac{1}{r}-\frac{1}{R}\right)+\frac{1}{2}(v^2-V^2)=m\left(\frac{1}{r}-\frac{1}{R}\right). \qquad ...(4)$$

From the equation of continuity, we have

$r^2v = R^2V = f(t)$

$2r^2dr = 2R^2\ dR = 2f\ (t)\ dt \qquad ...(5)$

or $-f'\ (t)\left(\frac{1}{r}-\frac{1}{R}\right)+\frac{1}{2}f^2(t)\left(\frac{1}{r^4}-\frac{1}{R^4}\right)$

Multiplying by 2f (t) dt both sides and integrating, we have

$$-2f\ (t)f'(t)\ dt\left(\frac{1}{r}-\frac{1}{R}\right)+\frac{1}{2}f^2(t)\left(\frac{2f(t)dt}{r^4}-\frac{2f(t)dt}{R^4}\right)$$

$$=\mu\left(\frac{2f(t)dt}{r}-\frac{2f(t)dt}{R}\right)$$

or $-2f\ (t)\ f'(t)\ dt\left(\frac{1}{r}-\frac{1}{R}\right)+f^2(t)\left(\frac{dr}{r^2}-\frac{dR}{R^2}\right)=\mu\ \{2r\ dr-2RdR\}$

or $-f^2\ (t)\left(\frac{1}{r}-\frac{1}{R}\right)=m\ (r^2-R^2)+B. \qquad ...(6)$

initially $r = a, R = A, v = 0; B = -\mu\ (a^2-A^2)$.

Equation (6) becomes

$$f^2(t)\left(\frac{1}{r}-\frac{1}{R}\right) = m\ (a^2 - r^2 - A^2 + R^2). \qquad ...(7)$$

Let $\bar{\omega}$ be the impulsive pressure at a distance r′ then

$$\overline{d\omega} = -\rho v'\,dr' = -\rho\,[f(t)/r'2]\,dr'.$$

By integrating, we have

$$\bar{\omega} = \frac{\rho f(t)}{r'} + C, \text{where C is an arbitrary constant.}$$

But $\bar{\omega} = 0$, $r' = R$, then $C = \dfrac{\rho f(t)}{R}$.

or $\bar{\omega} = r\,f(t)\left\{\dfrac{1}{r}, -\dfrac{1}{R}\right\}$,

which determines the impulsive pressure at any distance r′.

The impulsive pressure at any point of the ball where r′ = r is,

$$\bar{\omega} = \rho f(t)\left\{\frac{1}{r}, -\frac{1}{R}\right\}.$$

Substituting the value of f (t) from (7), we have

$$\bar{\omega} = \rho\sqrt{\left\{\frac{\mu(a^2 - r^2 - A^2 + R^2)}{\left(\frac{1}{r}-\frac{1}{R}\right)}\right\}}\left(\frac{1}{r}-\frac{1}{R}\right)$$

or $\bar{\omega} = \rho\,\sqrt{}\,\mu\sqrt{\left\{(a^2 - r^2 - A^2 - R^2)\left(\dfrac{1}{r}-\dfrac{1}{R}\right)\right\}}$

Hence the impulsive pressure varies as

$$\sqrt{\left\{(a^2 - r^2 - A^2 - R^2)\left(\frac{1}{r}-\frac{1}{R}\right)\right\}}.$$

Proved.

Example 18(a): *A sphere of radius a is surrounded by an infinite liquid of density ρ, the pressure at infinity being Π. The sphere suddenly annihilated. Show that the pressure at distance r from the centre immediately falls to P* $\left(I - \dfrac{a}{r}\right)$.

Show further that if the liquid is brought to rest by impinging on a concentric sphere of radius, $\frac{1}{2}a$, *the impulsive pressure sustained by the surface of the sphere is* $\sqrt{\left(\frac{7}{6}\Pi\rho a^2\right)}$.

Solution: Let v′ be the velocity at a distance r′ from the centre of the sphere at any time t and p be the pressure. The equation of continuity is

$$r'^2 v' = f(t). \qquad ...(1)$$

Equation of motion is

$$\frac{\partial v'}{\partial t} + v'\frac{\partial v'}{\partial r'} = -\frac{1}{\rho}\frac{\partial p}{\partial r'}$$

or $$\frac{f'(t)}{r'^2} + v'\frac{\partial v}{\partial r'} = -\frac{1}{\rho}\frac{\partial p}{\partial r'}.$$

Integrating with regard to r′, we have

$$\frac{f'(t)}{r'} + \frac{1}{2}v'^2 = -\frac{p}{\rho} + A,$$

where A is an arbitrary constant.

Since $r' = \infty$, $p = P$, $v' = 0$ so $A = \frac{\Pi}{\rho}$. ...(2)

When the sphere is suddenly annihilated *i.e.* $r' = a$, $v' = 0$, $p = 0$, then

$$-\frac{f'(t)}{a} = \frac{\Pi}{\rho} \Rightarrow f'(t) = \frac{\Pi a}{\rho}. \qquad ...(3)$$

The velocity v¢ vanishes just after annihilation, so from (2) and (3), we have

$$-\frac{\Pi a}{\rho}\frac{1}{r'} + 0 = \frac{\Pi - p}{\rho}\frac{1}{r'} \Rightarrow \frac{a\Pi}{r'} = \Pi - p.$$

Thus pressure at the time of annihilation ($r' = r$) is

$$\frac{a\Pi}{r} = -\Pi - p \Rightarrow p = \Pi\left(1 - \frac{a}{r}\right).$$ **Proved.**

Let $\bar{\omega}$ be the impulsive pressure at a distance r′, then

$\overline{d\omega} = -\rho v' \, dr'$.

From the equation of continuity, we have

$r^2 v = r'^2 v' = f(t)$

or $d\,\overline{\omega} = -\rho v \,(r^2/r'^2)\, dr'$,

where r be the radius of the inner surface and v bethe velocity there

By integrating, we have

$\overline{\omega} = B + \rho v\,(r^2/r')$

When $r' = \infty$, $\overline{\omega} = 0$ then $B = 0$. ...(4)

which determines the impulsive pressure $\overline{\omega}$ at a distance r'.

Since the liquid is brought to rest by impinging on a concentric sphere of radius a/2, then substituting $r = a/2$ in (4), we have

$$\overline{\omega} = \rho v\,(a^2/4r'). \qquad ...(5)$$

The velocity v at the inner surface of the sphere ($p = 0$) is obtained from (2)

$$-\frac{f'(t)}{r} + \frac{1}{2}v^2 = \frac{\Pi}{\rho}.$$

Using the relation (1), we have

$$\left(rv\frac{dv}{dr} + 2v^2\right) - \frac{1}{2}v^2 = \frac{\Pi}{\rho},$$

$$\text{or } rv\frac{dv}{dr}\frac{3}{2}v^2 = -\frac{\Pi}{\rho}.$$

Multiplying by $2r^2\,dr$ both the sides and integrating, we have

$$2r^3 v\,dv + 3v^2 r^2\,dr = -\frac{2\Pi}{\rho}r^2 dr$$

$r^3 v^2 = -(2\Pi/3\rho)\,r^3 + C$

Since $r = a$, $v = 0$ then $C = (2\Pi/2\rho)\,a^3$.

$$\text{or } r^3 v^2 = \frac{2\Pi}{3\rho}(a^3 - r^3).$$

The velocity v at the surface of the sphere of radius a/2 on which the liquid strikes is

$$v^2 = \frac{2\Pi}{3\rho}\frac{a^3 - r^3}{r^3} = \frac{2\Pi}{3\rho}.\frac{a^3 - a^3/8}{a^3/8} = \frac{14}{3}\frac{\Pi}{\rho}.$$

From the relation (4), we get

$$\bar{\omega} = \frac{1}{4}\rho\sqrt{\left(\frac{14}{3}\frac{\Pi}{\rho}\right)a^2\frac{1}{r'}}, \qquad ...(6)$$

determines the impulsive pressure at a distance r′.

Thus the impulsive pressure at the surface of the sphere of radius a/2 is given by

$$\bar{\omega} = \frac{1}{4}\rho\sqrt{\left(\frac{14}{3}\frac{\Pi}{\rho}\right)a^2\frac{2}{a}} = \sqrt{\left(\frac{7\Pi\rho a^2}{6}\right)}.$$ **Proved.**

Example 18(b): *If a jet issuing from an orifice under a constant head D travels a horizontal distance x from the plane of the vena contracta while it drops through a height h, obtain the coefficient of velocity.*

Solution: If a particle falls a vertical distance h under the influence of gravity, then

$$h = ut + \frac{1}{2}gt^2, \qquad ...(1)$$

where u is the initial vertical velocity and t is time taken by the particle to reach a distance h. Here initial velocity is zero, so

$$t = \sqrt{(2h/g)}.$$

Horizontal distance x = V × t, where V is the horizontal velocity at vena contracta.

$$V = \frac{x}{t} = \frac{x/\sqrt{(2h/g)}}{\sqrt{2gD}} = \frac{x}{\sqrt{4hD}}.$$ **Ans.**

Example 19: *A horizontal straight pipe gradually reduces in diameter from 24 in. to 12 in. Determine the total longitudinal thrust exerted on the pipe if the pressure at the larger end is 50 Ibf/in² and the velocity of the water is 8ft/sec.*

Solution: Let A_1 and A_2 be the cross-section of the larger and the smaller end. Let q_1 and q_2 be the velocity and p_1 and p_2 be the pressure at the larger and the smaller end of the pipe. From the equation of continuity, we have

$A_1q_1 = A_2q_2$

or $p\,(12)^2q_1 = p\,(6)^2q_2 \Rightarrow 4q_1 = q_2$. ...(1)

By bernoulli's equation, we have

$$\frac{p_1}{\rho}+\frac{1}{2}q_1^2 = \frac{p_2}{\rho}+\frac{1}{2}q_2^2$$

or $p_1 - p_2 = \frac{1}{2}\rho(q_2^2 - q_1^2) = \frac{1}{2}\rho \times 15 \times (96)^2$. ...(2)

Total longitudinal thrust exerted on the pipe

$= p_1A_1 - p_2A_2$

$= p\,(12)^2\,p_1 - p\,(6)^2p_2$ ...(3)

From (2), we have

$$p_2 = p_1 - \frac{1}{2}\rho \times 15 \times (96)^2 \quad ...(4)$$

From (3) and (4), we have

$$\text{Total thrust} = 36\pi\left(p_1 + \frac{1}{2}\rho \times 15 \times 96 \times 96\right)$$

$$= 36\pi\left(150 + \frac{1}{2} \times \frac{62.4 \times 15 \times 96 \times 96}{12 \times 12 \times 12}\right)$$

$= 36 \times 240\,\pi$ **Ans.**

Example 20: *A pitot static tube having a coefficient of 0.98 is used to measure the velocity of water in pipe. The stagnation pressure head is 6 metres and static pressure heads is 5 metres. Determine its velocity*

Solution: Velocity by pitot static tube is given as

$$q = c\sqrt{\left(2g\frac{p_1 - p_0}{w}\right)}$$

Here $c = 0.98$, $p_1/w = 6$m., $p_0/w = 5$m.

$\therefore q = 0.98\sqrt{\{2 \times 9.81(6-5)\}} = 4.35$ m/sec. **Ans.**

Example 21: *If the water jet is discharged from a nozzle, inclined at angle π/3 with the horizontal, at 20 ft/sec. Calculate the horizontal distance required for the jet striking the ground which is 3 ft below the horizontal line of the nozzle. What is the velocity of the jet just before reaching the ground?*

Solution: The horizontal and vertical components of the velocity at the nozzle exit are, respectively

$q_{1x} = q_1 \cos\alpha = 20 \cos \pi/3 = 10$ ft/sec.

$q_{1y} = q_2 \sin\alpha = 20 \cos \pi/3 = 17.32$ ft/sec.

We know that

$$y = \frac{q_{1y}}{q_{1x}} x - \frac{1}{2} g \frac{x^2}{q_1^2}$$

$$\Rightarrow -3 = \frac{17.32}{10} x - \frac{1}{2} g \frac{x^2}{(10)^2}.$$

or $x^2 - 10.74x - 18.61 = 0.$

The horizontal distance between the nozzle and jet striking the ground is

$$x = \frac{1}{2}[10.74 + \sqrt{\{10.74)^2 - 4 \times 18.62\}} = 12.27\text{ft}].$$

The jet velocity at $x = 12.\ 27$

and $y = -3$ is given by

$$q_2 = \sqrt{\{(Q/A_1)^2 - 2gy\}} = \sqrt{(20)^2 + 6g\}} = 24.24 \text{ ft./sec.} \qquad \textbf{Ans.}$$

Example 22(a): *Fluid is coming out from a small hole of cross-section s_1 in a tank. If the minimum cross-section of the stream coming out of the hole is s_2, then show that $s_2/s_1 = \frac{1}{2}$.*

Solution: Let PQ be the hole and P′ Q′ be its image on the opposite wall of the tank. Let p_1 be the pressure at PQ when the hole is closed. Let p_2 be the pressure and q_2 be the velocity at the minimum cross-section. The velocity of the fluid coming out from minimum cross-section is at right angles to the hole and the direction of velocity will be horizontal there. Equation of motion is

$$\sigma_1 (p_1 - p_2) = \sigma_2 \rho q_2^2$$

$$\Rightarrow (p_1 - p_2) = (\sigma_2/\sigma_1) \rho q_2^2 \qquad ...(1)$$

Bernoulli's equation for the stream line connecting a point of P′Q′ and a point of minimum cross-section of the jet, becomes

$$\frac{p_1}{\rho} = \frac{p_2}{\rho} + \frac{1}{2} q_2^2 \Rightarrow p_1 - p_2 = \frac{1}{2} \rho q_2^2 \qquad ...(2)$$

From (1) and (2), We have

$$\frac{\sigma_2}{\sigma_1} = \frac{1}{2}.$$ **Proved**

Example 22(b): *Liquid is discharged at the rate of 3.86 ft3/sec from a siphon in the reservoir. The siphon has a diameter of 6 in. Find the elevation z and the fluid pressure at the top of the siphon.*

Solution: Bernoulli's equation for the three points on the same streamline can be written as

$$\frac{q_0^2}{2g} + \frac{p_0}{\rho g} + z_0 = \frac{q_1^2}{2g} + \frac{p_1}{\rho g} + z_1 = \frac{q_2^2}{2g} + \frac{p_2}{\rho g} + z_2$$

Here $q_0 = 0$, $p_1 = p_2$, $z = z_0 - z_2$ (let)

$$q_2 = \frac{3.86}{\pi\left(\frac{1}{4}\right)^2} = \frac{3.86 \times 16 \times 7}{22} = 19.6 \text{ ft./sec.}$$

and $q_2^2 = 2gz$

$$\Rightarrow \quad z = \frac{19.62 \times 19.62}{2 \times 32} = 6 \text{ ft. (app).}$$

Since the velocity at the top is the same as that at the bottom. Bernoulli's equation written between these two level gives

$p_1/\rho g = -8$ ft. of liqid

i.e., below the atmospheric pressure. **Ans.**

Example 22(c): *Water flows through a pipe of length l which tapers from theentrance radius r_l to the exit radius r_2. If the entrance velocity is v_l, and the relation between r_l and r_2 is given by $r_2 = r_l \pm m_l$, where m is the slope, prove that the exit velocity v_2 is*

$$v_2 = v_l\left[1 - \frac{\pm 2m(l/r_1) + m^2(l/r_1)^2}{1 \pm 2m(l/r_1) + m^2(l/r_1)^2}\right].$$

Solution: Since r_1 and r_2 be the radius of the pipe of length l_1 at the entrance and the exit. Let A_1 and A_2 be the cross-section of the pipe at the entrance and the exit. From continuity equation, we have

$$A_1v_1 = A_2v_2 \Rightarrow (\pi r_1^2)\, v_2$$

$$\text{or } v_2 = \frac{r_1^2}{r_2^2} v_1 = \frac{r_1^2}{(r_1 \pm ml)^2} v_1$$

$$\text{or } v_2 = v_1 \left[\frac{1}{\{1 \pm (ml/r_1)\}^2}\right]$$

$$\text{or } v_2 = v_1 \left[\frac{1}{1 \pm 2m(l/r_1) + m_2(l^2/r_1^2)}\right]$$

$$\text{or } v_2 = v_1 \left[1 - \frac{\pm 2m(l/r_1) + m^2(l^2/r_1^2)}{1 + 2m(l/r_1) + m^2(l^2/r_1^2)}\right].$$ **Proved.**

Example 22(d): *A conical pipe has diameters of 10 cm. and 15 cm. at the two ends. If the velocity at the smaller end is 2m/sec, what is the velocity at the other end and the discharge through the pipe?*

Solution: Let q_1 and q_2 be the velocity at the smaller and larger ends. From continuity equation, we have

$q_1A_1 = q_2A_2$.

Here q_2 = 2 m/sec, $A_1 = (\pi/4)(0.1)^2$, $A_2 = (p/4)(0.15)^2$,

$$q_2 = q_1 \frac{A_1}{A_2} = 2 \frac{(0.1)^2}{(0.15)^2} = 0.89 \text{ m/sec.}$$ **Ans.**

Discharge through the pipe

$Q = q_1A_1 = 2(p/4)(0.1)^2 = 0.0157 \text{ m}^2\text{/sec.}$ **Ans.**

Example 23: *A horizontal conical pipe has diameters 25 cm. and 40 cm. at the two ends. (a) Calculate the pressure at the larger end if the pressure at the smaller end is 5m. of water and rate of flow is 0.3 m²/sec. (b) Calculate the discharge through the pipe if the manometer connected between the two ends reads 10 cm. of mercury.*

Solution: Let q_1, q_2 be the velocities and p_1, p_2 be the pressure at the larger and smaller ends of a conical pipe. Let Q be the discharge through the pipe then

$Q = A_1q_1$

$$q_1 = \frac{Q}{A_1} = \frac{0.3}{(\pi/4)(0.4)^2} \quad ...(1)$$

From the continuity equation, we have

$A1q_1 = A_2q_2$

$$\text{or } q_2 = \frac{A_1}{A_2} q_1 = z \frac{(0.4)^2}{(0.25)^2} \times 2.38 = 6.10 \text{m/sec.} \quad ...(2)$$

(a) Using Bernoulli's equation, we have

$$\frac{p_1}{\rho}+\frac{q_1^2}{2g}=\frac{p_2}{\rho}+\frac{q_2^2}{2g}. \quad \text{(Here } z_1 = z_2\text{)}$$

$$\text{or } 5+\frac{(2.38)^2}{2\times 9.81}=\frac{p_2}{\rho}+\frac{(6.10)^2}{2\times 9.81}$$

$$\text{or } \frac{p_2}{\rho}=5+\frac{(2.38)^2-(6.10)^2}{2\times 9.81}=3.4 \text{ m}=0.34 \text{ kg/cm}^2$$

Pressure at the larger end = 0.34 kg/cm².

(b) From manometer, we have

$$\frac{p_1}{\rho}-\frac{p_1}{\rho}=10\,(13.6-1)=126 \text{ cm.}=1.26 \text{ m.}$$

From manometer, we have

$A_1q_1 = A_2q_2$.

$$\Rightarrow q2=\frac{A_1}{A_2}q_1=\frac{(0.4)^2}{(0.25)^2}q_1=2.56q_1.$$

Using Bernoulli's equation, we have

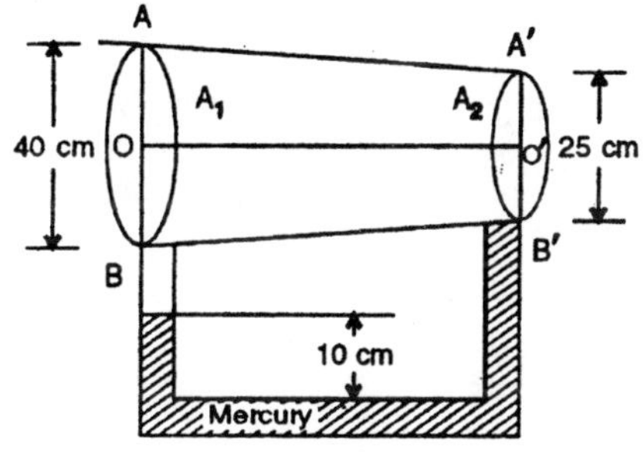

Fig. 2.21

$$\frac{p_1}{\rho.}+\frac{q_1^2}{2g}=\frac{p_2}{\rho}+\frac{q_2^2}{2g}$$

$$\text{or } \frac{q_1^2}{2g}\left(\frac{q_2^2}{q_1^2}-1\right)=\frac{p_1-p_2}{\rho}\Rightarrow\frac{q_1^2}{2g}[(2.56)^2-1]=1.26$$

$$\text{or } q_1=\sqrt{(1.26\times 2\times 9.81/5.55)}=2.11 \text{ m/sec.}$$

Hence discharge through the pipe is

$Q = A_1 q_1$

$Q = \frac{\pi}{4} \times (0.4)^2 \times 2.11 = 2.56 \text{ m}^3/\text{sec}.$ **Ans.**

Example 24: *A pipe of 10 cm. diameter is suddenly enlarged to 20 cm. diameter. Find the loss of head when 50 litres/sec. of water is flowing.*

Solution: Let q_1 and q_2 be the velocities at the smaller and larger ends of the pipe then

or $Q = A_1 q_1 = A_2 q_2$

or $q_1 = Q/A_1,\ q_2 = Q/A_2$

or $q_1 = \frac{0.05}{(\pi/4)(0.1)^2},\ q_2 = \frac{0.05}{(\pi/4)(0.2)^2}$

$q_1 = 6.36$ m/sec, $q_2 = 1.59$ m/sec.

Loss of head due to sudden enlargement

$$= \frac{(q_1 - q_2)^2}{2g} = \frac{(6.36 - 1.59)^2}{2 \times 9.81} = 1.6 \text{ m.}$$ **Ans.**

Example 25(a): *A jet of water 8 cm. in diameter impinges on a plate held normal to its axis. For a velocity of 4 m/sec, what force will keep the plate in equilibrium?*

Solution: Let q be the velocity and A be the area of the jet. Let F_x be the force on the plate then

Force on the plate

= change in momentum

$$F_x = \frac{WQV}{g} = \frac{WAV^2}{g}$$

$$= \frac{100 \times \frac{\pi}{4}(0.08)^2 \times 4^2}{9.81} = 32.8\text{kg.}$$

Example 25(b): *Show that*

$$\frac{x^2}{a^2} f(t) + \frac{y^2}{b^2}\phi(t) + \frac{z^2}{c^2}\psi(t) = 1$$

is a possible form of the boundary surface it

$f(t)\, \dot{\phi}(t)\, \psi(t) = 1.$

Solution: We know that a surface is a boundary surface if

$$\frac{\partial F}{\partial t} + u\frac{\partial F}{\partial x} + v\frac{\partial F}{\partial y} + w\frac{\partial F}{\partial z} = 0 \quad ...(1)$$

Here $F(x, y, z, t) \equiv \frac{x^2}{a^2} f(t) + \frac{y^2}{b^2}\phi(t) + \frac{z^2}{c^2}\psi(t) - 1 = 0$

$$\frac{\partial F}{\partial t} = \frac{x^2}{a^2} f'(t) + \frac{y^2}{b^2}\phi'(t) + \frac{z^2}{c^2}\psi'(t),$$

$$\frac{\partial F}{\partial x} = \frac{2x}{a^2} f(t),$$

$$\frac{\partial F}{\partial y} = \frac{2y}{b^2}\phi(t), \frac{\partial F}{\partial z} = \frac{2z}{c^2}\psi(t). \quad ...(2)$$

From the relations (1) and (2), we have

$$\frac{x^2}{a^2} f'(t) + \frac{y^2}{b^2}\phi'(t) + \frac{z^2}{c^2}\psi'(t) + \frac{2ux}{a^2} f(t) + \frac{2vy}{b^2}\phi(t) + \frac{2wz}{c^2}\psi(t) = 0$$

or $$\left[\frac{2ux}{a^2} f(t) + \frac{x^2}{a^2} f'(t)\right] + \left[\frac{2vy}{b^2}\phi(t) + \frac{y^2}{b^2}\phi'(t)\right] + \left[\frac{2wz}{c^2}\psi(t) + \frac{z^2}{c^2}\psi'(t)\right] = 0$$

This will be a boundary surface if

$$\frac{2ux}{a^2} f(t) + \frac{x^2}{a^2} f'(t) = 0$$

$$\Rightarrow u = -\frac{1}{2} x \frac{f'(t)}{f(t)},$$

$$\frac{2vy}{b^2}\phi(t) + \frac{y^2}{b^2}\phi'(t) = 0$$

$$\Rightarrow v = -\frac{1}{2} y \frac{\phi'(t)}{\phi(t)}$$

$$\frac{2wz}{c^2}\psi(t) + \frac{z^2}{c^2}\psi'(t) = 0$$

$$\Rightarrow w = -\frac{1}{2}z\frac{\psi'(t)}{\psi(t)},$$

where the velocity components u, v and w satisfy the continuity equation

$$\frac{\partial u}{\partial x} + \frac{\partial v}{\partial y} + \frac{\partial w}{\partial z} = 0$$

$$\Rightarrow -\frac{1}{2}\frac{f'(t)}{f(t)} - \frac{1}{2}\frac{\phi'(t)}{\phi(t)} - \frac{1}{2}\frac{\psi'(t)}{\psi(t)} = 0$$

$$\Rightarrow \phi(t)\,\psi(t)\,f'(t)\,\psi(t)\,\phi'(t) + f(t)\,\phi(t)\,\psi'(t) = 0$$

By integrating with regard to t, we have

$f(t)\,\phi(t)\,\psi(t)$ = constant = 1 (Let). **Proved.**

Example 26: *Show that*

$$\frac{x^2}{a^2}\tan^2 t + \frac{y^2}{b^2}\cot^2 t = 1,$$

is a possible form for the bounding surface of a liquid, and find an expression for the normal velocity.

Solution : The surface $F(x, y, z, t) = 0$ can be a possible boundary surface, if it satisfies the boundary condition

$$\frac{dF}{dt} = \frac{\partial F}{\partial t} + u\frac{\partial F}{\partial x} + v\frac{\partial F}{\partial y} + w\frac{\partial F}{\partial z} = 0,$$

where u, v, w satisfy the equation of continuity $\nabla . \mathbf{q} = 0$

Here $F(x, y, t) \equiv (x^2/a^2)\tan^2 t + (y^2/b^2)\cot^2 t - 1 = 0$

then $\dfrac{\partial F}{\partial t} = \dfrac{2x^2}{a^2}\tan t\sec^2 t - \dfrac{2y^2}{b^2}\cot t\,\mathrm{cosec}^2 t$,

and $\dfrac{\partial F}{\partial x} = \dfrac{2x}{a^2}\tan^2 t$ and $\dfrac{\partial F}{\partial y} = \dfrac{2y}{b^2}\cot^2 t$.

From the relation (1), we have

$$\frac{x\tan t}{a^2}(x\sec^2 t + u\tan t) + \frac{y\cot t}{b^2}$$

$(-y\,\mathrm{cosec}^2 t + v\cot t) = 0.$ ...(3)

Now (1) will be the boundary condition, if

$x\sec^2 t + u\tan t = 0,$

and $-y \operatorname{cosec}^2 t\ v \cot t = 0.$

or $u = -x \sec^2 t \cot t = -(x/\sin t \cos t)$,

and $v = y \operatorname{cosec}^2 \tan t = (y/\sin t \cos t)$,

Which satisfies the equation of continuity $\nabla . \mathbf{q} = 0$

Therefore the given surface is a possible form for the boundary surface of a liquid with velocity components

$u = -(x/\sin t \cos t)$ and $v = (y/\sin t \cos t)$.

Also, Normal velocity $= \dfrac{u(\partial F/\partial x)}{\sqrt{\{(\partial F/\partial x)^2 + (\partial F/\partial y)^2\}}}$

$$-\frac{\dfrac{x}{\sin t \cos}.\dfrac{2x}{a^2}\tan^2 t + \dfrac{y}{\sin t \cos t}.\dfrac{2y}{b^2}\cos^2 t}{\sqrt{\{(2x/a^2)\tan^2 t\}^2 + \{(2y/b^2)\cot^2 t)^2\}}}$$

$$= \frac{a^2y^2 \cot t \operatorname{cosec}^2 t + b^2x^2 \tan t \sec^2 t}{\sqrt{(x^2b^4 \tan^4 t + y^2a^4 \cot^4 t)}} \textbf{.Ans.}$$

Example 27: *Show the variable ellipsoid*

$$\frac{x^2}{a^2e^{-t}\cos(t+\pi/4)} + \frac{y^2}{b^2e^{t}\sin(t+\pi/4)} + \frac{z^2}{c^2\sec 2t} = 1,$$

is a possible form of boundary surface of a liquid for any, time t and determine the velocity ***q*** *of any particle on this boundary. Also, prove that the equation of continuity is satisfied.*

Solution : Since any boundary surface with equation F (x, y, z,t) = 0,is made up from a time-invariant set of liquid particles, we must have DF/Dt = 0 for all points on the boundary at time t. Therefore

$$\frac{DF}{Dt} = \frac{\partial F}{\partial t} + u\frac{\partial F}{\partial x} + v\frac{\partial F}{\partial y} + w\frac{\partial F}{\partial z} = 0 \ \forall\ t, (x\ y\ z) \in F. \qquad ...(1)$$

$$F(x, y, z, t) \equiv \frac{x^2}{a^2}e^{t}\sec\left(t+\frac{\pi}{4}\right) + \frac{y^2}{b^2}e^{-t}\operatorname{cosec}\left(t+\frac{\pi}{4}\right)$$

$$+ \frac{z^2}{c^2}\cos 2t - 1 = 0 \qquad ...(2)$$

From (1) and (2), we have

$$\frac{x^2}{a^2}e^t\left\{\sec\left(t+\frac{\pi}{4}\right)+\sec\left(t+\frac{\pi}{4}\right)\tan\left(t+\frac{\pi}{4}\right)\right\}$$

$$+\quad \frac{y^2}{b^2}e^{-t}\left\{-\operatorname{cosec}\left(t+\frac{\pi}{4}\right)-\operatorname{cosec}\left(t+\frac{\pi}{4}\right)\cot\left(t+\frac{\pi}{4}\right)\right\}$$

$$-\frac{2z^2}{c^2}\sin 2t+\frac{2ux}{a^2}e^t\sec\left(t+\frac{\pi}{4}\right)+\frac{2vy}{b^2}e^{-t}\operatorname{cosec}\left(t+\frac{\pi}{4}\right)$$

$$+\frac{2zw}{c^2}\cos 2t = 0,$$

or $$\frac{xe^t}{a^2}\sec\left(t+\frac{\pi}{4}\right)\left\{2u+x\left(1+\tan\left(t+\frac{\pi}{4}\right)\right)\right\}$$

$$+\quad \frac{ye^{-t}}{b^2}\operatorname{cosec}\left(t+\frac{\pi}{4}\right)\left\{2v-y\left(1+\cot\left(t+\frac{\pi}{4}\right)\right)\right\}$$

$$+\quad \frac{2z}{c^2}(w\cos 2t - z\sin 2t) = 0.$$

which will hold, if

$$u = -\ \frac{1}{2}x\left\{1+\tan\left(t+\frac{\pi}{4}\right)\right\},$$

$$v = -\ \frac{1}{2}y\left\{1+\cot\left(t+\frac{\pi}{4}\right)\right\},\ w = z\tan 2t.$$

For these components on the boundary for all t, we have

$$\frac{\partial u}{\partial x}+\frac{\partial v}{\partial y}+\frac{\partial w}{\partial z} = -\ \frac{1}{2}\left\{1+\tan\left(t+\frac{\pi}{4}\right)\right\}$$

$$-\ \frac{1}{2}\left\{1+\cot\left(t+\frac{\pi}{4}\right)\right\}\ +\tan 2t$$

$$= \frac{1-\tan^2\left(t+\frac{1}{4}\pi\right)}{2\tan\left(t+\frac{1}{4}\pi\right)}+\tan 2t$$

$$= \cot\left(2t+\frac{\pi}{2}\right)+\tan 2t = 0.$$

It follows that the equation of continuity is satisfied on the boundary. The total volume within this ellipsoid is V, where

$$V = \pi^2 a^2 e^{-t} \cos\left(t+\frac{\pi}{4}\right) b^2 e^t \sin\left(t+\frac{\pi}{4}\right) c^2 \sec 2t$$

$$= \frac{1}{2}\pi^2 a^2 b^2 c^2 = \text{constant.}$$

Thus the equation of continuity is satisfied. **Proved.**

Example 28: *Show that the ellipsoid*

$$\frac{x^2}{a^2k^2t^{2n}} + k^{tn}\left\{\left(\frac{y^2}{b^2}\right)+\left(\frac{z^2}{c^2}\right)\right\} = 1,$$

is a possible form of theboundary surface of a liquid.

Solution : The surface F (x, y, z, t) = 0 can be a possible boundary surface, if it satisfies the boundary condition

$$\frac{dF}{dt} = \frac{\partial F}{\partial t} + u\frac{\partial F}{\partial x} + v\frac{\partial F}{\partial y} + w\frac{\partial F}{\partial z} = 0. \qquad ...(1)$$

where u, v, w satisfy the equation of continuity $\nabla . \mathbf{q} = 0$

Here $F(x, y, z,t) \,^\circ\, \frac{x^2}{a^2k^2t^{2n}} + k^{tn}\left\{\left(\frac{y^2}{b^2}\right)+\left(\frac{z^2}{c^2}\right)\right\} - 1 = 0$

$$\frac{\partial F}{\partial t} = -\frac{x^2}{a^2k^2}\cdot\frac{2n}{t^{2n+1}} + nkt^{n-1}\left\{\left(\frac{y^2}{b^2}\right)+\left(\frac{z^2}{c^2}\right)\right\}$$

$$\frac{\partial F}{\partial x} = -\frac{2x}{a^2k^2t^{2n}},\ \frac{\partial F}{\partial y} = \frac{2kt^n y}{b^2} \text{ and } \frac{\partial F}{\partial z} = \frac{2kt^n z}{c^2}.$$

Now from the relation (1), we have

$$\frac{x^2}{a^2k^2}\frac{2n}{t^{2n+1}} + nkt^{n-1}\left\{\frac{y^2}{b^2}+\frac{z^2}{c^2}\right\}$$

$$+\frac{2xu}{a^2k^2t^{2n}} + \frac{2kt^n vy}{b^2} + \frac{2kt^n zw}{c^2} = 0$$

or $\left(u-\frac{nx}{t}\right)\frac{2x}{a^2k^2t^{2n}} + \left(v+\frac{ny}{2t}\right)\frac{2kyt^n}{b^2} + \left(w+\frac{nz}{2t}\right)\frac{2kzt^n}{c^2} = 0,$

which will hold, if

$$u - \frac{nx}{t} = 0,\ v + \frac{ny}{2t} = 0 \text{ and } w + \frac{nz}{2t} = 0$$

or $u = nx/t, v = -ny/2t$ and $w = -nz/2t$, ...(2)

which shows that the velocity components satisfy the continuity equation.

Thus the equation (1) is a possible form for the boundary surface of a liquid with velocity components

$u = nx/t,\ v = -ny/2t$ and $w = nz/2t$. **Proved.**

Example 29: *Show that all necessary conditions can be satisfied by a velocity potential of the form*

$$\phi = \alpha x^2 + \beta y^2 + \gamma z^2$$

and a bounding surface of the form

$$F \equiv ax^4 + by^4 + cz^4 - \chi(t) = 0$$

where $\chi(t)$ is a given function of the time, and $\alpha, \beta, \gamma, a, b, c$ are the suitable functions of the time.

Solution: The necessary conditions are :

(i) ϕ satisfies the Laplace's equation *i.e.*, $\nabla^2\phi = 0$ for incompressible fluid flow

(ii) F satisfies the condition for bounding surface

$$\frac{\partial F}{\partial t} + u\frac{\partial F}{\partial x} + v\frac{\partial F}{\partial y} + w\frac{\partial F}{\partial z} = 0.$$

The velocity potential f is given as

$$\phi = \alpha x^2 + \beta y^2 + \gamma z^2 \qquad ...(1)$$

The Laplace's equation $\nabla^2\phi = 0$ will be satisfied if $\alpha + \beta + \gamma = 0$ where α, β, γ are some suitable functions of the time.

Again $F \equiv ax^4 + by^4 + cz^4 - \chi(t)$, ...(2)

can be a possible form for the bounding surface of a liquid, if

$$\frac{\partial F}{\partial t} + u\frac{\partial F}{\partial x} + v\frac{\partial F}{\partial y} + w\frac{\partial F}{\partial z} = 0$$

or $x^4\dfrac{\partial a}{\partial t} + y^4\dfrac{\partial b}{\partial t} + z^4\dfrac{\partial c}{\partial t} - \chi'(t) + 4ax^3u + 4by^4v + 4cz^3w = 0$ ...(3)

But $u = -(\partial\phi/\partial x) = -2\alpha x,\ v = -(\partial\phi/\partial y) = -2\beta y$,

and $w = -(\partial\phi/\partial x) = -2\gamma z$.

Substituting the value of u,v, w in the (3), we have

$$x^4 \frac{\partial a}{\partial t} + y^4 \frac{\partial b}{\partial t} + z^4 \frac{\partial c}{\partial t} - \chi'(t) - 8a\alpha x^4 - 8b\beta y^4 - 8czy^4 = 0$$

$$\text{or } x^4 \left(\frac{\partial a}{\partial t} - 8a\alpha\right) + v^4 \left(\frac{\partial b}{\partial t} - 8a\beta\right) + z^4 \left(\frac{\partial c}{\partial t} - 8x\gamma\right) - \chi'(t) = 0$$

Comparing this with the equation of the bounding surface,

$F \equiv ax^4 + by^4 + cz^4 - \chi(t) = 0$, we have

$$\frac{\partial a/\partial t - 8a\alpha}{a} = \frac{\partial b/\partial t - 8a\beta}{b} = \frac{\partial c/\partial t - 8c\gamma}{c}$$

$$= \frac{\chi'(t)}{\chi(t)}.$$

The condition will hold if a, b, c, α, β, γ are some suitable functions of time.

Hence the velocity potential f and the boundary surface F = 0 satisfy the necessary condition for velocity potential and boundary surface if a, b, g,a,b, c are some suitable function of time. **Ans.**

EXERCISES

1. A mass of fluid moves in such a way that each particle describes a circle in one plane about a fixed axis, show that the equation of continuity is

   $\partial r/\partial t + \partial(\rho\omega)/\partial\theta = 0.$

   where ω is the angular velocity of a particle whose azimuthal angle is q at time t.

2. Each particle of a mass of liquid moves in a plane through the axis of Z; show that the equation of continuity is

   $$\frac{\partial \rho}{\partial t} + \frac{1}{r^2}(\rho u r^2) + \frac{1}{r\sin\theta}\frac{\partial}{\partial\theta}(\rho v \sin\theta) = 0$$

3. A spherical hollow of radius a initially exists in an infinite fluid subject to constant pressure at infinity. Shew that the pressure at distance r from the centre when the radius of the cavity is x to pressure at infinity as

   $$\{3x^2r^4 + (a^3 - 4x^3)\, r^3 - (a^3 - x^3\}; 3x^2r^4.$$

4. A sphere whose radius at time t is b + a cos nt is surrounded by liquid extending to infinity under no forces. Prove that the pressure at distance r from the centre is less than the pressure at an infinite distance by

$$\frac{\rho n^2 a}{r}(b+a)\cos nt\left\{a(1-3\sin^2 nt)+b\cos nt + \frac{a}{2r^3}\sin^2 nt(b+a\cos nt)^3\right\},$$

5. A sphere of radius a is alone in an unbounded liquid which is at rest at a great distance from the sphere and is subject to no external forces. The sphere is forced to vibrate radially keeping its spherical shape, the radius r at any time being by r = a + b cos nt. Show that if Π is the pressure in the liquid at a great distance from the sphere, the least pressure (assumed positive) at the surface of the sphere during the motion is

$$\Pi - n^2\rho b\ (a + b).$$

6. A homogeneous liquid is contained between two concentric spherical the radius of the inner being a and that of the outer indefinitely great. The fluid is attracted to the centre of these surfaces by a force ϕ (r), and constant pressure Π is exerted at the outer surface. Suppose

$$\int \phi(r)dr = \chi\ (r),$$

and that χ (r) vanishes when r is infinite. Show that if the inner surface is suddenly removed, the pressure at the distance r is suddenly diminished by

$$\Pi\left(\frac{a}{r}\right)-\left(\frac{\sigma\rho}{r}\right)\chi(a).$$

Find ϕ (r) so that the pressure immediately after the inner surface is removed may be the same as it would be if no attractive force existed. Also, with this value of f (r), find the velocity of the inner boundary of the fluid at any period of the motion.

7. A mass of uniform liquid is in the form of a thick spherical shell bounded by concentric spheres of radii a and b (a < b). The cavity is filled with gas the pressure of which varies according to Boyle's law, and is initially equal to the atmospheric pressure

Π, and the mass of which may be neglected. The outer surface of the shell is exposed to atmospheric pressure. Prove that if the system is symmetrically disturbed, so that each particle moves along particle moves along the line joining it to the centre, the time of an oscillation is

$$2pa\left\{\rho\frac{b-a}{3\Pi b}\right\}^{1/2},$$

where ρ is the density of the liquid.

8. Homogeneous liquid moves so that the path of any particle P lies in the plane POX where OX is fixed axes. Prove that if OP = r, and the angle ∠ XOP = θ, the equation of continuity may be written as

$$\frac{\partial}{\partial r}(ur^2) - \frac{\partial}{\partial \mu}(vr\sin\theta) = 0,$$

where u, v, are the component velocities along and perpendicular to OP in the plane POX and μ = cos θ.

9. (a) Streamlines are drawn for depicting the flow around a bridge pier. The streamlines have a spacing of 4 cm. where the stream flowsat uniform velocity of 3m/s. Determine the velocity indicated by a spacing of 2.5 cm. between two adjacent streamlines near the nose of the pier?

Hint : $dq = q_1\, dn_1 = q_2\, dn_2$

$\Rightarrow 300 \times 4\ q_2 \times 2.5$

$\Rightarrow q_2 = 480 = 4.8$ m/s.

(b) Determine an expression for the mean velocity along axes of the pipe if a pipe converges from 7.5 cm. to 3 cm. diameter at a reach, the disharge is 1200 litres/min.

10. A mass of perfect in compressible fluid, of density ρ, is bounded by concentric surface. The outer surface i contained by a flexible envelope which exerts continuously a uniform pressure Π and contracts from radius R_1 to radius R_2. The hollow is filled with a gas obeying Boyle's law, its radius contracts from C_1 to C_2, and the pressure of the gas is initially p_1. Initially the whole mass is at rest. Prove that, neglecting the mass of the gas, the velocity

v of the inner surface when the configuration (R_2, C_2) is reached is given by

$$\frac{1}{2}v^2 = \frac{C_1^3}{C_2^1}\left\{\frac{1}{3}\left(1-\frac{C_2^3}{C_1^3}\right)\frac{\Pi}{\rho}-\frac{p_1}{p}\log\frac{C_1}{C_2}\right\}\Big/\left(1-\frac{C_2}{R_2}\right),$$

11. A mass of homogeneous liquid is moving so that the velocity at any point is proportional to the time, and that the pressure is given by

$$\frac{p}{\rho} = \mu xyz - \frac{1}{2}t^2(y^2z^2 + z^2x^2 + x^2y^2).$$

Prove that this motion may have been generated from rest by finite natural forces independent of the time, and show that if the direction of motion at every point coincide with the direction of the acting force, each particle of the liquid describe a curve which is the intersection of the two hyperbolic cylinders.

Hint : Equation of motion is

$$\frac{p}{\rho}-\frac{\partial\phi}{\partial t}+\frac{1}{2}q^2 + V = f(t). \qquad \text{...(i)}$$

Here $q = \lambda t$

and $p = \rho\left\{\mu xyz-\frac{1}{2}t^2(y^2z^2 + z^2x^2 + x^2y^2)\right\}$. ...(ii)

From (i) and (ii) $\lambda^2 = y^2z^2 + z^2x^2 + x^2y^2$. ...(iii)

So $q^2 = t^2(y^2z^2 + z^2x^2 + x^2y^2)$.

Let $\phi = txyz$.

From (i) we have

$$\frac{p}{\rho} = xyz - \frac{1}{2}t^2(y^2z^2 + z^2x^2 + x^2y^2) - V + f(t). \qquad \text{...(iv)}$$

From (i) and (iv), we have

$$f(t) = 0,$$

$$V = xyz(1-\mu).$$

Let, u, v, are the component velocities and X, Y, Z be the component forces, then

$$u = -\frac{\partial \phi}{\partial x} = -\ tyz \text{ etc.}$$

and $$X = -\frac{\partial V}{\partial x} = (\mu - 1)\ yz \text{ etc.}$$

Since the direction of motion coincides with the direction of acting forces, so

$$\frac{u}{X} = \frac{v}{Y} = \frac{w}{Z}.$$

Thus the equation to the path is given by

$$\frac{dx}{X} = \frac{dy}{Y} = \frac{dz}{Z}.$$

3

Motion in Two Dimensions

STRENGTH OF A SOURCE

Thus it $2\pi m$ is the volume of the fluid flowing out per unit time, then m is called the strength of the source. The strength of a source is defined in terms of the amount of liquid out from it.

Sink is a source of negative strength.

If gr is the radial velocity at a distance r from the source, then amount of liquid flowing of circle of radius r $= 2\ \pi m$

i.e., $$2\pi r q^r = 2\pi m$$

or $$q^r = \frac{m}{r}$$

At source there is creation of the fluid and at a sink the annihilation of the fluid.

COMPLEX POTENTIAL FOR A SOURCE

Let a source of strength m be situated at origin.

If q_r is the velocity at P (r, θ), then

$$2\pi r q_r = 2pm, \text{ i.e., } q_r = \frac{m}{r} \quad ...(1)$$

If u and v are the component velocity at P (r, θ), then

$$u = \frac{m}{r}\cos\theta \text{ and } v = \frac{m}{r}\sin\theta.$$

Again let w be the complex potential at P; then

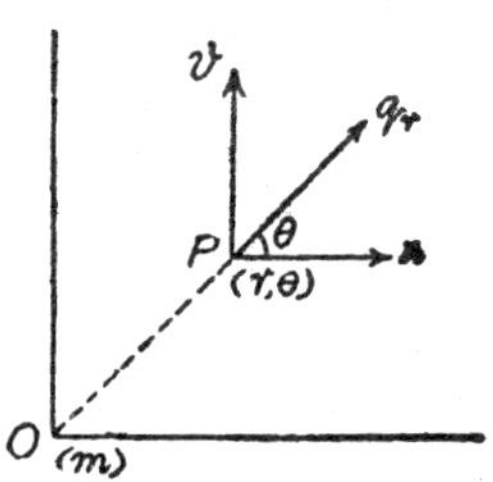

Fig. 3.1

$$w = \phi + i\psi,$$

so that $\dfrac{dw}{dz}.\dfrac{\partial z}{\partial x} = \dfrac{\partial \phi}{\partial x} + i\dfrac{\partial \psi}{\partial x}.$

i.e., $\dfrac{dw}{dz} = \dfrac{\partial \phi}{\partial x} - i\dfrac{\partial \phi}{\partial y} \text{ as } \dfrac{\partial z}{\partial x} = 1$

and $\dfrac{\partial \psi}{\partial x} = \dfrac{\partial \phi}{\partial y}$

$= -u + iv$ as $u = -\dfrac{\partial \phi}{\partial x}$ and $v = -\dfrac{\partial \phi}{\partial y}$

$$= -\frac{m}{r}(\cos\theta - i\sin\theta)$$

$$= \frac{m}{r}e^{-i\theta} = -\frac{m}{re^{i\theta}}$$

$= -\dfrac{m}{z}$ as $z = re^{i\theta}$.

Integrating, $w = -m \log z$.

This gives complex potential due to a source at origin. If the source is situated at z_1, then the complex potential is given by

Several Sources

If sources of strengths m_1, m_2, m_3... are situated at points $z = z_1$, z_2, z_3,..., then the complex potential is given by

$$w = -m_1 \log(z - z_1) - m_2 (z - z_2) - m_3 \log(z - z_3)...$$

since $w = \phi + i\psi$.

Therefore equating real and imaginary parts, we get

$$\phi = -m_1 \log r_1 - m_2 \log r_2 - m_3 \log r_3 - ...$$

and $\psi = -m_1\theta_1 - m_2\theta_2 - m_3\theta_3 - ...,$

where $r_i = |z - z_i|$ and $\theta_i = \arg(z - z_i)$.

Aliter. Complex Potential of a Source

Consider irrotational motion due to a source of strength m placed at the origin.If q_r and q_θ be radial and transverse velocities at P (r, θ), then

$$2\pi r q_r = 1\pi m \text{ and } q\theta = 0.$$

$$-\frac{\partial\phi}{\partial r} = \frac{m}{r} \text{ and } -\frac{\partial\phi}{r\partial\theta} = 0,$$

which yield $\phi = -m \log r$.

Also $\frac{\partial\phi}{\partial r} = \frac{\partial\psi}{\partial\theta}$ and $\frac{\partial\phi}{r\partial_\theta} = -\frac{\partial\psi}{\partial r}$, Cauchy Riemann equation

$$-m = \frac{\partial\psi}{\partial\theta} \text{ and } 0 = \frac{\partial\psi}{\partial r},$$

which yield $\psi = -m\theta$.

The complex potendial w at P (r, θ) is

$$w = \phi + i\psi$$

$$= -m(\log r + i\theta)$$

$$= -m \log (re^{i\theta})$$

$$= -m \log z.$$

If the source is located at the point $z = z_1$, then the complex potential is

$$w = -m \log (z - z_1).$$

TWO-DIMENSIONAL SOURCES AND SINKS

A point is called a two-dimensional simple Source if liquid flows out from this point radially and symmetrically in all directions in the xy-plane.

We know that the two-dimensional motion is essentially a three-dimensional motion with motion similar in all planes parallel to xy-plane. Therefore a simple two dimensional source is a line source parallel to z-axis from which liquid is flowing and from each point and flows in all directions normally to the line source. On the other hand a point is called a sink if there is an inward radial liquid flow towards that point from all directions.

DOUBLETS IN TWO DIMENSIONS

A doublet is a combination of a source of strength m and a sink of strength $-m$ at a small distance δs apart such that $\mu\, \delta s$ is finite.

Strength of a doublet : If $m\, \delta s = \mu$, a finite quantity, where m is taken infinitely large and δs infinitely small, then μ is called the strength of the doublet.

The line from – m to + m (i.e., from sink to source) is called the axis of the doublet.

Complex Potential for a Doublet : Let doublet be formed of a sink – m at z = a and source + m at z = a + δa, AB making an angle α with x-axis.

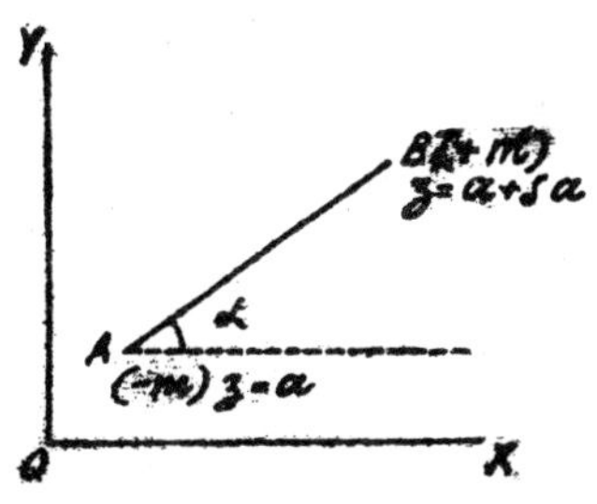

Fig. 3.2

Then the complex potential is given by

$$w = m \log (z - a) - m \log (z - a - \delta a)$$

$$= - m\, \delta a \frac{\log[z - \overline{a + \delta a}] - \log[z - a]}{\delta a}$$

$$= - m\, \delta a \frac{\partial}{\partial a} [\log (z - a)]$$

$$\text{as } \frac{\log[z - \overline{a + \delta a}] - \log[z - a]}{\delta a} = \frac{\partial}{\partial a} \log(z - a) = \frac{m \delta a}{z - a}$$

$$= \frac{m.AB.e^{i\alpha}}{z - a} \text{ as da = } ABe^{i\alpha}$$

$$= \frac{ae^{i\alpha}}{z - a} \text{ as m. AB = } \mu\text{, the strength of the doublet.}$$

SEVERAL DOUBLETS

If doublets of strengths a_1, a_2, a_3,...are situated at Z = a_1, a_2, a_3,... and their axes making angles α_1, α_2, α_3, ... with x-axis, then the complex potential due to the above system is given by

$$w = \frac{a_1 e^{i\alpha_1}}{Z - a_1} = \frac{a_2 e^{i\alpha_2}}{z - a_2} + \frac{a_3 e^{i\alpha_3}}{z - a_3} + \ldots$$

COMPLEX POTENTIAL OF DOUBLET

Let A and B be the position of the sink and source of equal strength m and P by any point in the space. Let AP = r, BP = r + δr and PB makes an angle θ with the axis of the doublet. The velocity potential is given by

$$\phi = m \log r - m \log (r + \delta r)$$

$$\Rightarrow \phi = -m \log \left(\frac{r+\delta r}{r}\right)$$

$$= -m \log \left(1+\frac{\delta r}{r}\right)$$

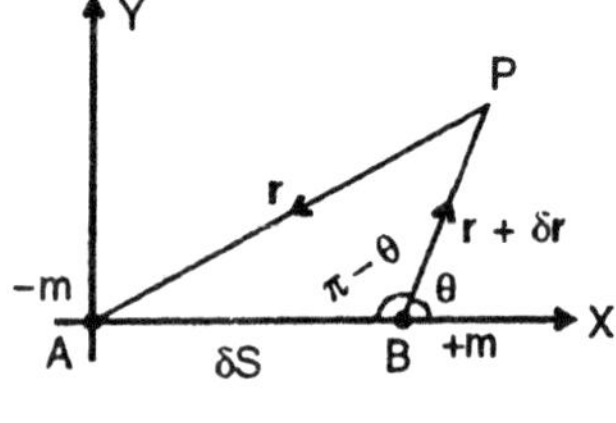

Fig. 3.3

$$\Rightarrow \phi = -\frac{m\delta r}{r}$$

(to a first approximation)

$$\Rightarrow \phi = \frac{m\delta s \cos\theta}{r} = \frac{\mu \cos\theta}{r},$$

where μ (= m δs) is the strength of the doublet.

Let $\frac{\mu\cos\theta}{r} = C$, where C is a constant.

$$\Rightarrow \mu r \cos\theta = Cr^2$$

$$\Rightarrow x^2 + y^2 = \frac{\mu}{C} x.$$

The curves ϕ = constant represent the circles touching the Y-axis at the origin.

We know that $\frac{1}{r}\frac{\partial \psi}{\partial \theta} = \frac{\partial \phi}{\partial r}$

$$\Rightarrow \frac{1}{r}\frac{\partial \psi}{\partial \theta} = \frac{\mu\cos\theta}{r^2}$$

$$\Rightarrow \frac{\partial \psi}{\partial \phi} = -\frac{\mu\cos\theta}{r} \Rightarrow \psi = -\frac{\mu \sin\theta}{r},$$

constant of integration vanishes as θ and ψ, both vanish.

Thus, we have

$$\phi + i\psi = \frac{\mu\cos\theta}{r} - i\frac{\mu\sin\theta}{r} = \frac{\mu}{r}(\cos\theta - i\sin\theta) = \frac{\mu e^{-\theta i}}{r}$$

$$\Rightarrow w = \phi + i\psi = \frac{\mu}{re^{\theta i}} = \frac{\mu}{z},$$

this determines the complex potential for a doublet of strength μ at the origin directed along the X-axis.

The complex potential of a two-dimensional doublet of strength μ at the origin making an angle α with X-axis is given by

$$w = \frac{\mu}{r}e^{-(\theta-a)i} = \frac{\mu e^{ai}}{re^{\theta i}} = \frac{\mu e^{ai}}{z} \qquad ...(3)$$

Similarly, the complex potential of two-dimensional doublets at z = a_1, c_2, a_3,... of strengths μ1, μ2, μ3,... αvδ direction a_1, a2, a3,... with the X-axis is determine as

$$w = \frac{\mu_1^{a_1^i}}{z-a_1} + \frac{\mu_2 e^{a_2^j}}{z-a_2} + \frac{\mu_e e^{\alpha_3^i}}{z-a_3} + ... \qquad ...(4)$$

Example 1: *Prove that for the complex potential tan^{-1} z the stream lines and equi-potentials are circles. Find the velocity at any points and examine the singularities at z = ± i.*

Solution: The complex potential is given by

$$w = \phi + i\psi = \tan^{-1}z, \qquad ...(1)$$

$$\text{Also } \overline{w} = \phi + i\psi = \tan^{-1}\overline{z}, \qquad ...(2)$$

By subtracting (1) and (2), we have

$$2i\psi = \tan^{-1}z - \tan^{-1}\overline{z} = \tan{-1}\frac{z-\overline{z}}{1+z\overline{\overline{z}}}$$

$$\text{or } \tan 2i\,\psi = \frac{2iy}{1+x^2+y^2}$$

$\Rightarrow x^2 + y^2 + 1 = 2y \coth 2\psi.$

The stream lines ψ = constant represent the circles

$x^2 + y^2 + 1 = 2y \coth 2\psi.$

Similarly, by adding (1) and (2), we have

$$2\phi = \tan^{-1}z + \tan^{-1}\overline{z} = \tan^{-1}\frac{z-\overline{z}}{1+z\overline{\overline{z}}}$$

or $1 - x^2 - y^2 = 2x \cot 2\phi.$

The equi-potentials ϕ = const. also represent circles which are orthogonal to the streamlines ψ = const. and form a co-axial system with limit points at z = ± i. The velocity component (u, v) are given by

$$\frac{dw}{dz} = -m + iv = \frac{1}{z^2 + 1},$$

The denominator vanishes at $z = \pm i$, therefore, it represents the singularities at these points.

At $z = + i$, substitute $z = i + z_1$, where $|z_1|$ is very small

$$-u + iv = \frac{dw}{dz} = \frac{dw}{dz_1} = \frac{1}{1 + (-1 + 2iz_1)} = \frac{1}{2iz_1}.$$

by integrating, we have $w = -\frac{1}{2}i \log z_1$

$\Rightarrow$ that the singularity at $z = i$ is a vortex of strength $k = -1/2$ with circulation $-\pi k$.

Similarly, the singularity at $z = -i$ is a vortex of strength $k = 1/2$ with circulation πk.

IMAGES IN TWO DIMENSION

If a surface S can be drawn in a moving fluid in such a way that there is no transport of fluid across that surface then any system of sources, sinks and doublets on one side of the surface is said to be the *image system* of sources, sinks and doublets on the side with regard to the surface S. The fluid flows tangentially to the surface.

As there is no flow across the surface S, the surface S is necessarily a streamline. If we introduce a rigid boundary in place of the surface then the fluid motion will remain unaltered and the fluid velocity at any point, normal to the rigid boundary must vanish.

IMAGE OF A SOURCE WITH REGARD TO A PLANE

Consider two source of equal strength + m at A (a, 0) and B (– a, 0) at an equidistant from the plane OY. The complex potential is given by

$$w = -m \log (z - a) - m \log (z + a),$$

$$w = -m \log (r_1 e^{i\theta_1}) - m \log (r_2 e^{i\theta_2}),$$

$$\Rightarrow w = -m \log \{r_1 r_2 ei(\theta_1 + \theta_2)\}$$

$$\Rightarrow w = -m \log r_1 r_2 - im (\theta_1 + \theta_2).$$

$$\Rightarrow \phi + i\psi = -m \log r_1 r_2 - im (\theta_1 + \theta_2).$$

$$\Rightarrow \psi = -m (\theta_1 + \theta_2).$$

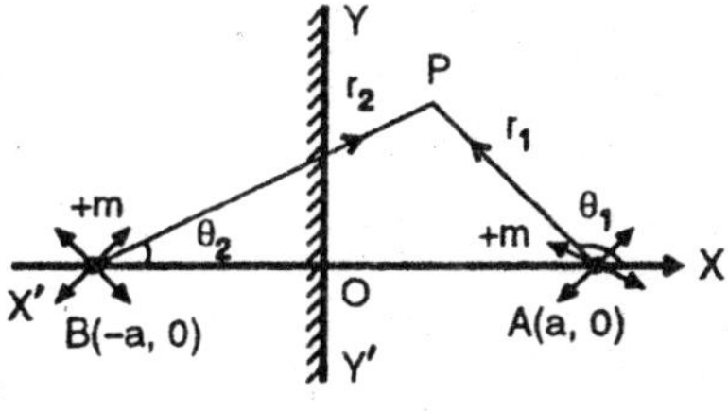

Fig. 3.4

The stream lines ψ = const, are given by $\theta_1 + \theta_2$ = const. Let the point P be on Y axis which is a streamline the $\theta_1 + \theta_2 = p$. Thus there is no fluid flow across the line YOY′.

Therefore, the source of strength + m at B (–a, 0) is the image of the source of strength + m of A (a, 0) with regard to the line YOY′.

IMAGE OF A DOUBLET WITH REGARD TO A PLANE

Let the axis of the doublet PQ makes an angle θ with the real axis. Then the image of a doublet at A (z = a) with its axis PQ inclined at an angle θ with X-axis consists an equal doublet at A′ (z = a) inclined with its axis P′ Q′ at an angle (π –θ) with X-axis. The complex potential is given by

$$w = \frac{\mu e^{\theta i}}{z-a} + \frac{\mu e^{(\pi-\theta)i}}{z+a},$$

$$w = \frac{\mu e^{\theta i}}{z-a} + \frac{\mu e^{\theta i}}{z+a}.$$

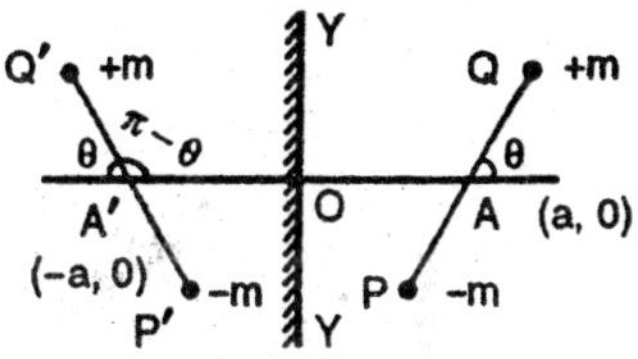

Fig. 3.5

THE CIRCLE THEOREM

Let f (z) be the complex potential of motion of a two-dimensional irrotational flow of an incompressible fluid with no rigid boundaries and f (z) has no singularities with in the circle | z | = a. If a circular cylinder, classified by its cross-section, the circle | z | = a, be introduced in to the field of flow, the complex potential becomes

$$w = f(z) + \bar{f}(a^2/z), \ |z| \geq a$$

where $\bar{f}$ is the complex conjugate form $\overline{zz} = a^2$, where $\bar{z}$ is the complex conjugate function of z. The complex potential $w = f(z) + \bar{f}(\bar{z})$ will be a real quantity on the circle, and hence y = 0. Therefore the circle is a stream line.

Again, if the point z lies outside the circle then the point $\bar{z}$ (= a^2/z) will lie inside the circle and vice-versa, since all the sigularities of f (z) and $\bar{f}\,\bar{z}$ lie in the domain $|z| > a$ and $|z| < a$, respectively. In particular $\bar{f}(\bar{z})$ has no singularity at infinity and f (z) has no singularity with in the circle. It follows that the hydrodynamic image of f (z) in $|z| = a$ is the complex potential $\bar{f}(\bar{z})$ such that $f(z) + \bar{f}(\bar{z})$ is real on $|z| = a$.

IMAGE OF A SOURCE WITH REGARD TO A CIRCLE

Let Q is an inverse point of P (z = f) with regard to the circle such that

$$OP.\ OQ = a^2 \Rightarrow OQ = a^2/f.$$

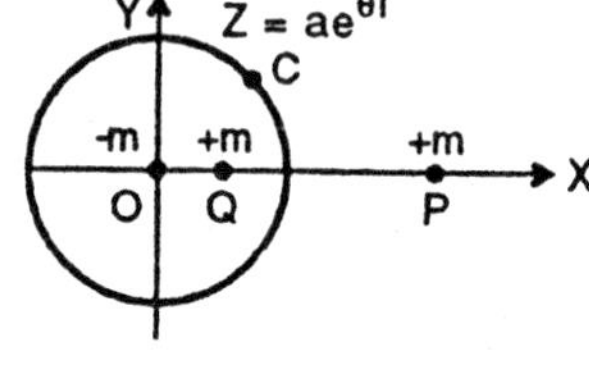

Fig. 3.6

By using the circle theorem, the complex potential is given by

$$w = -m \log (z - f) - m \log \left(\frac{a^2}{z} - f\right). \qquad ...(1)$$

By adding a constant terms {m log (–f)}, the relation (1) becomes

$$w = -m \log (z - f) - m \log \left(\frac{a^2}{z} - f\right) + m \log (-f)$$

$$\text{or } w = -m \log (z - f) - m \log \left\{(-f)\left(z - \frac{a^2}{z} - f\right)\right\} + m \log (-f) + m \log z$$

IMAGE OF A SOURCE WITH RESPECT TO A LINE

Let OY be the line with respect to which the image of the source m at A on X- axis is to be found. Consider an equal source at B such as that

OA = OB.

Let us now take any point P and OY.

Velocity at P due to source at

$A = \dfrac{m}{r}$ along AP and velocity at P due to source at $B = \dfrac{m}{r}$ along BP.

The velocity at P along the normal to OY.

$$= \frac{m}{r} \cos\theta\ \theta - \frac{m}{r} \cos\theta$$

i.e., there is no flow across OY.

Therefore the image of a simple source with respect to a line in two dimensions is an equal source equidisatrint from the line opposite to the source.

Aliter. Consider two equal sources of strength m at points

A (a, o) and B (– a, 0).

Complex potential w at p is given by

$$w = - m \log (z - a) - m (z + a)$$

$$= - m \log (z - a) (z + a)$$

$$= - m \log (r_1 e^{i\theta} a\ r_2 e^{i\theta}{}_2)$$

$$= - m \log r_1 r_2 - i\, m (\theta_1 + \theta_2).$$

Streamlines are given by

$$\theta_1 + \theta_2 = \text{Constant}.$$

In particularly y-axis is a streamline for if P is on Y-axis

$$\theta_1 + \theta_2 = \pi,$$

and thus there is no flow across OY.

∴ Source m at B is the image of source m at A w. r. t OY, or image of a source in a line is an equal source at the optical image point. The same is true for a sink.

IMAGE OF A SOURCE IN A CIRCLE

Let us determine the image of a source m place data with respect to a circle with O as centre.

Let B be the inverse point of A with respect to the circle, so that

$$OA.\ OB = (\text{radius})^2. \qquad ...(1)$$

place an equal source m at B and take a point P on the circle. Velocity at P due to source m at A = $\frac{m}{aP}$ along AP and velocity at P due to source m at B = $\frac{m}{aP}$ along BP.

Hence velocity at p along OP (normally to the circle)

$$= \frac{-m}{aP} \cos OPQ + \frac{m}{BP} \cos BPO \qquad ...(2)$$

$$= -\frac{m}{BP} \cos \theta + \frac{m}{BP} \cos \alpha,$$

where $\angle OP\theta = \theta$ and $\angle BPO = \alpha$.

Triangle OBP and OPA are similar {since angle AOP is common and $\frac{OA}{OP} = \frac{OP}{OB}$ from (1)}.

$\therefore \angle OAP = \alpha$.

Now $\cos \alpha = \frac{AL}{OA} = \frac{AP + PL}{OA}$

$$= \frac{AP + OP\cos\theta}{OA}$$

$$= \frac{AP}{OA} + \frac{OP}{OA}\cos\theta$$

$$= \frac{BP}{OP} + \frac{BP}{AP} \cos \theta \text{ as } \frac{OA}{OP} = \frac{OP}{OB} = \frac{AP}{BP}$$

Hence (2) gives

velocity at P along OP

$$= \frac{m}{AP}\cos\theta + \frac{m}{BP}\left(\frac{BP}{OP} + \frac{BP}{AP}\cos\theta\right)$$

$$= \frac{m}{OP}.$$

and if we place a sink – m at 0, the velocity due to it at P will be $\frac{-m}{OP}$ along OP and then the velocity of the system along OP will reduced to zero.

Thus the image of the source m at A consists of

(i) an equal source m at B, the inverse point of A with regard to the circle, and

(ii) an equal sink - m at the centre 0 of the circle.

CIRCLE THEOREM OF MILNE-THOMSON

Let f (z) be the complex potential of a two-dimensional irrotational motion of an incompressible inviscid liquid with no rigid boundaries.

Then if a circular cylinder r | z | = a is inserted in the flow field, the complex potential of the resulting motion is given by

$$w = f(z) + \overline{f(a^2 1\bar{z})}$$

provided f (z) has no singularities inside $|z| \le a$.

ALTERNATIVE METHOD OF

Let a 2- dimensional source of strength m be placed at the point z = b. Then the complex potential at any point z is

$$f(z) = -m \log(z - b)$$

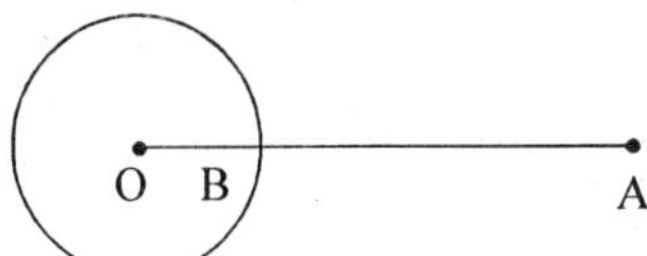

Fig. 3.7

Let a circular cylinder | z | –a (a < b) be inserted in the flow field. The complex potential w (z) in the region | z | ³ a, is by the circle theorem of Milne Thomson,

$$w(z) = f(z) + \overline{f(a^2/\bar{z})}$$

$$= -m \log(z - b) - m \log \overline{(a^2/\bar{z} - b)}$$

$$= -m \log(z - b) - m \log\left(\frac{a^2}{z} - b\right)$$

$$= -m \log(z - b) - m \log\left(z - \frac{a^2}{b}\right)$$

$$+ m \log z - m \log(-b)$$

Ignoring the complex constant [– m log (–b)] which is immaterial

$$w(z) = -m \log(z - b) - m \log\left(z - \frac{a^2}{b}\right) + m \log z.$$

This is the complex potential due to

(i) a source of strength m at z = b,

(ii) a source of strength m at $z = \dfrac{a^2}{b}$,

(iii) a sink of strength m at z = 0.

Then the image of a source in a circle is an equal source at the inverse point together with an equal sink at the centre of the circle.

ALTERNATIVE PROOF OF USING CIRCLE THEOREM

Let a two-dimensional double of strength μ be placed at z = b and making an angle a with the axis of X. Then complex potential due to this doublet at any point z is

$$f(z) = \frac{\mu e^{ix}}{z-b}.$$

If a circular cylinder $|z| = a$ $(a < b)$ is inserted, then by the circle theorem, the complex potential w (z) of the flow field in the region $|z| \geq a$ is

$$w(z) = f(z) + \overline{f(a^2/\bar{z})}$$

$$= \frac{\mu e^{i\alpha}}{z-b} + \left[\frac{\mu e^{i\alpha}}{(a^2/\bar{z}-b)}\right]$$

$$= \frac{\mu e^{2\alpha}}{z-b} + \frac{\mu e^{i\alpha}}{\frac{a^2}{z}-b}$$

$$= \frac{\mu e^{i\alpha}}{z-b} + \frac{\mu z e^{i(\pi-\alpha)}}{b\left(z-\frac{a^2}{b}\right)}$$

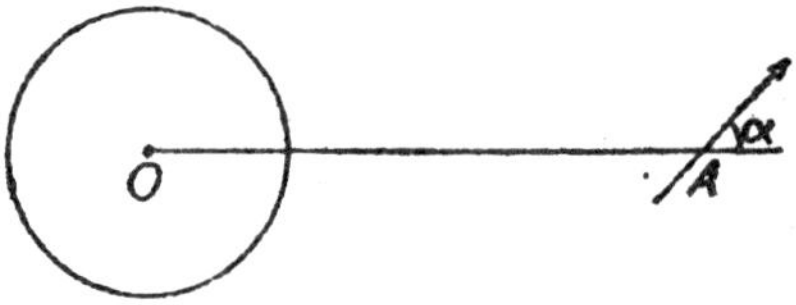

Fig. 3.8

$$= \frac{\mu e^{i\alpha}}{z-b} + \frac{\mu e^{i(\pi-\alpha)}}{b}\frac{[z-a^2/b+a^2/b]}{z-a^2/b}$$

$$= \frac{\mu e^{i\alpha}}{z-b} + \frac{\mu a^2}{b^2}\frac{e^{i(\pi-\alpha)}}{(z-a^2/b)} + \frac{\mu}{b}e^{i(\pi-\alpha)}.$$

Ignoring the last term which is constant,

$$w(z) = \frac{\mu e^{i\alpha}}{z-a} + \frac{\mu a^2}{b^2}\frac{e^{i(\pi-\alpha)}}{z-a^2/b}.$$

This is the complex potential due to

(i) a doublet of strength m at z = b, inclined at an angle α to the x-axis.

(ii) a doublet of strength $\frac{\mu a^2}{b^2}$ at $z = \frac{a^2}{b}$ and inclined at an angle $(\pi - \alpha)$ to the x-axis.

The doublet (ii) is the image of doublet (i) in the circle $|z| = a$.

Example 1: *Find the lines of flow in two-dimensional fluid motion given by* $\phi + i\psi = -\frac{1}{2}\pi\ (x + iy)^2 e^2 int.$

Prove or verify that the paths of the particles of the fluid (in polar co-ordinates) may be obtained by eliminating t from the equations $r \cos (nt + \theta) - x_0 = r \sin (nt + \theta) - y_0 = nt\,(x_0 - y_0)$.

Solution: We have $\phi + i\psi = -\frac{1}{2} n\ (x + iy)^2\ e^{2int}$.

Putting $x = r \cos \theta$, $y = r \sin \theta$, i.e., $x + iy = re^{i\theta}$,

$$\phi + i\psi = -\frac{1}{2}nr^2e^{2i\theta}\ e^{2int} = -\frac{1}{2}nr^2e^{2i(\theta + nt)}.$$

Equating real and imaginary parts,

$$\phi = \frac{1}{2}nr^2 \cos (2\theta + 2nt)$$

$$\text{and } \psi = -\frac{1}{2}nr^2 \sin (2\theta + 2nt) \qquad ...(1)$$

Now the lines of flow are given by ψ = const.

i.e. $r^2 \sin (2\theta + 2nt)$ = const.

For the second par, we have

$$r = n\ \frac{dr}{dt} = -\frac{\partial\phi}{\partial r} = nr \cos (2\theta + 2nt)$$

$$\text{and } r\theta = r\ \frac{d\theta}{dt} = \frac{1}{r}\frac{\partial\phi}{\partial r} = -\ nr \sin (2\theta + 2nt) \qquad ...(2)$$

Now by actual differentiation,

$$\frac{d}{dt}\{r \cos (nt + \theta) = r \cos (nt + \theta) - r\theta \sin (nt + \theta) - rn \sin (nt + \theta)$$

$= nr\ [\cos (nt + \theta) \cos (2nt + 2\theta) + \sin (nt + \theta) \sin (2nt + 2\theta) - \sin (nt + \theta)]$

putting values of r and rq from (2)

$= nr\ [\cos (nt + \theta) - \sin (nt + \theta)]$. ...(3)

Similarly, $\frac{d}{dt}$ [r sin (nt + θ)]

= nr [cos (nt + q) – sin (nt + q)]. ...(4)

Thus from (3) and (4),

$\frac{d}{dt}$ [r cos (nt + θ) = $\frac{d}{dt}$ [r sin (nt + θ)],

i.e. r cos (nt + θ) = r sin (nt + θ) + A,

where A is the constant of integration

i.e., r [cos (nt + θ) – sin (nt + θ) = A. ...(5)

To determine A, we have when t = 0,

x = r cos θ = x_0, y = r sin θ = y_0.

∴ [$x_0 - y_0$] = A.

Thus (5) becomes

r [cos (nt + θ) – sin (nt + 0)] = $x_0 - y_0$.

Putting this value on the right hand side of (3), it becomes

$\frac{d}{dt}$ {r cos (nt + θ)} = n (x0 – y0).

Integrating, r cos (nt + q) + C1 = nt (x0 – y0).

Again when t = 0, x = r cos q = x0.

∴ $x_0 + C_1 = 0$, i.e., $C_1 = - x_0$.

Thus r cos (nt + θ) – x_0 = nt ($x_0 - y_0$).

Similarly from (4), r sin (nt + θ) – y_0 = nt ($x_0 - y_0$); so we have finally r cos (nt + θ) – x_0 = r sin (nt + θ) – y_0 = nt ($x_0 - y_0$).

Example 2: *Use the method of images to prove that if there be a source m at the point (z_0) in a fluid bounded by the lines θ = 0 and θ* = $\frac{1}{3}\pi$, *the solution is*

$$\phi + i\psi = - m \log \{(z^3 - z_0^3)(z^3 - z_0'^3)\}$$

where $z_0 = x_0 + iy_0$ *and* $z'_0 = x_0 - iy_0$.

Solution: Let us transform the z- plane to ζ-plane by the transformation ζ = z^3, where = $re^{i\theta}$ and ζ = $Re^{i\phi}$, so that

$Re^{i\phi} = r^3 e^{3i\theta}$, *i.e.*, R= r^3 and φ = 3θ.

Therefore the boundaries $\theta = 0$, $\theta = \frac{1}{3}\pi$ in z-plane transform to $\phi = 0$ and $\phi = \pi$, *i.e.* real axis in ζ-plane.

The image system of source m at the point z_0 in z-plane, ie., $\zeta = z_0^3$ in ζ- plane with respect to real axis ($\phi = 0$, $\phi = \pi$) consists of

(i) a source m at $\zeta_0 = z_0^3$, and

(ii) a source m at $\zeta'_0 = z'^3_0$.

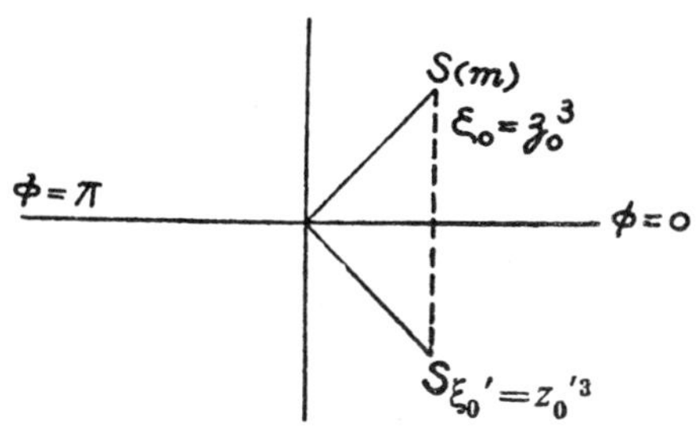

Fig. 3.9

Hence the complex potential is given by

$$w = -m \log(\zeta - \zeta_0^3) - m \log(\zeta - \zeta'_0)$$

$$= -m \log(z^3 - z^3_0) - m \log(z^3 - z_0'^{\,3}\},$$

i.e., $\phi + i\psi = -m\{(z^3 - z_0^3)(z^3 - z'^3_0)\}$.

This proves the result.

Example 3: *In the case of the motion of liquid in a part of a plane bounded by a straight line due to a source in the plane prove that if mρ is the mass of fluid (of density ρ) generated at the source per unit of time the pressure on the length 2l of the boundary immediately opposite to the source is less than that on an equal length at a great distance by*

$$\frac{1}{2}\frac{m^2\rho}{\pi^2}\left\{\frac{1}{c}\tan^{-1}\frac{l}{c} - \frac{l}{l^2 + c^2}\right\},$$

where c is the distance of the source from the boundary.

Solution: Let the bounding line oy be taken as y- axis and $\frac{m}{2\pi}$ be the source at A (c, 0).

The equivalent image system consists of

(i) a source $\frac{m}{2\pi}$ at A (c, 0),

(ii) a source $\frac{m}{2\pi}$ at A′ (– c, 0).

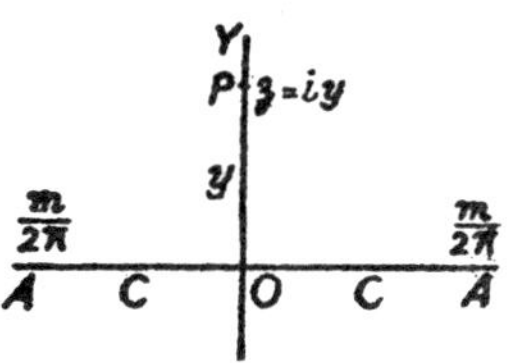

Fig. 3.10

Hence the complex potential is given by

$$w = -\frac{m}{2\pi}\log(z-c) - \frac{m}{2\pi}\log(z+c)$$

$$= -\frac{m}{2\pi}\log(z^2 - c^2).$$

Speed, $q = \left|\frac{dw}{dz}\right| = \frac{m}{2\pi}\left|\frac{2z}{z^2 - c^2}\right|.$

If P (z = iy) is a point on y-axis, then speed at this point is given by

$$q = \frac{m}{\pi}\frac{y}{y^2 + c^2},$$

Also by Bernoulli's theorem,

$$\frac{p}{\rho} = \frac{p_0}{\rho} - \frac{1}{2}q^2,$$

where p_0 is the pressure on OY when y is the infinite, where velocity is zero.

$$\therefore \quad \frac{p_0 - p}{\rho}\ \frac{1}{2}q^2 = \frac{1}{2}\frac{m^2}{\pi^2}\frac{y^2}{(y^2+c^2)^2}.$$

Therefore the required difference in pressure

$$= \int_{-1}^{1}(p_0 - p)dy = \frac{1}{2}\frac{m^2\rho}{\pi^2}\int_{-1}^{1}\frac{y^2}{(y^2+c^2)}dy$$

$$\frac{m^2}{\pi^2}\rho\int_0^1\frac{y^2}{(y^2+c^2)^2}dy$$

$$\frac{m^2\rho}{\pi^2 c}\int_0^1\frac{c^2\tan^2\theta}{c^4\sec^4\theta.}c\sec^2\theta d\theta, \text{ putting } y = c\tan\theta$$

$$= \frac{m^2\rho}{\pi^2 c}\int_0^1\sin^2\theta d\theta = \frac{m^2\rho}{2\pi^2 c}\int_0^1(1-\cos 2\theta)\,d\theta \text{ putting } y = c\tan\theta$$

$$= \frac{m^2\rho}{\pi^2 c}\left[\theta - \frac{\sin 2\theta}{2}\right] = \frac{1}{2}\frac{m^2\rho}{\pi^2}\left[\frac{1}{c}\theta - \frac{1}{c}\sin\theta\cos\theta\right]_0^l$$

$$= \frac{1}{2}\frac{m^2\rho}{\pi^2}\left[\frac{1}{c}\tan^{-1}\frac{y}{c} - \frac{y}{y^2 + c^2}\right]_0^l \quad \text{as } y = c\tan\theta$$

$$= \frac{1}{2}\frac{m^2\rho}{\pi^2}\left[\frac{1}{c}\tan^{-1}\frac{l}{c} - \frac{l}{l^2 - c^2}\right]_0^l.$$

This proves the result.

Example 4: *What arrangement of sources and sinks will give rise to the function $w = \log (z - a^2/z)$?*

Draw a rough sketch of the stream lines in this case and prove that two of them subdivide into the circle $r = a$ and the axis of y.

Solution: We have

$$w = \log\left(z - \frac{a^2}{z}\right) = \log\left(\frac{z^2 - a^2}{z}\right)$$

$$= \log (z - a) + \log (z + a) - \log z.$$

This shows that there are two sinks of units strength $m = -1$ at $z = a$ and $z = -a$ and a source of unit strength at origin.

Further since $w = \phi + i\psi$,

$$\phi + i\psi = \log (z - a) + \log (z + a) - \log z$$

$$= \log (x + iy - a) + \log (x + iy + a) - \log (x + iy).$$

Equating imaginary parts,

$$\psi = \tan{-1}\ \frac{y}{x-a} + \tan^{-1}\frac{y}{x+a}\tan^{-1}\frac{y}{x}$$

$$= \tan^{-1}\frac{\frac{y}{x-a} + \frac{y}{x+a}}{1 - \frac{y^2}{x^2 - a^2}} - \tan^{-1}\frac{y}{x}$$

$$= \tan^{-1}\frac{2xy}{x^2 - y^2 - a^2} - \tan^{-1}\frac{y}{x} = \frac{\frac{2xy}{x^2 - y^2 - a^2} - \frac{y}{x}}{1 - \frac{2xy^2}{x(x^2 - y^2 - a^2)}}$$

$$= \tan^{-m1} \frac{y(x^2 + y^2 + a^2)}{x(x^2 + y^2 - a^2)}.$$

For stream lines, ψ = const.,

i.e., $$\frac{y(x^2 + y^2 + a^2)}{x(x^2 + y^2 - a^2)} = \text{const.}$$

or $$\frac{x(x^2 + y^2 - a^2)}{y(x^2 + y^2 + a^2)} = \text{const.} \quad ...(1)$$

When const. is taken to be zero, the corresponding stream lines are given by $x\ (x2 + y^2 - a^2) = 0$,

i.e., $x = 0$ and $x^2 + y^2 = a^2$,

i.e., y-axis and the circle $x^2 + y^2 = a^2$ or $r = a$.

Also from (1) $y = 0$, *i.e.* x-axis is a stream line. Hence the rough sketch of the stream lines is as given in the figure with sources at the origin and sinks at

(a, 0) and (–a, 0).

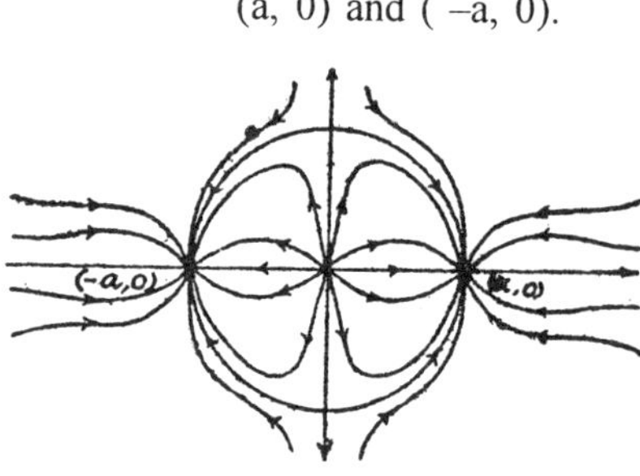

Fig. 3.11

Example 5(a): *Find the stream function of two-dimensional motion due to two equal sources and an equal sink midway between them; sketch the stream lines and find the velocity at any point.*

In a region bounded by a fixed quadrantal arc and its radii, deduce the motion due to a source and an equal sink situated at the ends of one of the bounding radii. Show that the stream line leaving either end at an angle α with the radius is

$$r^2 \sin(\alpha + \theta) = a^2 \sin(\alpha - \theta).$$

Solution: Let S and S′ be the two sources of equal strength m and 2a be the distance between them. Take the middle point of SS¢ as real axis. There being a sink – m at O, the complex potential w at any point P (z = x + iy) is given by $w = \phi + i\psi$

$$= -m \log (z - a)$$

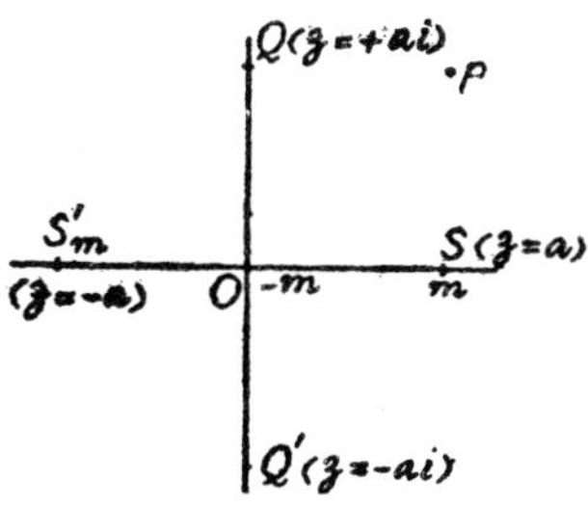

Fig. 3.12

$$- m \log (z + a) + m \log z.$$

Equating imaginary parts, the stream function ψ is given by

$$\psi = -m \tan^{-1}\frac{y}{x-a} - m\tan^{-1}\frac{y}{x+a} + m\tan^{-1}\frac{y}{x}$$

$$= -m \tan^{-1}\left[\frac{y/(x-a) + y/(x+a)}{1-\{y^2/(x^2-a^2)\}}\right] + m\tan^{-1}\frac{y}{x}$$

$$= -m \tan^{-1}\frac{2xy}{x^2-y^2-a^2} + m\tan^{-1}\frac{y}{x}$$

$$= -m \tan^{-1}\frac{y(x^2+y^2+a^2)}{x(x^2+y^2-a^2)}.$$

The stream lines are given by ψ = constant.

When $\psi = 0$, $y = 0$ is the stream line and when $\psi = \infty$, $x = 0$ and $x^2 + y^2 = a^2$ are the stream lines.

Now the stream lines can be sketched as in the previous example except that the arrowheads are now reversed.

Velocity : Let q be the magnitude of velocity at any point P (z).

$$q = \left|\frac{dw}{dz}\right|$$

$$= \left|-\frac{m}{z-a} - \frac{m}{z+a} + \frac{m}{z}\right|$$

$$= m \left| \frac{z^2 + a^2}{z(z-a)(z+a)} \right|$$

2nd part : The boundaries of the quadrantal are viz. x = 0, y = 0 and x2 + y2 = a2 are streamlines.

Now streamlines are given by

$$\frac{y(x^2 + y^2 + a^2)}{x(x^2 + y^2 - a^2)} = C \text{ (= constant)}$$

$$x^2 y + y^3 + a^2y = C (x^3 + xy^2 - a^2x)$$

whence $$\frac{dy}{dx} = \frac{2xy - 3Cx^2 - Cy^2 + Ca^2}{x^2 - 3y^2 + a^2 - zCxy}.$$

For a stream line which leaves the ends (± a, 0) at an angle α, we must have

$$\left(\frac{dy}{dx}\right)_{(\pm a,0)} = \tan\alpha \text{ , } i.e., C = -\tan\alpha,$$

The stream line leaving at an angle α the end S is given by

$$-\tan\alpha = \frac{y(x^2 + y^2 + a^2)}{x(x^2 + y^2 - a^2)},$$

$$i.e., -\tan\alpha = \frac{r\sin\theta(r^2 + a^2)}{r\cos\theta(r^2 - a^2)}$$

or $r^2 \sin(\alpha + \theta) = a^2 \sin(\alpha - \theta)$,

which is the required equation of the streamlines.

Example 5(b): An area A is bounded by that part of the x-axis for which x > a and by that branch of $x^2 - y^2 = a^2$ which is in the positive quadrant. There is a two-dimensional unit source at (a, 0) which sends out liquid uniformly in all directions. Shew by means of the transformation $w = \log(z^2 - a^2)$ that in steady motion the stream lines of the liquid within the area A are portions of rectangular hyperbolas. Draw the stream lines corresponding to

Solution: Equating imaginary parts,

$$y = -m\tan^{-1}\frac{2xy}{x^2 - y^2 - a^2} + 2m\tan^{-1}\frac{y}{x}$$

$$= -\tan^{-1}\frac{2xy}{x^2-y^2-a^2} + m\tan^{-1}\frac{2xy}{x^2-y^2}$$

as $2\tan^{-1}A = \tan^{-1}\dfrac{2A}{1-A^2}$

$$= -m\tan^{-1}\frac{2a^2xy}{(x^2+y^2)^2-a^2(x^2-y^2)}.$$

The streamlines are given by y = const.

i.e., $\dfrac{2a^2xy}{(x^2+y^2)^2-a^2(x^2-y^2)} = \dfrac{2}{\lambda}$ (say)

or $(x^2 + y^2)^2 = a^2 (x^2 - y^2 + \lambda xy)$;

by properly adjusting the constant, which is the parameter in this case and for different values of λ, we get different streamlines.

$$\text{speed} = \left|\frac{dw}{dz}\right| = \left|\frac{2m}{z^2-a^2} - \frac{2m}{z}\right|$$

$$= \frac{2ma^2}{|z||z-a||z+a|} = \frac{2ma^2}{r_3 r_1 r_2}.$$

This proves the result.

Example 5(c): *Between the fixed boundaries* $\theta = \frac{1}{6}\pi$ *and* $\theta = -\frac{1}{6}\pi$, *there is a two-dimensional liquid motion due to a source at the point* ($r = c$, $\theta = \alpha$) *and a sink at the origin, absorbing water at the same rate as the source produces it. Find the stream function, and show that one of the stream lines is a part of the curve*

$$r^3 \sin 3\alpha = c^3 \sin 3\theta.$$

Solution: Let us transform the z-plane (xy-plane) to ζ-plane ($\zeta\eta$-plane) by the transformation

$$\zeta = z^3,$$

where $z = re^{i\theta}$ and $\zeta = Re^{i\phi}$,

so that $Re^{i\phi} = r^3e^{3i\theta}$, *i.e.*, $R = r^3$ and $\phi = 3\theta$.

Therefore the boundaries $\theta = \pm\frac{1}{6}\pi$ in z-plane transform to $f = \pm\frac{1}{2}\pi$, i.e., imaginary axis in ζ-plane.

The point ($r = c$, $\theta = a$) transforms to ($R = C^3$, $\phi = 3\alpha$),

$\therefore$ The image system with regard to imaginary axis ($\phi = \pm \frac{1}{2}\pi$) in ζ-plane consists of

(i) a source of strength m at $(c^3, 3\alpha)$,

(ii) a sink of strength $-$ m at (0, 0),

(iii) a source of strength m at $(c^3, \pi - 3\alpha)$,

(iv) a sink of strength $-$ m at (0, 0)

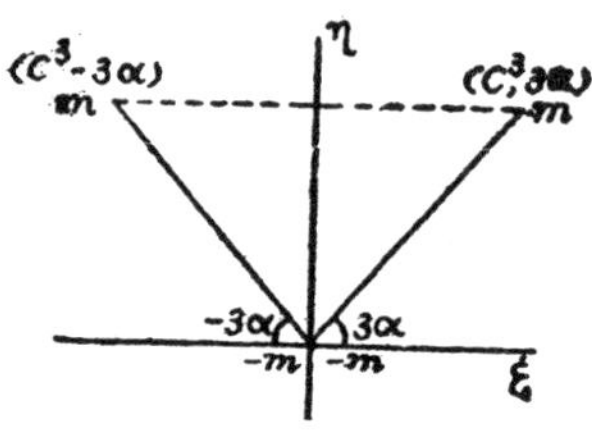

Fig. 3.13

Therefore the complex potential is given by

$w = -m \log(\zeta - c^3 e^{3i\alpha}) + m \log(\zeta - 0)$

$-m \log\{-c^3 e^{i(\pi - 3\alpha)}\} + m \log(\zeta - 0)$

$= -m \log(z^3 - c^3 e^{3i\alpha})(z^3 + c^3 e^{-3i\alpha}) + 2m \log z^3$ as $z^3 = \zeta$

$= -m \log(z^6 - c^6 - 2ic^3 z^3 \sin 3\alpha) + 6m \log z$

$= -m \log(r^6 e^{6i\theta} - c^6 - 2ic^3 r^3 e^{3i\theta} \sin 3\alpha) + 6m \log re^{i\theta}$ as $z = re^{i\theta}$

$= -m \log\{r^6 \cos 6\theta + 2c^3 r^3 \sin 3\alpha \sin 3\theta - c^6)$

$+ i(r^6 \sin 6\theta - 2c^3 r^3 \sin 3\alpha \cos 3\theta)\}$

$+ 6m \log(r \cos\theta + ri \sin\theta)$.

But $w = \phi + i\psi$. Equating imaginary parts,

$$\mu\psi = -m \tan^{-1}\left[\frac{r^6 \sin 6\theta - 2c^3 r^3}{r^6 \cos 6\theta + 2c^3 r^3 \sin 3\alpha \sin 3\theta - c^6}\right] + 6m\theta,$$

which is the stream function.

Now puting $\psi = 0$, corresponding stream line is given by

$$\tan 6\theta = \frac{r^6 \sin 6\theta - 2c^3 r^3 \sin 3\alpha \cos 3\theta}{r^6 \cos 6\theta + 2c^3 r^3 \sin 3\alpha \sin 3\theta - c^6}$$

or $\sin 6\theta\,[r^6 \cos 6\theta + 2c^3 r^3 \sin 3\alpha \sin 3\theta - c^6]$

$= \cos 6\theta\,[r^6 \sin 6\theta - 2c^3 r^3 \sin 3\alpha \cos 3\theta]$.

i.e. $2r^3c^3 \sin 3\alpha\, [\sin 3\theta \sin 6\theta + \cos 3\theta \cos 6\theta] - c^6 \sin 6\theta = 0$,

i.e. $2r^3c^3 \sin 3\alpha \cos 3\theta - c^6 .\, 2 \sin 3\theta \cos 3\theta = 0$,

i.e. $\cos 3\theta\, (r^3 \sin 3\alpha - c^3 \sin 3\theta) = 0$.

when $\cos 3\theta = 0$, $\theta = \pm \dfrac{\pi}{6}$ which are the given boundaries.

$\therefore$ other stream line for $\psi = 0$ is part of the curve

$r^3 \sin 3\alpha = c^3 \sin 3\theta$.

Example 6: *Between the fixed boundaries* $\theta = \dfrac{1}{4}\pi$ *and* $\theta = -\dfrac{1}{4}\pi$, *there is a two-dimensional liquid due to a source of strength m at the point* ($r = a$, $\theta = 0$). *Show that the stream function is*

$$-m \tan^{-1}\left\{\frac{r^4(a^4 - b^4)\sin 4\theta}{r^8 - r^4(a^4 + b^4)\cos 4\theta + a^4b^4}\right\}.$$

Show also that the velocity at (r, θ) *is*

$$\frac{4m(a^4 - b^4)r^3}{(r^8 - 2a^4r^4 \cos 4\theta + a^8)^{\frac{1}{2}}(r^8 - 2b^4r^4 \cos 4\theta + b^8)^{\frac{1}{2}}}$$

Solution: Let us transform the z-plane to ζ-plane by the transformation $\zeta = z^2$,

where $z = re^{i\theta}$ and $\zeta = Re^{i\phi}$,

so that $Re^{i\phi} = r^2e^{2i\theta}$,

i.e., $R = r^2$ and $\phi = 2\theta$.

Therefore the boundaries $\theta = \pm \dfrac{1}{4}\pi$ in z-plane (xy-plane) transform to $\phi = \pm \dfrac{1}{2}\pi$ in ζ-plane ($\xi\eta$-plane).

The point $r = a$, $\theta = 0$ becomes $R = a^2$, $\phi = 0$ and the point $r = b$, $\theta = 0$ becomes $R = b^2$, $\phi = 0$.

The equivalent image system with regard to $\phi = \pm \dfrac{1}{2}\pi$, *i.e.* line perpendicular to real axis in -ζ plane consists of

(i) a source of strength m at $(a^2, 0)$,

(ii) a sink of strength $-$ m at $(b^2, 0)$,

(iii) a source of strength m at $(-a^2, 0)$.

(iv) a sink of strength $-$ m at $(-b^2, 0)$.

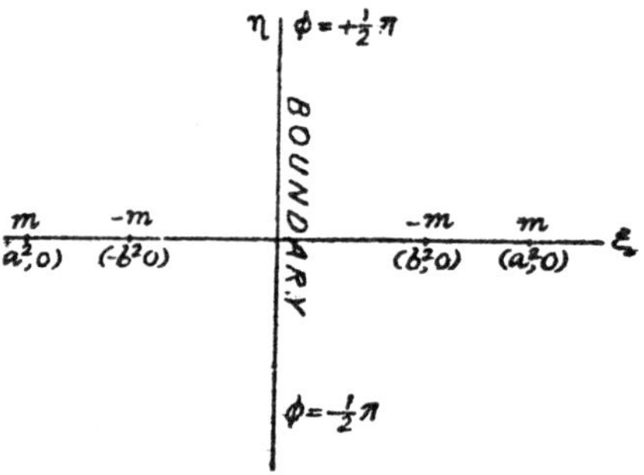

Fig. 3.14

And therefore the complex potential is given by

$$w = -m\log(\zeta - a^2) + m\log(\zeta - b^2)$$
$$- m\log(\zeta + a^2) + m\log(\zeta + b^2)$$
$$= m\log(\zeta^2 - a^4) + m\log(\zeta^2 - b^4)$$
$$= -m\log(z^4 - a^4) + m\log(z^4 - b^4) \text{ as } \zeta = z^2.$$

i.e., $\phi + i\psi = -m\log[r^4e^{4i\theta} - a^4] + m\log[r^4e^{4i\theta} - b^4]$,

so that $\psi = -m\left[\tan^{-1}\dfrac{r^4\sin 4\theta}{r^4\cos 4\theta - a^4}\tan^{-1}\dfrac{r^4\sin 4\theta}{r^4\cos 4\theta - b^4}\right]$

$$= m\tan^{-1}\frac{r^4(a^4-b^4)\sin 4\theta}{r^8 - r^4(a^4-b^4)\cos 4\theta + a^4b^4}$$

This proves the first part of the result.

If q be the velocity at (r, θ),

$$q = \left|\frac{dw}{dz}\right| = \left|-m\frac{4z^3}{z^4-a^4} + m\frac{4z^3}{z^4-b^4}\right|$$

$$= 4m\left|\frac{z^3(a^4-b^4)}{(z^4-a^4)(z^4-b^4)}\right|$$

$$= 4\,m\,\frac{\left|r^3e^{3i\theta}\right|(a^4-b^4)}{\left|r^4e^{4i\theta}-a^4\right|\left|r^4e^{4i\theta}-b^4\right|}$$

$$= \frac{4mr^3(a^4-b^4)}{(r^8-2a^4r^4\cos 4\theta + a^8)^{1/2}(r^8-2b^4r^4\cos 4\theta+b^8)^{1/4}},$$

which proves the second part of the result.

Example 7: *In the case of the two-dimensional fluid motion produced by a source of strength m placed at a point S outside a rigid circular disc of radius a whose centre is O, show that the velocity of slip of the fluid in contact with the disc is greatest at the points where the lines joining S to the ends of the diameter at right angles to OS cut the circle; and prove that its magnitude at these points is*

$$\frac{2m.OS}{(OS^2 - a^2)}.$$

Solution: Let S′ be the inverse point of S with regard to the circular disc, so that if

OS = c,

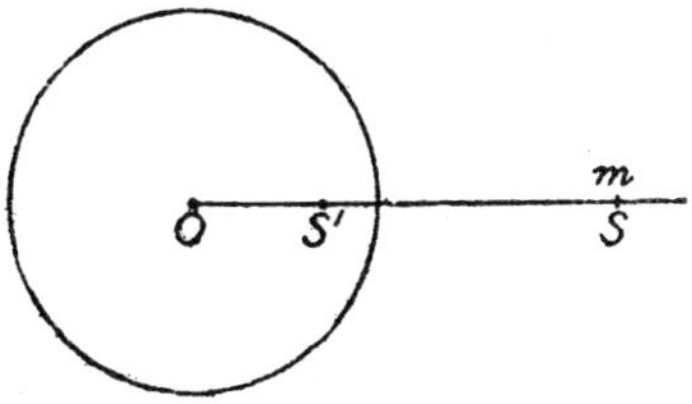

Fig. 3.15

os. $OS' = a^2$, i.e. $OS' = \dfrac{a^2}{c}$.

The equivalent image system consists of

(i) a source of strength m at S,

(ii) a source of strength m at S′,

(iii) a sink of strength – m at O.

Taking O as origin and OS as real axis, the complex potential for the motion of the fluid at any point z = (z + iy) is given by

$$w = -m \log (z - c) - m \log \left(z - \frac{a^2}{c} \right) + m \log z,$$

so that $$\frac{dw}{dz} = -m\left(\frac{1}{z-c}\right) - m\left(\frac{1}{z - a^2/c}\right) + \frac{m}{z}.$$

If q is the velocity at point z, we have

$$q = \left|\frac{dw}{dz}\right| = \left| -\frac{m}{z-c} - \frac{m}{z-a^2/c} + \frac{m}{z} \right|$$

$$= m\left|\frac{z^2 - a^2}{(z-c)(z-a^2/c)z}\right|$$

$$= m\left|\frac{(z-a)(z+a)}{(z-c)(z-a^2/c)z}\right|.$$

If θ is the vectorial angle of any point on the boundary of the disc, then

$$z = a^{i\theta}.$$

∴ velocity at any point $ae^{i\theta}$ on the boundary of the disc is given by

$$q = m\left|\frac{(ae^{i\theta} - a)(ae^{i\theta} + a)}{(ae^{i\theta} - c)(ae^{i\theta} - a^2/c)ae^{i\theta}}\right|$$

$$= mc\left|\frac{(1-e^{-i\theta})(1+e^{i\theta})}{(a - ce^{-i\theta})(ce^{i\theta} - a)}\right|$$

$$= \frac{2mc\sin\theta}{a^2 + c^2 - 2ac\cos\theta} \qquad ...(1)$$

For q to be maximum, we have

$$\frac{dq}{d\theta} = 2mc\ \frac{(a^2 + c^2 - 2ac\cos\theta)\cos\theta - \sin\theta(2ac\sin\theta)}{(a^2 + c^2 - 2ac\cos\theta)^2} = 0,$$

i.e., $(a^2 + c^2)\cos\theta - 2ac = 0$

$$\text{or} \quad \cos\theta = \frac{2ac}{a^2 + c^2}. \qquad ...(2)$$

As θ = 0 gives maximum (zero) velocity, value of θ given by (2) corresponds to the maximum value of q.

Putting from (2) in (1),

$$q_{\text{maxi.}} = 2mc\frac{a^2 - c^2}{a^2 + c^2}.$$

$$\text{as } \sin\theta = \frac{a^2 - c^2}{a^2 + c^2}$$

$$= \frac{2mc}{c^2 - a^2} = \frac{2m.OS}{OS^2 - a^2}.$$

The boundary being stream line, the velocity on the boundary is the velocity of the slip.

Example 8: *If a homogeneous liquid is acted on by a repulsive force from the origin, them magnitude of which at distance r from the origin is μr per unit mass, shew that it is possible for the liquid to move steadily, without being constrained by any boundaries, in the space between one branch of the hyperbola $x^2 - y^2 = a^2$ and the asymptotes, and find the velocity potential.*

Solution: The liquid moves steadily between the space given by one branch of

$$x^2 - y^2 = a^2 \qquad ...(1)$$

and its asymptotes given by $x^2 - y^2 = 0$. ...(2)

Thus $y = A(x^2 - y^2) = Ar^2 \cos^2 2\theta = Ar^2 \sin\left(\frac{\pi}{2} + 2\theta\right)$,

where A is some constant.

The harmonic conjugate of ψ which is given by ϕ is clearly

$$f = Ar^2\cos\left(\frac{\pi}{2} + 2\theta\right),$$

so that $w = \phi + i\psi = Ar^2 e^{i(\frac{1}{2}\pi + 2\theta)}$

$$= Ae^{\frac{1}{2}i\pi r^2 e^2 i\theta} = Aiz^2 \text{ as } z = re^{i\theta}.$$

$$\therefore \quad \frac{dw}{dz} = Ai.\ 2z.$$

Since the only (repulsive) force is along the radius vector, the velocity is a function of r alone.

So if v is the velocity at a distance r, $v = 2\,Air$.

The motion is steady ; hence the equation of motion becomes

$$\frac{p}{p} + \frac{1}{2}v^2 + V = \text{const.},$$

where as given $-\frac{\partial V}{\partial r} = \mu r$ or $V = -\frac{1}{2}\mu r^2$,

so that $\frac{p}{p} - 2A^2r^2 - \frac{1}{2}\mu r^2 = \text{const.}$

On the free surface p = const. $\therefore\ 2A^2 - \frac{1}{2}\mu = 0.$

Hence $v = 2Ai.\ r = -\sqrt{\mu r}$

and velocity potential,

$$\phi = Ar^{2\frac{1}{2}i\pi} \cos 2\theta = Air^2 \cos 2\theta$$

$$= -\frac{1}{2}\sqrt{\mu.r^2} \cos 2\theta.$$

Example 9: *In the part of an infinite plane bounded by a circular quadrant AB and the productions of the radii OA, OB, there is a two-dimensional motion due to the production of liquid at A, and its absorption at B, at the uniform rate m. Find the velocity potential of the motion; and show that the fluid which issues from A in the direction making an angle μ with OA follows the path whose polar equation is*

$$r = a \sin^{1/2} 2\theta \left[\cot \mu + \sqrt{(\cot^2 \mu + \operatorname{cosec}^2 2\theta)^{1/2}}\right],$$

the positive sign being taken for all the square roots.

Solution: The equivalent image system of source $\frac{m}{2\pi}$ at A w.r.t. circular boundary consists of

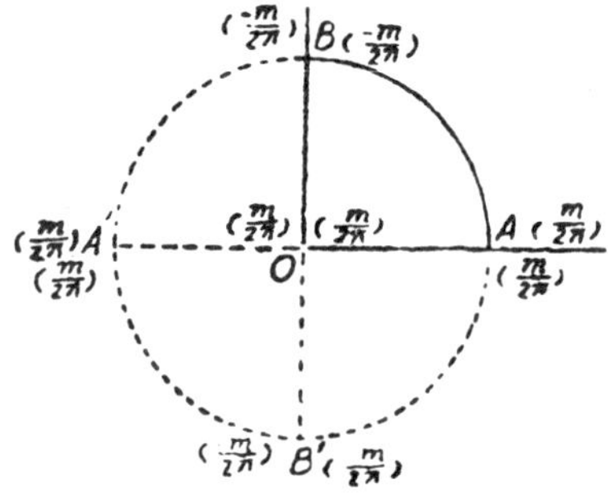

Fig. 3.16

a source $\frac{m}{2\pi}$ at A.

a source $\frac{m}{2\pi}$ at A,as A is the inverse pt. of itself and

a sink $-\frac{m}{2\pi}$ at O.

Of the above system the image system with respect to the lines OA and OB is

(i) sources $\frac{m}{2\pi} + \frac{m}{2\pi}$ at A,

(ii) source $\frac{m}{2\pi} + \frac{m}{2\pi}$ at A′ (image of sources at A),

(iii) sink $-\frac{m}{2\pi}$ at O.

Similarly considering the image system of sink $-\frac{m}{2\pi}$ at B with respect to circular boundary and OA and OB, we get

(i) sink $-\frac{m}{2\pi} - \frac{m}{2\pi}$ at B,

(ii) sink $-\frac{m}{2\pi} - \frac{m}{2\pi}$ at B′,

(iii) source $\frac{m}{2\pi}$ at O.

Thus if P (z = z + iy) is any point in the fluid, the complex potential at P due to the above system is given by

$$w = -\frac{m}{\pi}\log(z-a) - \frac{m}{\pi}\log(z+a) + \frac{m}{\pi}\log(z-ai) + \frac{m}{\pi}\log(z+ai)$$

as source and sink at origin cancel each other,

i.e., $$\phi + i\psi = -\frac{m}{\pi}\log(z^2-a^2) + \frac{m}{\pi}\log(z^2+a^2). \qquad ...(1)$$

$$\therefore \phi = -\frac{m}{\pi}\log|z-a| - \frac{m}{\pi}\log|z+a| + \frac{m}{\pi}\log|z-ai| + \frac{m}{\pi}\log|z+ai|$$

$$= -\frac{m}{\pi}\log AP - \frac{m}{\pi}\log A'P + \frac{m}{\pi}\log BP + \frac{m}{\pi}\log B'P$$

$$= \frac{m}{\pi}\log\frac{.B'P}{AP.A'P}.$$

Also putting $z = re^{i\theta}$ in (1) and then equating imaginary parts,

$$\psi = -\frac{m}{\pi}\tan^{-1}\left(\frac{r^2\sin 2\theta}{r^2\cos 2\theta - a^2}\right) + \frac{m}{\pi}\tan^{-1}\left(\frac{r^2\sin 2\theta}{r^2\cos 2\theta + a^2}\right)$$

$$= -\frac{m}{\pi}\tan^{-1}\frac{\dfrac{r^2\sin 2\theta}{r^2\cos 2\theta - a^2} - \dfrac{r^2\sin 2\theta}{r^2\cos 2\theta + a^2}}{1+\dfrac{r^4\sin^4 2\theta}{r^4\cos^2 2\theta - a^4}}$$

$$= -\frac{m}{\pi}\tan^{-1}\left(\frac{2a^2r^2\sin 2\theta}{r^4 - a^4}\right).$$

The stream line that leaves A at an inclination μ is obtained by putting $y = -\frac{m}{\pi}\mu$ and is given by

$$\tan\mu = \frac{2a^2r^2\sin 2\theta}{r^4 - a^4} \text{ or } r^4 - 2a^2r^2\sin 2\theta\cot\mu - a^4 = 0,$$

i.e., $r^2 = a^2\sin 2\theta\cot\mu + \sqrt{(a^4\sin^2 2\theta\cot^2\mu + a^4)}$,

taking + Ve sign before the square root

or $\quad r = a\sin^{1/2} 2\theta\,[\cot\mu + \sqrt{(\cot^2\mu + \operatorname{cosec}^2 2\theta)}]^{1/2}$.

This proves the result.

Example 10(A): *Prove that in the two-dimensional liquid motion due to any number of sources at points on a circle, the circle is a stream line provided that there is no boundary and that the algebraic sum of the strengths of the sources is zero.*

Show that the same is true if the region of flow is bounded by a circle which cuts orthogonally the circle in question.

Solution: Let $m_1, m_2, m_3,\ldots$ be the strengths of the sources at $A_1, A_2, A_3,\ldots$ Also let P be any point on the circle. Take the diameter through P as initial line.

If $\angle A_1PO = \theta$, $\angle A_2PA_1 = \alpha_1$, $A_3PA_2 = \alpha_2,\ldots$etc.,

then $\psi = m_1\theta - m_2(\theta + \alpha_1) - m_3(\theta + \alpha_1 + \alpha_2)\ldots$

$= -\theta(m_1 + m_2 + m_3 + \ldots) - [m_2\alpha_1 + m_3(\alpha_1 + \alpha_2) + \ldots]$

$= -\theta(m_1 + m_2 + m_3 + \ldots) - \text{const.}$

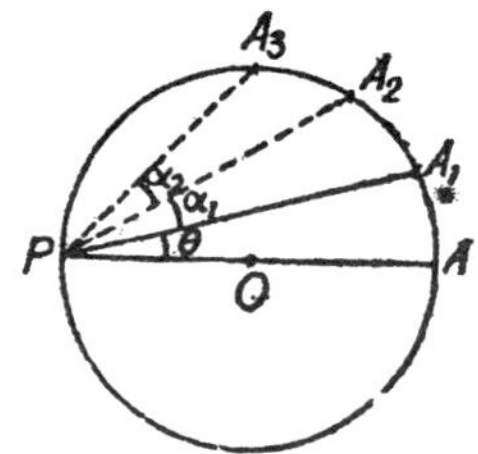

Fig. 3.17

as $m_2\alpha_1 + m_3(\alpha_1 + \alpha_2) + ...$ is a constant, $\alpha_1, \alpha_2,...$ being constants for all positions of P (since the angle subtended at the circumference by an arc is always the same).

Again if the algebraic sum of the strengths of sources is zero,

i.e., if $m_1 + m_2 + m_3 + ... = 0$,

then ψ = constant, i.e., the circle is a stream line.

Again let O′ be the centre of a circle that cuts the given circle orthogonally. The image of m_1 at A_1 is m_1 at A′, the inverse pt. of A_1 and –m at O′; the point A′ is where A_1O' meets the given circle.

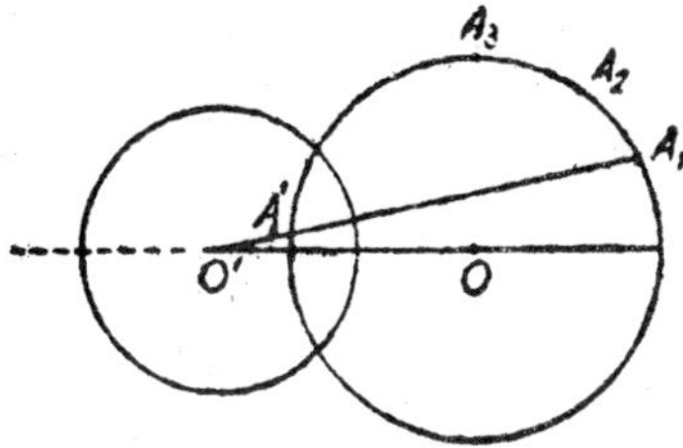

Fig. 3.18

If the barriers are omitted, then we are left with system 2 $(m_1 + m_2 + ...)$ on the boundary of the given circle and $(-m_1 + m_2 + ...)$ at O′.

And since $m_1 + m_2 + m_3 + ... = 0$, the result follows.

Example 10(b): *In a two-dimensional liquid motion ϕ and ψ are the velocity potential and current function; show that a second fluid motion exists in which ψ is the velocity potential and $-\phi$ the current function; and prove that if the first motion be due to sources and sinks, the second motion can be built up by replacing a source and an equal sink by a line of doublets uniformly distributed along any curve joining them.*

Solution: Since ϕ and ψ are the velocity and stream function for the two-dimensional motion, therefore we have the conditions

$$\frac{\partial\phi}{\partial x} = \frac{\partial\psi}{\partial y} \frac{\partial\phi}{\partial y} = -\frac{\partial\psi}{\partial x}. \quad ...(1)$$

Now the fluid motion with ψ as velocity potential and $-\phi$ as current function will exist if conditions of the of (1) are satisfied when we put ψ for ϕ and $-\phi$ for ψ,

$$i.e., \text{ if } \frac{\partial\psi}{\partial x} = \frac{\partial(-\phi)}{\partial y} \text{ and } \frac{\partial\psi}{\partial y} = \frac{\partial(-\phi)}{\partial x},$$

i.e., if $\frac{\partial \psi}{\partial x} = -\frac{\partial \phi}{\partial y}$ and $\frac{\partial \psi}{\partial y} = \frac{\partial \phi}{\partial x}$.

which hold because of (1). Hence such a motion exists.

Thus if $w = \phi + i\psi$ exists,

$w' = +\psi - i\phi = iw$ also exists.

Second part : Let there be a source m at B (a, 0) and a sink – m at (–a, 0). The complex potential for this is

$w = -m \log (z - a) + m \log (z + a)$

$= m \log \left(\frac{z+a}{z-a}\right).$ ...(1)

Now let us join AB by any curve. The axis of the doublet on this curve is normal to AB.

If w_1 is the complex potential due to this line of doublets,

then $w_1 = \int_A^B \frac{me^{i\pi/2}}{z-u} du = me^{i\pi/2} \log \frac{z-a}{z+a}$

$= mi \log \frac{z-a}{z+a} = -iw.$

The result follows from the first part now.

Example 10(c): *Prove that for liquid circulating irrotationally in part of the plane between two-intersecting circles the curves of constant velocity are Cassini's ovals.*

Solution: Let OO′ be the line of centres of the circles. Two points A and B can be found such that these are inverse points with respect to both the circles.

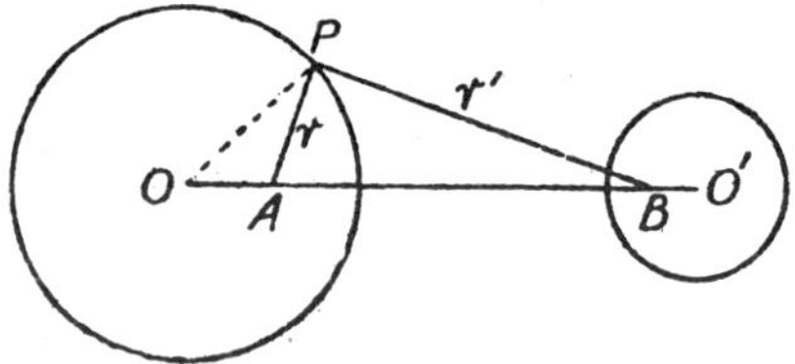

Fig. 3.19

Take a point P on the circle at a distance r from A and r¢ from B. Since OA. OB = OP^2, triangles OAP and OPB are similar i.e.

$$\frac{r}{r'} = \frac{OP}{OB} = \text{const.}$$

Thus the equations of two circles can be taken as

$$\frac{r}{r'} = c_1 \text{ and } \frac{r}{r'} = c_2,$$

and since these are the two streamlines, the stream function ψ must be the form $f\left(\frac{r}{r'}\right)$ and it being a plane harmonic, we have

$$\psi = C \log \frac{r}{r'} \qquad \text{(as log r is the only plane-harmonic of r)}$$

Again if ϕ is the conjugate harmonic of ψ, τηεν ϕ + iψ or ψ – iϕ must be an analytic function of z (see previous example).

So $\phi = -C(\theta - \theta')$

and then $w = \psi - i\phi = C \log \frac{r}{r'} + iC(\theta - \theta')$

$$= C[\log r + i\theta] - C[\log r' + i\theta']$$

$$= C \log \frac{re^{i\theta}}{r'e^{i\theta}} = C \log \frac{z+a}{z-a}$$

choosing A to be (–a, 0) and B to be (a, 0).

Now $q = \left|\frac{dw}{dz}\right| = 2a\left|\frac{1}{(z+a)(z-a)}\right|$

$$= \frac{2Ca}{rr'}.$$

Thus curves of equal velocity are given by

q = constant,

i.e. rr′ = constant,

which are clearly Cassini-ovals.

Example 11: *Show that velocity potential*

$$\phi = \frac{1}{2}\log\frac{(x+a)^2 + y^2}{(x-a)^2 + y^2},$$

gives a possible motion. Determine the form of streamlines and the curves of equal speed.

Solution: We have

$$\phi = \frac{1}{2}\log\,[(x + a)^2 + y^2] - \frac{1}{2}\,\log\,[(x-a)^2 + y^2],$$

so that $$\frac{\partial \phi}{\partial x} = \frac{x+a}{(x+a)^2+y^2} - \frac{x-a}{(x-a)^2+y^2},$$

and $$\frac{\partial \phi}{\partial y} = \frac{y}{(x+a)^2+y^2} - \frac{y}{(x-a)^2+y^2}.$$

Now $$\frac{\partial u}{\partial x} = \frac{\partial}{\partial x}\left(-\frac{\partial \phi}{\partial x}\right) = -\frac{y^2-(x+a)^2}{[(x+a)^2+y^2]^2} + \frac{y^2-(x-a)^2}{[(x-a)^2+y^2]^2}$$

and $$\frac{\partial v}{\partial y} = \frac{\partial}{\partial y}\left(-\frac{\partial \phi}{\partial y}\right) = -\frac{(x+a)^2-y^2}{[(x+a)^2+y^2)]^2} + \frac{(x-a)^2-y^2}{[(x-a)^2+y^2]^2}.$$

Clearly the equation of continuity, *i.e.*,

$\frac{\partial u}{\partial x} + \frac{\partial v}{\partial y} = 0$ is satisfied.

Therefore $\phi = \frac{1}{2}\log\frac{(x+a)^2+y^2}{(x-a)^2+y^2}$ gives a possible motion.

Now to find streamlines, we should determine the stream function ψ, which holds :

$$\frac{\partial \phi}{\partial x} = \frac{\partial \psi}{\partial y} \text{ and } \frac{\partial \phi}{\partial y} = -\frac{\partial \psi}{\partial x}. \qquad ...(3)$$

so $$\frac{\partial \psi}{\partial y} = \frac{(x+a)}{(x+a)^2+y^2} - \frac{x-a}{(x-a)^2+y^2}.$$

Integrating w.r.t. y, we get

$$\psi = \tan{-1}\frac{y}{x+a} - \tan^{-1}\frac{y}{x-a} + f(x)$$

to determine f (x), we have

$$\frac{\partial \psi}{\partial x} = \frac{y}{(x+a)^2+y^2} + \frac{y}{(x-a)^2+y^2} + f'(x)$$

$$= -\frac{\partial \phi}{\partial y} = \frac{y}{(x+a)^2+y^2} + \frac{y}{(x-a)^2+y^2} \text{ from (2).}$$

$\therefore$ f′ (x) = 0, i.e. f (x) = absolute constant and may be omitted.

$$\therefore\ \psi = \tan^{-1}\frac{y}{x+a}\tan^{-1}\frac{y}{x-0}$$

$$= \tan^{-1}\left[\frac{-2ay}{x^2+y^2-a^2}\right].$$

Thus streamlines are given by ψ = const.

when $\psi = 0$ and infinity, the streamlines are

$y = 0$ and $x^2 + y^2 = a^2$.

Now $w = \phi \; i\psi = \frac{1}{2}\log\{(x + a)^2 + y^2\} - \frac{1}{2}\log\{(x - a)^2 + y^2\}$

$$+ i \tan^{-1} x \frac{y}{x+a} - i\tan^{-1}\frac{y}{x-a}$$

$$= \log\{(x + a) + iy\} - \log\{(x - a) + iy\}$$

$= \log(z + a) - \log(z - a)$ where $z = x + iy$.

Speed $= \left|\frac{dw}{dz}\right| = \left|\frac{1}{z+a} - \frac{1}{z-a}\right| = \frac{2a}{|z+a||z-a|}$.

If $|z + a| = r'$ and $|z-a| = r$, then

Speed $= \frac{2a}{rr'}$

and putting speed equal to constant, the curves of equal speed are given by

$\frac{2a}{rr'}$ = constant, *i.e.* rr' = const.

which are Cassini ovals.

Example 12: *In irrotational motion in two dimensions, prove that*

$$\left(\frac{\partial q}{\partial x}\right)^2 + \left(\frac{\partial q}{\partial y}\right)^2 = q\nabla^2 q.$$

Solution: Since the motion is irrotational, the velocity potential ϕ exists and satisfies the equation

$$\frac{\partial^2\phi}{\partial x^2} + \frac{\partial^2\phi}{\partial y^2} = 0.$$

Also $q^2 = \left(\frac{\partial\phi}{\partial x}\right)^2 + \left(\frac{\partial\phi}{\partial y}\right)^2$.

Differentiating it partially w.r.t. x and y respectively, we get

$$q\frac{\partial q}{\partial x} = \frac{\partial\phi}{\partial x}\frac{\partial^2\phi}{\partial x^2}+\frac{\partial\phi}{\partial y}\frac{\partial^2\phi}{\partial x\partial y} \qquad ...(2)$$

$$\text{and } q\frac{\partial q}{\partial y} = \frac{\partial\phi}{\partial x}\frac{\partial^2\phi}{\partial x\partial y}+\frac{\partial\phi}{\partial y}\frac{\partial^2\phi}{\partial y^2}. \qquad ...(3)$$

Differentiating these again partially w.r.t. x and y respectively, we get

$$q\frac{\partial^2 q}{\partial x^2}+\left(\frac{\partial q}{\partial x}\right)^2 = \left(\frac{\partial^2\phi}{\partial x^2}\right)^2+\frac{\partial\phi}{\partial x}\frac{\partial^3\phi}{\partial x^3}+\left(\frac{\partial^2\phi}{\partial x\partial y}\right)^2+\frac{\partial\phi}{\partial y}\cdot\frac{\partial^3\phi}{\partial x^2\partial y} \qquad ...(4)$$

$$q\frac{\partial^2 q}{\partial y^2}+\left(\frac{\partial q}{\partial y}\right)^2 = \left(\frac{\partial^2\phi}{\partial x\partial y}\right)^2+\frac{\partial\phi}{\partial x}\frac{\partial^3\phi}{\partial y^2\partial x}+\left(\frac{\partial^2\phi}{\partial y^2}\right)^2+\frac{\partial\phi}{\partial y}\cdot\frac{\partial^3\phi}{\partial y^3}. \qquad ...(5)$$

Adding (4) and (5), we get

$$q\,\nabla^2 q+\left(\frac{\partial q}{\partial x}\right)^2+\left(\frac{\partial q}{\partial y}\right)^2 = \left(\frac{\partial^2\phi}{\partial x^2}\right)^2+2\left(\frac{\partial^2\phi}{\partial x\partial y}\right)^2+\left(\frac{\partial^2\phi}{\partial y^2}\right)^2$$

$$+\frac{\partial\phi}{\partial x}\frac{\partial}{\partial x}\left\{\frac{\partial^2\phi}{\partial x^2}+\frac{\partial^2\phi}{\partial y^2}\right\}+\frac{\partial\phi}{\partial y}\frac{\partial}{\partial y}\left\{\frac{\partial^2\phi}{\partial x^2}+\frac{\partial^2\phi}{\partial y^2}\right\}$$

$$=\left(\frac{\partial^2\phi}{\partial x^2}\right)^2+\left(\frac{\partial^2\phi}{\partial x\partial y}\right)^2+\left(\frac{\partial^2\phi}{\partial y^2}\right)^2 \quad \text{from (1)}$$

$$=2\left(\frac{\partial^2\phi}{\partial x^2}\right)^2+\left(\frac{\partial^2\phi}{\partial x\partial y}\right)^2 \text{ as } \frac{\partial^2\phi}{\partial x^2}=-\frac{\partial^2\phi}{\partial y^2}. \qquad ...(6)$$

Also squaring and adding (2) and (3), we get

$$q^2\left[\left(\frac{\partial q}{\partial x}\right)^2+\left(\frac{\partial q}{\partial y}\right)^2\right] = \left(\frac{\partial\phi}{\partial x}\right)^2\left[\left(\frac{\partial^2\phi}{\partial x^2}\right)^2+\left(\frac{\partial^2\phi}{\partial x\partial y}\right)^2\right]$$

$$+\left(\frac{\partial\phi}{\partial y}\right)^2\left[\left(\frac{\partial^2\phi}{\partial x\partial y}\right)^2+\left(\frac{\partial^2\phi}{\partial y^2}\right)^2\right]+2\frac{\partial\phi}{\partial x}\frac{\partial\phi}{\partial y}\frac{\partial^2\phi}{\partial x\partial y}\left[\frac{\partial^2\phi}{\partial x^2}+\frac{\partial^2\phi}{\partial y^2}\right]$$

$$=\left[\left(\frac{\partial\phi}{\partial x}\right)^2+\left(\frac{\partial\phi}{\partial y}\right)^2\right]\left[\left(\frac{\partial^2\phi}{\partial x^2}\right)^2+\left(\frac{\partial^2\phi}{\partial x\partial y}\right)^2\right] \qquad \text{using (1)}$$

$$= q^2\left[\left(\frac{\partial^2\phi}{\partial x^2}\right)^2 + \left(\frac{\partial^2\phi}{\partial x\partial y}\right)^2\right],$$

$$i.e. \left(\frac{\partial^2\phi}{\partial x^2}\right)^2 + \left(\frac{\partial^2\phi}{\partial x\partial y}\right)^2 = \left(\frac{\partial q}{\partial x}\right)^2 + \left(\frac{\partial q}{\partial y}\right)^2.$$

Putting this in (6), we get

$$q\nabla^2 q + \left(\frac{\partial q}{\partial x}\right)^2 + \left(\frac{\partial q}{\partial y}\right)^2 = 2\left[\left(\frac{\partial q}{\partial x}\right)^2 + \left(\frac{\partial q}{\partial y}\right)^2\right]$$

$$\text{or } q\nabla^2 q = \left(\frac{\partial q}{\partial x}\right)^2 + \left(\frac{\partial q}{\partial y}\right)^2.$$

This proves the result.

Example 13: *If the fluid fill the region of space on the positive side of x-axis, which is a rigid boundary, and if there be a source m at the point (0, a) and an equal sink at (0, b), and if the pressure on the negatives side of the boundary be the same as the pressure of the fluid at infinity, show that the resultant pressure on the boundary y is* $\pi\rho m^2 (a-b)^2/ab\,(a+b)$, *where* ρ *is the density of the fluid.*

Solution: The equivalent image system consists of

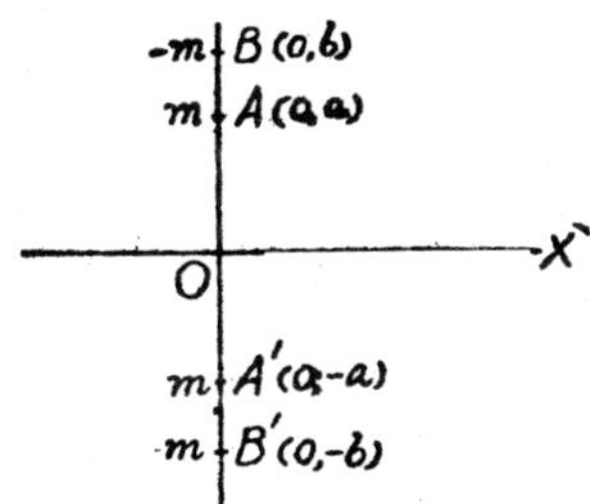

Fig. 3.20

(i) a source + m at A (0, a), i.e., at z = ai,

(ii) a sink – m at B (0, b), *i.e.*, at z = bi,

(iii) a source + m at A′ (0, – a), i.e., at z = – ai.

(iv) a sink – m at B′ (0, – b), i.e., at z = –bi.

Therefore the complex potential is given by

$w = -m \log(z - ai) - m \log(z + ai) + m \log(z - bi) + m \log(z + bi)$

$= -m \log(z^2 + a^2) + m \log(z^2 + b^2).$

$$\therefore \text{ Speed } = \left|\frac{dw}{dz}\right| = m\left|\frac{2z}{z^2 + a^2} - \frac{2z}{z^2 + b^2}\right|.$$

Let us consider a point (x, 0) on x-axis; if q is the speed at this point, then putting z = x,

$$q = m\left|\frac{2x(a^2 - b^2)}{(x^2 + a^2)(x^2 + b^2)}\right| = \frac{2xm(a^2 - b^2)}{(x^2 - a^2)(x^2 + b^2)}.$$

If p_0 is the pressure at infinity, then by Beronoulli's theorem, the pressure p at any point is given by

$$\frac{p}{p} = \frac{p_0}{p} - \frac{1}{2}q^2, \textit{ i.e.,} \frac{p_0 - p}{\rho} = \frac{1}{2}q^2.$$

Therefore the resultant pressure on the boundary

$$= \int_{-\infty}^{\infty} (p_0 - p)dx = \frac{1}{2}\rho \int_{-\infty}^{\infty} q^2 dx$$

$$= \frac{1}{2}\rho.4m^2(a^2 - b^2)^2 \int_{-\infty}^{\infty} \frac{x^2 dx}{(x^2 + a^2)^2 (x^2 + b^2)^2}$$

$$= 4m^2\rho \int_{-\infty}^{\infty} \left[\frac{(a^2 + b^2)}{b^2 - a^2}\left(\frac{1}{x^2 + a^2} - \frac{1}{x^2 + b^2}\right) - \frac{a^2}{(x^2 + a^2)} - \frac{b^2}{(x^2 + b^2)^2}\right] dx$$

resolving into partial fractions

$$= 4m^2\rho\left[\frac{a^2 + b^2}{b^2 - a^2}\left(\frac{\pi}{2a} - \frac{\pi}{2b}\right) - \frac{\pi}{4a} - \frac{\pi}{4b}\right]$$

$$= \pi \rho m^2 \frac{(a - b)^2}{ab(a + b)}.$$ This proves the result.

Example 14: *A single source is placed in an infinite perfectly elastic fluid, which is also a perfect conductor of heat ; shew that if the motion be steady, the velocity V at a distance r from the source satisfies the equation*

$$\left(V - \frac{k}{V}\right)\frac{\partial V}{\partial r} = \frac{2k}{r};$$

and hence that $r = \frac{1}{\sqrt{V}} e^{V^{2/4k}}.$

Solution: The fluid will obey Boyle's law as the liquid extends to infinity and is perfectly elastic, there being hardly any change in temperature.

$\therefore$ $p = k\rho$.

Again since the motion is steady and is due to a single source, the flow is readial. The equation of continuity is

$$\frac{\partial}{\partial r}(\rho r^2 V) = 0.$$

or $Vr^2 \frac{\partial \rho}{\partial r} + \rho\left(r^2 \frac{\partial V}{\partial r} + 2rV\right) = 0$ ...(2)

and the equation of motion is

$$V\frac{\partial V}{\partial r} = \frac{1}{\rho}\frac{\partial p}{\partial r},$$

i.e., $V\frac{\partial V}{\partial r} + \frac{K}{\rho}\frac{\partial \rho}{\partial r} = 0$ by (1). ...(3)

Eliminating $\frac{\partial \rho}{\partial r}$ from (2) and (3), we get

$$Vr^2\left(-\frac{\rho V}{K}\frac{\partial V}{\partial r}\right) + \rho\left(r^2\frac{\partial V}{\partial r} + 2rV\right) = 0$$

or $\frac{\partial V}{\partial r}.(k - V^2) + 2k\frac{V}{r} = 0$

or $\left(V - \frac{k}{V}\right)\frac{\partial V}{\partial r} = \frac{2k}{r}.$

This proves the first result.

The above equation after separating the variables can be written as

$$V\,dV - \frac{k}{V}dV = \frac{2k}{r}dr.$$

Integrating $\frac{1}{2}V^2 - k\log V = 2k\log r + \log C'$

or $\frac{V^2}{2k} = \log(Vr^2C)$

or $Vr^2C = eV^2/2k,$

where $\log C' = k\log C$

or $r = \frac{1}{\sqrt{(VC)}} eV^2/4k$

Giving C the particular value 1, this becomes

$$r = \frac{1}{\sqrt{V}} eV^2/4k.$$

This proves the second result.

Example 15(a): *An infinite mass of liquid is moving irrotationally and steadily under the influence of a source of strength μ and an equal sink at a distance 2a from it. Prove that the kinetic energy of the liquid which passes in unit time across the plane which bisects at right angles the line joining the source and sink is $\frac{8}{7}\pi\rho\mu^3/a^4$, ρ being the density of the liquid.*

Solution: Let A and B be two points each on either side of yz-plane at a distance a from it.

Take a point P on the yz-plane at a distance r from O.

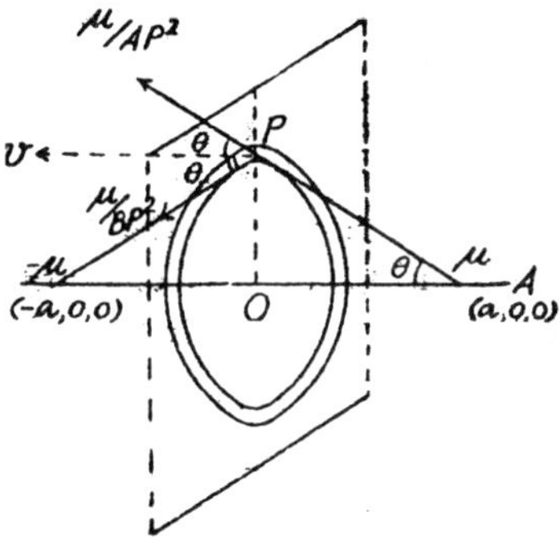

Fig. 3.21

The velocity at P due to source μ at A

$$= \frac{\mu}{AP^2}.$$

The velocity at P due to sink $-\mu$ at B $= \frac{\mu}{BP^2}$ along PB.

If $\angle PAO = \theta$ (also we have AP = BP),

$\therefore$ velocity at P,

$$v = 2\frac{\mu}{AP^2}\cos\theta \text{ parallel to AB}$$

$$= \frac{2\mu}{a^2+r^2}\frac{a}{\sqrt{(r^2+a^2)}}$$

as OP = r, OA = a.

Consider an elementary ring of radius r and breadth dr.

The amount of liquid passing through this circular strip

$= 2\pi r\,\delta r\,\rho v$ per unit time.

The K. E. of the liquid passing through the yz-plane in unit time

$$= \frac{1}{2}\int_0^\infty (2\pi r\,\delta r\,\rho v)v^2$$

$$= \pi\rho\int_0^\infty r.\frac{8\mu^3a^3}{(a^2+r^2)^{9/2}}dr$$

$$= 8\,\pi\rho\mu^3a^3\int_0^{\pi/2}\frac{1}{a^2}\cos^6\theta\sin\theta\,d\theta, \text{ where } a\tan\theta = r$$

$$= \frac{8}{7}\frac{\pi\rho\mu^3}{a^4}.$$ This proves the result.

Example 15(b): *A source and a sink, each of strength m, exist in an infinitive liquid on opposite sides of, and at equal distances c from, the centre of a rigid sphere of radius a. Show that the velocity potential V may be expressed in the form.*

$$V = \frac{2\mu}{c}\sum_{n=0}^{n=\infty}\left\{\left(\frac{r}{c}\right)^{2n+1} + \frac{2n+1}{2n+2}.\frac{c}{a}.\left(\frac{a^2}{rc}\right)^{2n+2}\right\}P_{2n+1}(\cos\theta),$$

θ being the vectorial angle measured from the diameter of the sphere on which the source and sink lie, and $r < c$; and find an expression for V when $r < c$.

Solution: If ϕ_1 be the velocity potential due to source μ at A and sink $-\mu$ at B when the sphere is not present and ϕ_2 the contribution to the contribution to the velocity potential due to the presence of the sphere, then

$$V = \phi_1 + \phi_2.$$

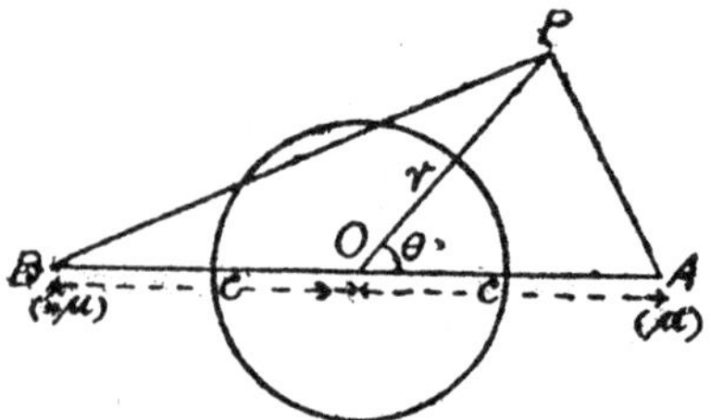

Fig. 3.22

Now $\phi_1 = \dfrac{\mu}{Ap} - \dfrac{\mu}{BP}$ ...(1)

$= \mu\,(r^2 + c^2 - 2rc\cos\theta)^{-1/2} - \mu\,(r^2 + c^2 + 2rc\cos\theta)^{-1/2}$

$$= \frac{\mu}{c}\left[\left(1 - \frac{2r}{c}\cos\theta + \frac{r^2}{c^2}\right)^{-1/2} - \left(1 + \frac{2r}{c}\cos\theta\frac{r^2}{c^2}\right)^{-1/2}\right]$$

$$= \left[\left\{1 + \sum_1^{\infty}\frac{r^n}{c^n}P_n(\cos\theta)\right\} - \left\{1 + \sum_1^{\infty}\left(-\frac{r}{c}\right)^n P_n(\cos\theta)\right\}\right]$$

(for r < c)

$$= \frac{2\mu}{c}\sum_0^{\infty}\left(\frac{r}{c}\right)^{2n+1} P_{2n+1}(\cos\theta)$$

Again let $\sum_{n=0}^{\infty}\dfrac{A_n}{r^{n+1}}P_n(\cos\theta)$,

this form being suitable, so that f2 vanishes at infinity and satisfies Laplace's equation.

The constants A_n etc. are to be determined with the condition that velocity normal to the sphere is zero.

i.e., $\dfrac{\partial V}{\partial r} = 0$ when r = a

or $\left(\dfrac{\partial\phi_1}{\partial r} + \dfrac{\partial\phi_2}{\partial r}\right)_{r=a} = 0,$

$$\frac{2\mu}{c}\sum_0^{\infty}\frac{(2n+1)a^{2n}}{c^{2n+1}}P_{2n+1}(\cos\theta) - \sum_{n=0}^{\infty}\frac{A_n(n+1)}{a^{n+2}}P_n(\cos\theta) = 0.$$

Equating to zero, coefficients of various Pn's, we get

$A_{2n} = 0$ for all values of n

$$\text{and } A_{2n+1}\frac{(2n+2)}{a^{2n+3}} - \frac{2\mu}{c}\frac{(2n+1)a^{2n}}{c^{2n+1}} = 0,$$

$$i.e.,\ A_{2n+1} = \frac{\mu(2n+1)a^{4n+3}}{c^{2n+2}(n+1)} = \frac{\mu}{a}\cdot\frac{2n+1}{n+1}\left(\frac{a^2}{c}\right)^{2n+2}.$$

$$\therefore\ \phi_2 = \sum_0^{\infty}\frac{\mu}{a}\cdot\frac{2n+1}{n+1}\left(\frac{a^2}{rc}\right)^{2n+2} P_{2n+1}(\cos\theta).$$

$$\therefore\ V = \phi_1 + \phi_2 = \frac{2\mu}{c}\sum_{n=0}^{\infty}\left[\left(\frac{r}{c}\right)^{2n+1} + \frac{2n+1}{2n+2}\cdot\frac{c}{a}\left(\frac{a^2}{rc}\right)^{2n+2}\right]P_{2n+1}(\cos\theta).$$

This proves the result.

Second Part : When r > c, we take r common in ϕ_1 instead of c ; thus in this case

$$\phi_1 = \frac{2\mu}{r}\sum_0^{\infty}\left(\frac{c}{r}\right)^{2n+1} P_{2n+1}(\cos\theta)$$

$$\text{and } \phi_2 = \sum_0^{\infty}\frac{B_n}{r^{n+1}}P_n(\cos\theta), \text{ as before;}$$

the value of B_n's can be determined as before.

Example 15(c): *The space on one side of an infinite plane wall, y = 0, is filled with inviscid, incompressible fluid, moving at infinity with velocity U in the direction of the axis of x. The motion of the fluid is wholly two-dimensional, in the (x, y) plane. A doublet of strength μ is at a distance of a from the wall, and points in the negative direction of the axis of x. Shew that if μ is less than $4a^2U$, the pressure of the fluid on the wall is a maximum at points distant $\sqrt{3}a$ from O, the foot of the perpendicular from the doublet on the wall, and is a minimum at O.*

If μ is equal to $4a^2U$, find the points where the velocity of the fluid is zero, and shew that the stream lines include the circle

$$x^2 + (y-a)^2 = 4a^2,$$

where the origin is taken at O.

Solution: Since the doublet points in the negative direction of x-axis, it makes an angle π with real axis, the image of it being an equal doublet on the other side of real axis.

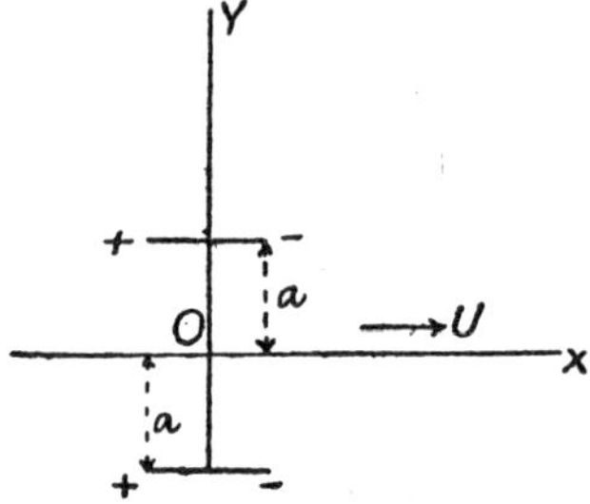

Fig. 3.23

Hence the complex potential is given by

$$w = \frac{\mu_e i\pi}{z - ai} + \frac{\mu_e i\pi}{z + ai} - Uz$$

$$= -\frac{2\mu z}{z^2 a + a^2} - Uz.$$

$$\text{Velocity} = \left|\frac{dw}{dz}\right| = \left|2\mu\frac{a^2 - z^2}{(z^2 + a^2)^2} + U\right|.$$

If q is the velocity at a point (x, 0) on x-axis,

$$q = \left[2\mu\frac{a^2 - x^2}{(x^2 + a^2)^2} + U\right].$$

Now by Bernoulli's theorem, we have

$$\frac{p}{\rho} + \frac{1}{2}q^2 = \frac{\Pi}{\rho} + \frac{1}{2}U^2,$$

i.e.,
$$\frac{\Pi - p}{\rho} = \frac{1}{2}(q^2 - U^2) = \frac{1}{2}\left[2\mu\frac{a^2 - x^2}{(x^2 + a^2)} + U\right]^2 - \frac{1}{2}U^2$$

$$= \frac{2\mu^2(a^2 + x^2)^2}{(x^2 + a^2)^4} + 2\mu U\frac{a^2 - x^2}{(x^2 + a^2)^2} \qquad ...(2)$$

For pressure to be maximum or minimum,

$$\frac{dp}{dx} = 0,$$

$$i.e., \; 2\mu^2\left[\frac{4x(a^2 - x^2)}{(x^2 - a^2)^4} + \frac{8x(a^2 - x^2)^2}{(x^2 + a^2)^5}\right]$$

$$+ 2\mu U\left[\frac{2x}{(x^2 - a^2)^2} + \frac{4x(a^2 - x^2)^2}{(x^2 + a^2)^3}\right] = 0$$

or $4\mu x \dfrac{(3a^2 - x^2)}{(x^2 + a^2)^5}$ [2μ $(a^2 - x^2) + U$ $(a^2 + x^2) = 0$.

This gives $x = 0$ and $3a^2 - x^2 = 0$, i.e. $x = \pm \sqrt{3a}$,

$$\text{Again} \quad \left[\frac{dw}{dz}\right]_{x = \pm 3a} = \left[U - 2\mu\frac{2a^2}{16a^4}\right] \text{from (1)}$$

$$= U - \frac{\mu}{4a^2}.$$

Thus if $\mu < 4a^2U$, $\dfrac{d^2p}{dx^2}$ is negative at $x = \sqrt{3a}$,

so that the pressure at the wall is maximum.

Also $\dfrac{d^2p}{dx^2}$ is + ve, when $x = 0$; so there the pressure is minimum.

But if $\mu = 4a^2U$, we get $\dfrac{dw}{dz} = 0$ and then (1) gives

$$8a^2U\frac{a^2 - z^2}{(a^2 + z^2)^2} + U = 0,$$

i.e. $z^4 - 6a^2z^2 + 9a^4 = 0$ or $(z^2 - 3a^2)^2 = 0$

or $z = \pm \sqrt{3a}$.

Thus $(\sqrt{3a}\,,0)$ and $(-\sqrt{3a}\,, 0)$ are the stagnation points.

We have, as already found, $w = -\dfrac{2\mu z}{z^2 + a^2} - Uz$.

Equating imaginary parts,

$$\psi = \frac{2\mu(x^2 + y^2 - a^2)}{(x^2 + y^2)^2 + 2a^2(x^2 - y^2) + a^4} - Uy.$$

The stream lines when $\psi = 0$, $\mu = 4a^2U$ are given by

$(x^2 + y^2)^2 - 6a^2x^2 - 10a^2y^2 + 9a^4 = 0,$

i.e., $(x^2 + y^2 - 2ay - 3a^2)(x^2 + y^2 + 2ay - 3a^2) = 0,$

which includes $x^2 + y^2 - 2ay = 3a^2,$

i.e., $x^2 + (y - a)^2 = 4a^2.$

This proves the result.

Example 16: *Parallel line -sources (perpendicular to the xy-plane) of equal strength m are place at the points z = nia, where n = ...–2, –1, 0, 1, 2, 3,...; prove that the complex potential is* $w = -m \log \sin h \dfrac{\pi z}{a}$.

Hence show that the complex potential for two-dimensional doublets (line doublets), with their axes parallel to the x-axis, of strength μ at the some points, is given by

$$w = \mu \coth\left(\frac{\pi z}{a}\right).$$

Solution: Hence we have source at points

$z = 0;\ ai, -ai;\ 2ai, -2ai;\ 3ai, -3ai;$ ...etc.

Hence the complex potential of the system at any point z is given by

$$w = -m \log z - m \log (z - ai) - m \log (z + ai)$$

$$- m \log (z - 2ai) - m\log (z + 2ai)$$

$$- m \log (z - 2ai) - m\log (z + 3ai)$$

$$= - m \log z - m \log (z^2 + a^2) - m \log (z^2 + 2^2 a^2) + \ldots$$

$$= - m\log \left\{\frac{\pi z}{a}\left(1 + \frac{z^2}{a^2}\right)\left(1 + \frac{z^2}{2^2 a^2}\right)\right\}$$

$$- m \log \left\{\frac{a}{\pi} a^2 (2^2 a^2)(3^2 a^2)\ldots\right\}$$

$$= - m \log \sin h \frac{\pi z}{a} + \text{const.}$$

$$= - m \log \sin h \frac{\pi z}{a}, \text{ leaving the constant} \qquad \ldots(1)$$

Since $\sin h\, x = x\left(1 + \dfrac{x^2}{\pi^2}\right)\left(1 + \dfrac{x^2}{2^2 \pi^2}\right)\left(1 + \dfrac{x^2}{3^2 \pi^2}\right) - \ldots$

This proves the first part of the result.

The complex potential due to doublets at these points can be obtained by differentiating (1) which is therfore given by

$$w = -\frac{m\pi}{a}\coth\left(\frac{\pi z}{a}\right)$$

$$= \mu \coth\left(\frac{\pi z}{a}\right) \text{numerically.}$$

This proves the result.

Example 17: *λ denoting a variable parameter, and f a given function, find the condition that f (x, y, l) = 0 should be a possible system of stream lines for steady irrotational motion in two dimension.*

Solution : We know that stream lines are given by

$$\psi = \text{const.} \quad \text{...(1)}$$

If $$f(x, y, l) = 0 \quad \text{...(2)}$$

is a system of stream lines for different values of λ, then for $\lambda = \lambda_1$, (2) must represent a stream line which must be given by (1), also for some value of the constant say c_1. Thus ψ is a function of λ alone (λ of course depending on x and y),

$$\text{i.e. } y = y(\lambda) \quad \text{...(3)}$$

Now for irrotational steady motion, ψ satisfies the equation

$$\frac{\partial^2\psi}{\partial x^2} + \frac{\partial^2\psi}{\partial y^2} = 0. \quad \text{...(4)}$$

The required condition will be obtained with the help of (3) and (4). We have

$$\frac{\partial\psi}{\partial x} = \frac{d\psi}{d\lambda}\cdot\frac{\partial\lambda}{\partial y}.$$

$$\text{Again } \frac{\partial^2\psi}{\partial x^2} = \frac{\partial}{\partial x}\left[\frac{\partial\psi}{\partial x}\right] = \frac{\partial}{\partial x}\left[\frac{d\psi}{d\lambda}\cdot\frac{\partial\lambda}{\partial x}\right]$$

$$= \frac{d^2\psi}{d\lambda^2}\cdot\left(\frac{\partial\lambda}{\partial x}\right)^2 + \frac{d\psi}{d\lambda}\cdot\frac{\partial^2\lambda}{\partial x^2}$$

$$\text{and } \frac{\partial^2\psi}{\partial y^2} = \frac{\partial}{\partial y}\left[\frac{d\psi}{d\lambda}\cdot\frac{\partial\lambda}{\partial y}\right]$$

$$= \frac{d^2\psi}{d\lambda^2} \cdot \left(\frac{\partial\lambda}{\partial y}\right)^2 + \frac{d\psi}{d\lambda} \cdot \frac{\partial^2\lambda}{\partial y^2}.$$

Putting these in (4), we get

$$\frac{d^2\psi}{d\lambda^2} \cdot \left(\frac{\partial\lambda}{\partial x}\right)^2 + \left(\frac{\partial\lambda}{\partial y}\right)^2 + \frac{d\psi}{d\lambda} \cdot \left[\frac{\partial^2\lambda}{\partial x^2} + \frac{\partial^2\lambda}{\partial y^2}\right] = 0,$$

which is the required condition.

Example 18: *In two-dimensional irrotational fluid motion show that, if the stream lines are confocal ellipses,*

$$\frac{x^2}{a^2+\lambda} + \frac{y^2}{b^2+\lambda} = 1,$$

$$\psi = A \log \left[\sqrt{(a^2+\lambda)} + \sqrt{(b^2+\lambda)}\right] + B,$$

and the velocity at any point is inversely proportional to the square root of the rectangle under the focal radii of the point.

Solution: In example 13, we have shown that if f (x,y,λ) = 0 is a possible stream line, ψ is a function of λ alone and we have the condition

$$\frac{\left[\frac{\partial^2\lambda}{\partial x^2} + \frac{\partial^2\lambda}{\partial y^2}\right]}{\left(\frac{\partial\lambda}{\partial x}\right)^2 + \left(\frac{\partial\lambda}{\partial y}\right)^2} = \frac{\frac{d^2\psi}{d\lambda^2}}{\frac{d\psi}{d\lambda}}. \quad \text{....(1)}$$

Here f (x, y, λ) = 0 is

$$\frac{x^2}{a^2+\lambda} + \frac{y^2}{b^2+\lambda} - 1 = 0.$$

Differentiating it partially w.r.t. x, we get

$$\frac{2x}{a^2+\lambda} - \left[\frac{x^2}{(a^2+\lambda)^2} + \frac{y^2}{(b^2+\lambda)^2}\right]\frac{\partial\lambda}{\partial x} = 0, \quad \text{...(2)}$$

$$\text{i.e. } 2\xi - [\xi^2 + \eta^2]\frac{\partial\lambda}{\partial x} = 0,$$

$$\text{where } \xi = \frac{x}{a^2+\lambda}, \eta = \frac{y}{b^2+\lambda}$$

or $\quad \dfrac{\partial\lambda}{\partial x} = \dfrac{2\xi}{\xi^2+\eta^2}$ ...(3)

Similarly, $\dfrac{\partial\lambda}{\partial y} = \dfrac{2\eta}{\xi^2+\eta^2}$.

Again differentiating (2), w.r.t.x, we get

$$\frac{2}{a^2+\lambda} - \frac{4x}{(a^2+\lambda)^2}\left(\frac{\partial\lambda}{\partial x}\right) + 2\left[\frac{x^2}{(a^2+\lambda)^3} + \frac{y^2}{(b^2+\lambda)^3}\right]\left(\frac{\partial\lambda}{\partial x}\right)^2$$

$$= \frac{\partial^2\lambda}{\partial x^2}[\xi^2+\eta^2] \quad \text{i.e.,} \quad \frac{\partial^2\lambda}{\partial x^2}(\xi^2+\eta^2)$$

$$= \frac{2}{a^2+\lambda} - \frac{4x}{(a^2+\lambda)^2}\left(\frac{\partial\lambda}{\partial x}\right) + 2\left[\frac{\xi^2}{a^2+\lambda} + \frac{\eta^2}{b^2+\lambda}\right]\left(\frac{\partial\lambda}{\partial x}\right)^2.$$

Similarly differentiating twice w.r.t. y, we get

$$\frac{\partial^2\lambda}{\partial y^2}(\xi^2+\eta^2)$$

$$= \frac{2}{b^2+\lambda} - \frac{4y}{(b^2+\lambda)^2}\left(\frac{\partial\lambda}{\partial y}\right) + 2\left[\frac{\xi^2}{a^2+\lambda} + \frac{\eta^2}{b^2+\lambda}\right]\left(\frac{\partial\lambda}{\partial y}\right)^2.$$

Adding these, we get

$$\left(\frac{\partial^2\lambda}{\partial x^2} + \frac{\partial^2\lambda}{\partial y^2}\right)(\xi^2+\eta^2) = \frac{2}{a^2+\lambda} - \frac{2}{b^2+\lambda}$$

$$-4\left[\frac{x}{(a^2+\lambda)^2}\frac{\partial\lambda}{\partial x} + \frac{y}{(b^2+\lambda)^2}\frac{\partial\lambda}{\partial y}\right]$$

$$+2\left[\frac{\xi^2}{a^2+\lambda} + \frac{\eta^2}{b^2+\lambda}\right]\left[\left(\frac{\partial\lambda}{\partial x}\right)^2 + \left(\frac{\partial\lambda}{\partial y}\right)^2\right]$$

$$= \frac{2}{a^2+\lambda} + \frac{2}{b^2+\lambda};$$

other terms cancel when we put values of $\dfrac{\partial\lambda}{\partial x}\,\dfrac{\partial\lambda}{\partial y}$ and from (3).

Thus $\left(\dfrac{\partial^2\lambda}{\partial x^2} + \dfrac{\partial^2\lambda}{\partial y^2}\right) = \dfrac{2}{\xi^2+\eta^2}\left[\dfrac{1}{a^2+\lambda} + \dfrac{1}{b^2+\lambda}\right]$.

Also from (3), $\left(\frac{\partial\lambda}{\partial x}\right)^2+\left(\frac{\partial\lambda}{\partial y}\right)^2=\frac{4}{\xi^2+\eta^2}$.

Thus, (1) gives $\frac{1}{2}\left[\frac{1}{a^2+\lambda}+\frac{1}{b^2+\lambda}\right]=-\frac{\psi''(\lambda)}{\psi'(\lambda)}$.

Integrating $.\frac{1}{2}[\log(a^2+\lambda)+\log(b^2+\lambda)]-\log A=-\log\psi'(\lambda)$

or $\psi'(\lambda)=\frac{A}{\sqrt{[(a^2+\lambda)(b^2+\lambda)]}}$ ...(4)

Integrating again,

$$\psi=A,\int\frac{1}{\sqrt{\{(a^2+\lambda)(b^2+\lambda)}}d\lambda+B$$

$$=A\log\left[\sqrt{(a^2+\lambda)}+\sqrt{(b^2+\lambda)}\right]+B.$$

This proves the result

Second part : If q is the velocity, then

$$q2=\left(\frac{\partial\psi}{\partial x}\right)^2+\left(\frac{\partial\psi}{\partial y}\right)^2$$

$$=\left(\frac{d\psi\partial\psi}{d\lambda\partial x}\right)^2+\left(\frac{d\psi}{d\lambda}.\frac{\partial\psi}{\partial y}\right)^2$$

$$=[\psi'(\lambda)]^2\left[\left(\frac{\partial\lambda}{\partial x}\right)^2+\left(\frac{\partial\lambda}{\partial y}\right)^2\right]$$

$$=\frac{4A^2}{(a^2+\lambda)(b^2+\lambda)}\left[\frac{1}{\frac{x^2}{(a^2+\lambda)^2}+\frac{y^2}{(b^2+\lambda)^2}}\right]$$

$$=\frac{4A^2}{a^2+\lambda}\left[\frac{1}{\frac{(b^2+\lambda)}{(a^2+\lambda)^2}\left(1-\frac{x^2}{a^2+\lambda}\right)}\right]$$

from the given equation

$$= \frac{4A^2}{(a^2+\lambda) - \frac{x^2(a^2-b^2)}{a^2+\lambda}}.$$

But from the property of the ellipse,

$a^2 - b^2 = (a^2 + \lambda) - (b^2 + \lambda) = (a^2 + \lambda)\ e^2.$

$$\therefore\ q^2 = \frac{4A^2}{(a^2+\lambda)-e^2x^2} = \frac{4A^2}{[\sqrt{(a^2+\lambda)}-ex][\sqrt{(a^2+\lambda)}+ex]}$$

$$= \frac{4A^2}{r_1 r_2},$$

where $r_1 = \sqrt{(a^2+\lambda)-ex}$ and $r2 = \sqrt{(a^2+\lambda)-ex}$ are the focal distances of the point.

Thus $\quad q = \dfrac{2A}{\sqrt{(r_1)}\sqrt{(r_2)}}$

in inversely proportional to square - root of rectangle formed of focal radii to the point.

Example 19: *Find the lines of flow in the two dimensional fluid motion given by*

$\phi + i\psi = -\ (1/2)\ n\ (x + iy)^2 e^{2int}.$

Prove or verify that the paths of the particles of the fluid (in polar coordinates) may be obtained by eliminating t from the equations.

$r \cos (nt + \theta) - x_0 = r \sin (nt + \theta) - y_0 = nt\ (x_0 - y_0)$

Solution: Since $\phi + i\psi = -\ \frac{1}{2} n\ (x + iy)^2\ e^{2int}$ ...(1)

Substituting $x = r \cos \theta$ and $y = r \sin \theta$, we have

$$\phi + i\psi = -\ \frac{1}{2} n\ (r \cos \theta + ir \sin \theta)^2\ e^{2int}$$

$$\phi + i\psi = -\frac{1}{2} nr^2\ e^{2i\theta}\ e^{2int} = -\ \frac{1}{2} nr^2\ e^{2i}(\theta + nt)$$

$$\phi + i\psi = -\frac{1}{2} nr^2\ \{\cos 2\ (\theta + nt) + i \sin 2\ (\theta + nt)\ \}.$$

Separating into real and imaginary parts, we get

$\phi = -\ \dfrac{1}{2}nr^2 \cos 2\ (\theta + nt),\ \psi = -\dfrac{1}{2}nr^2 \sin 2\ (\theta + nt).$

The lines of flow are given by y = const. i.e.,

$r^2 \sin 2\ (\theta + nt) = \text{const.}$

Now, we shall determine the path of the particles of fluid

$$\frac{\partial r}{\partial t} = \frac{\partial \phi}{\partial r} = nr \cos (2\theta + 2nt) = nr \cos 2\lambda,\ \lambda = \theta + nt \qquad ...(2)$$

$$\text{and } r\frac{\partial \theta}{\partial t} = -\ \frac{1}{r}\frac{\partial \phi}{\partial \theta} = -\ nr \sin (2\theta + 2nt) = -\ nr \sin 2\lambda. \qquad ...(3)$$

From (2), we have

$$nr \cos 2\lambda = \frac{dr}{dt} = \frac{dr}{d\lambda}\frac{d\lambda}{dt} = \frac{dr}{d\lambda}\left(\frac{d\theta}{dt} + n\right)$$

$$\text{or } nr \cos 2\lambda = \frac{dr}{d\lambda}\ (n - n \sin 2\lambda)$$

$$\text{or } \frac{2dr}{r} = \frac{2 \cos 2\lambda}{1 - \sin 2\lambda} d\lambda$$

By integrating, we have

$2 \log r + \log (1 - \sin 2\lambda, = \log C$

or $r^2 (1 - \sin 2\lambda\ ) = C$

or $r (\cos \lambda - \sin \lambda) = D,$ ...(4)

where C and D are arbitrary constants.

At $t = 0;\ \lambda = \theta = \theta_1$ (initially), $r\ r_0$, so $D = x_0 - y_0$. ...(5)

Equation (4) reduces to

$r (\cos \lambda - \sin \lambda) = x_0 - y_0$

or $r \cos (\theta + nt) - x_0 = r \sin (\theta + nt) - y_0$ ...(6)

$$\text{Again } \frac{d\lambda}{dt} = n - n \sin 2\lambda$$

$$\text{or } \int \frac{d\lambda}{1 - \sin 2\lambda} = n \int dt$$

$$\text{or } \int \frac{d\lambda}{(\cos \lambda - \sin \lambda)^2} = n \int dt$$

$$\text{or } \int \frac{\sec^2 \lambda d\lambda}{(1 - \tan \lambda)^2} = n \int dt$$

or $\dfrac{1}{1-\tan\lambda} = nt + A.$ ...(7)

or $\dfrac{\cos\lambda}{\cos\lambda - \sin\lambda} = nt + A$

Using (5), the value of the constant becomes

$$A = \frac{\cos\theta_0}{\cos\theta_0 - \sin\theta_0} = \frac{x_0}{x_0 - y_0}$$

$$\Rightarrow \frac{r\cos\lambda}{r\cos\lambda - r\sin\lambda} = nt + \frac{x_0}{x_0 - y_0}$$

$$\Rightarrow r\cos\lambda = nt\,(x_0 - y_0) + x_0\,. \qquad ...(8)$$

From (6) and (8), we have

$r\cos(\theta + nt) - x_0 = r\sin(\theta + nt) - y_0 = nt\,(x_0 - y_0)$. **Proved.**

Example 20(a): *In the case of motion of liquid in a part of a place bounded by a straight line due to source in the plane, prove that if mρ is the mass of fluid (of density ρ) generated at the source per unit of time the pressure on the length 2l of the boundary immediately opposite to the source is less than that on an equal length at a great distance by*

$$\frac{1}{2}\frac{m^2\rho}{\pi^2}\left[\frac{1}{c}\tan^{-1}\frac{1}{c} - \frac{1}{1^2 + c^2}\right],$$

where c is the distance of source to the boundary.

Solution: Let Y-axis be the bounding line and a source of strength $(m/2\pi)$ is placed at S (c, 0). The image system consists of :

(i) a source of strength $(m/2\pi)$ at S (c, 0).

(ii) a source of strength $(m/2\pi)$ at S (– c, 0).

The complex potential w is given by

$$w = -\frac{m}{2\pi}\log(z - c) - \frac{m}{2\pi}\log(z + c)$$

$$= -\frac{m}{2\pi}\log(z^2 + c^2)$$

The velocity q is given by

$$q = \left|\frac{dw}{dz}\right| = \frac{m}{2\pi}\left|\frac{2z}{z^2 - c^2}\right|$$

Velocity at the point P (i.e., z = iy) becomes

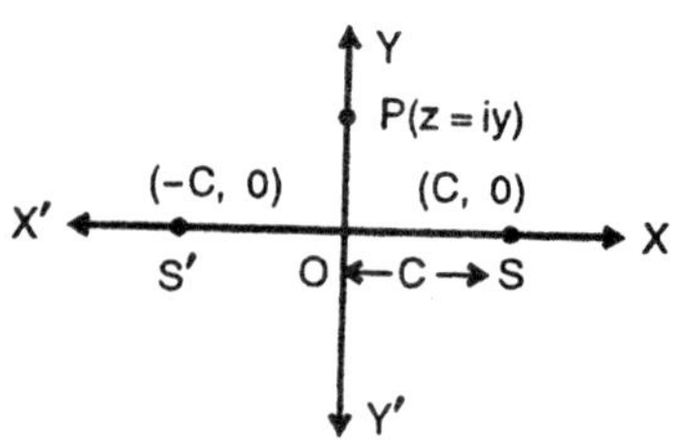

Fig. 3.24

$$= \frac{m}{2\pi}\left|\frac{2iy}{-y^2 - c^2}\right| = \frac{m}{\pi}\frac{y}{y^2 + c^2}.$$

Let p_0 be the pressure on Y-axis

at a great distance from O then by Bernoulli's theorem, we have

$$\frac{p_0 - p}{\rho} = \frac{1}{2}q^2 = \frac{1}{2}\frac{m^2}{\pi^2}\frac{y^2}{(y^2 + c^2)^2}$$

Let P be the pressure on the length 2l of the boundary then

$$P = \int_{-l}^{l}(p_0 - p)dy = \frac{1}{2}\frac{m^2\rho}{\pi^2}\int_{-l}^{l}\frac{y^2 dy}{(y^2 + c^2)^2}.$$

Put $y = c \tan\theta$, $dy = c \sec^2\theta \, d\theta$

$$\text{or } P = \frac{1}{2}\frac{m^2\rho}{\pi^2}\int\frac{c^2 \tan^2\theta c \sec^4\theta}{c^4 \sec^2\theta}d\theta = \frac{1}{2}\frac{m^2\rho}{\pi^2 c}\int \sin^2\theta d\theta$$

$$\text{or } P = \frac{1}{2}\frac{m^2\rho}{\pi^2 c}(\theta - \sin\theta\cos\theta)$$

$$\text{or } P = \frac{1}{4}\frac{m^2\rho}{\pi^2 c}\left\{\tan^{-1}\left(\frac{y}{c}\right) - \frac{yc}{\sqrt{(c^2 + y^2)}}\right\}_{-l}^{l}$$

$$\text{or } P = \frac{1}{2}\frac{m^2\rho}{\pi^2}\left\{\frac{1}{c}\tan^{-1}\left(\frac{l}{c}\right) - \frac{l}{\sqrt{(c^2 + l^2)}}\right\}.$$ **Proved.**

Example 20(b): *What arrangement of sources and sinks will give rise to the function.*

$$w = \log\left(z - \frac{a^2}{z}\right)$$

Draw a rough sketch of the stream lines to this curve and prove that two of them sub-divide into the circle r = a and the axis of Y.

Solution: The complex potential is given by

$$w = \log\left(z - \frac{a^2}{z}\right) = \log\left\{\frac{(z-a)(z+a)}{z}\right\}$$

or $w = \log(z - a) + \log(z + a) - \log z$, ...(1)

which shows that there are two sinks of unit strength at distance z = a and z = – a and a source of unit strength at an origin.

The relation (1) can be expressed as

$$\phi + i\psi = \log\{(x - a) + iy\} + \log\{(x + a) + iy\} - \log(x + iy).$$

Equating the imaginary parts, we have

$$\psi = \tan^{-1}\frac{y}{x-a} + \tan^{-1}\frac{y}{x+a} - \tan^{-1}\frac{y}{x}$$

$$\text{or } \psi = \tan^{-1}\left(\frac{\dfrac{y}{(x-a)} + \dfrac{y}{(x+a)}}{1 - \dfrac{y}{(x-a)(x+a)}}\right) - \tan^{-1}\frac{y}{x}$$

$$\text{or } \psi = \tan^{-1}\left(\frac{2xy}{x^2 - y^2 - a^2}\right) - \tan^{-1}\frac{y}{x}$$

$$\text{or } \psi = \tan^{-1}\left(\frac{\dfrac{2xt}{x^2 - y^2 - a^2} - \dfrac{y}{x}}{1 + \dfrac{2xy}{x^2 - y^2 - a^2}\dfrac{y}{x}}\right),$$

$$\text{or } \psi = \tan^{-1}\frac{y\,(x^2 + y^2 + a^2)}{x\,(x^2 + y^2 - a^2)}.$$

The streamlines are given by y = const.

$$\textit{i.e.}, \tan^{-1}\frac{y(x^2 + y^2 + a^2)}{x(x^2 + y^2 - a^2)} = \text{const.}$$

$$\text{or } \frac{y(x^2 + y^2 + a^2)}{x(x^2 + y^2 - a^2)} = k. \qquad ...(2)$$

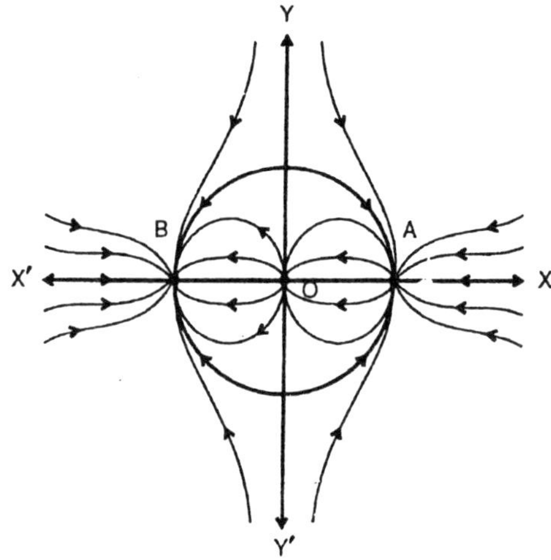

Fig. 3.25

I. If the constant k is infinite then

$$\frac{y(x^2 + y^2 + a^2)}{x(x^2 + y^2 - a^2)} = \infty$$

$\Rightarrow x\ (x^2 + y^2 - a^2) = 0,$

$\Rightarrow x = 0$ and $x^2 + y^2 = a^2$.

Thus $x = 0$ shows that Y-axis is a streamline and the equation $x^2 + y^2 = a^2$ (i. e., $r = a$) shows that the circle is a streamline with centre as origin.

II. If the constant k is zero, then $y = 0$ implies that axis of X is a streamline. Therefore the rough sketch of the lines is with a source of unit strength at origin O (0, 0) and two sinks of unit strength at A (a, 0) and B (– a, 0).

Example 20(c): *(A) There is a source of strength m at an origin and equa. sinks at (1, 0) and (–1, 0). Discuss two dimensional motion. Also, draw a sketch of the streamlines.*

Solution: Equation (1) of Ex. 12 becomes

$w = m \log (z - 1) + m \log (z + 1) - m \log (z - 0)$

$\Rightarrow \phi + i\psi = m \log \log [z^2 - 1] - m \log (z)$

$\Rightarrow \phi + i\ \psi = m \log [\ (x^2 - y^2 - 1) + 2ixy] - m \log (x + iy)$

$$\Rightarrow y = m\left[\tan^{-1}\frac{2xy}{x^2-y^2-1} - \tan^{-1}\frac{y}{x}\right].$$

Put a = 1 in Example and proceed accordingly.

Example 21: *(B) Find the stream function of the two dimensional motion due to two equal sources of strength m at a distance 2a apart and an equal sink of strength 2m between them, Determine the stream lines. Also find the fluid speed at any point.*

Solution: The complex potential at any point P is given by

$$w = -m\log(z-a) - m\log(z+a) + 2m\log z$$

$$\Rightarrow w = -m\log(z^2-a^2) + m\log z^2$$

$$\Rightarrow \phi + i\psi = -m\log\{(x^2-y^2-a^2) + 2ixy\}.$$

$$+ m\log\{(x^2-y^2+2ixy\}.$$

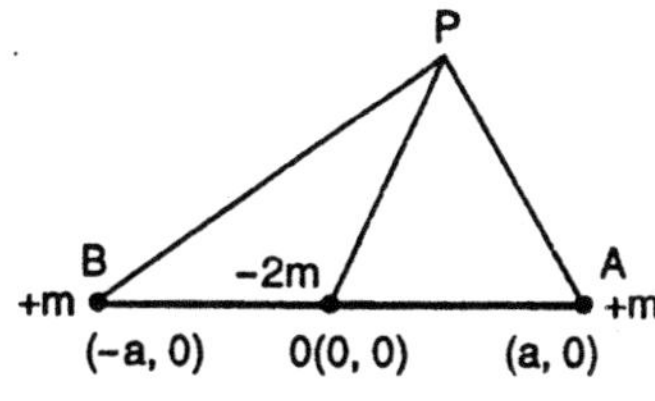

Fig. 3.26

Equating imaginary parts, we have

$$\psi = -m\tan^{-1}\frac{2xy}{x^2-y^2-a^2} + m\tan^{-1}\frac{2xy}{x^2-y^2}$$

$$\Rightarrow \psi = -m\tan^{-1}\left\{\frac{\dfrac{2xy}{x^2-y^2-a^2} - \dfrac{2xy}{x^2-y^2}}{1+\dfrac{2xy}{x^2-y^2-a^2}\cdot\dfrac{2xy}{x^2-y^2}}\right\}$$

$$\Rightarrow \psi = -m\tan^{-1}\left\{\frac{2a^2xy}{(x^2-y^2)(x^2-y^2-a^2)+4x^2y^2}\right\}$$

The streamlines are given by y = constant *i.e.*,

$$-m\tan^{-1}\frac{2a^2xy}{(x^2-y^2)(x^2-y^2-a^2)+4x^2y^2} = -m\tan^{-1}\left(\frac{2}{\lambda}\right)\text{(say)}$$

$\Rightarrow \lambda a^2xy = (x^2 - y^2)(x^2 - y^2)(x^2 - y^2 - a^2) + 4x^2y^2$

$\Rightarrow \lambda a^2xy = (x^2 - y^2)^2 - a^2(x^2 - y^2) + 4x^2y^2$

$\Rightarrow \lambda a^2xy = (x^2 - y^2)^2 - a^2(x^2 - y^2)$

$\Rightarrow (x^2 + y^2)^2 = a^2[x^2 - y^2 + \lambda xy]$,

where λ is a variable parameter.

The fluid speed q at any point is determined as

$$q = \left|\frac{dw}{dz}\right|$$

$$q = \left|-\frac{2mz}{z^2 - a^2} + \frac{2mz}{z^2}\right|$$

$$q = \frac{2ma^2}{|z(z^2 - a^2)|}$$

$$q = \frac{2ma^2}{|z||z-a|\ |z+a|} = \frac{2ma^2}{PO.PA.PB},$$

where $PO = |z|$, $PA = |z + a|$ and $PB = |z - a|$. **Ans.**

Example 22: *In this case of the two dimensional fluid motion produced by a source of strength m placed at a point S outside a rigid circular disc of radius a whose centre is O, show that the velocity of slip of the fluid in contact with the disc is the greatest at the point where the line joining prove that its magnitude at these point is*

$$\frac{2m.OS}{(OS^2 - a^2)}.$$

Solution: Let S′ be an inverse point of S with regard to the circular disc such that

$OS\ OS' = a^2$

$OS' = a^2/c,\ OS = c.$

The image system of a source of strength + m placed at a point S (c, 0) outside the circular disc consists of

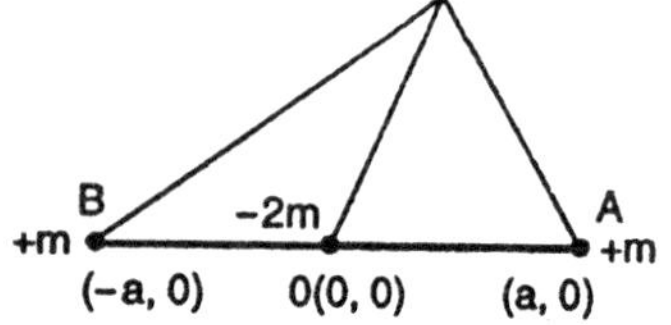

Fig. 3.27

(i) a source of strength + m at S (c, 0),

(ii) a source of strength + m at S′ $(a^2/c, 0)$,

(iii) a sink of strength – m at O (0, 0).

The complex potential w for the motion of the fluid element at any point z is given by

$$w = -m \log(z - c) - m \log(z - (a^2/c)\} + m \log z.$$

Differentiating with regard to z, we have

$$\frac{dw}{dz} = -m\,\frac{1}{z-c} - m\frac{1}{z-(a^2/c)} + m\frac{1}{z}$$

Let q be the velocity given by

$$q = \left|\frac{dw}{dz}\right| = \left|-\frac{m}{z-c}\frac{m}{z-(a^2/c)} + \frac{m}{z}\right|$$

$$\text{or } q = m\left|\frac{z\{z-(a^2/c)\} + z(z-c) - (z-c)(z-(a^2/c)\}}{z(z-c)(z-a^2/c)\}}\right|$$

$$\text{or } q = m\left|\frac{(z-a)(z+a)}{z(z-c)\{z-(a^2/c)\}}\right|$$

The velocity at any point P, $z = ae^{\theta i}$, on the boundary of the circular disc reduces to

$$q = m\left|\frac{(ae^{\theta i}-a)(ae^{\theta i}+a)}{ae^{\theta i}(ae^{\theta i}-c)(ae^{\theta i}-(a^2/c)\}}\right|$$

$$\Rightarrow q = mc\left|\frac{(e^{\theta i}-1)(e^{\theta i}+1)}{e^{\theta i}(ae^{\theta i}-c)(ce^{\theta i}-a)}\right|$$

$$\Rightarrow q = mc\left|\frac{(e^{2\theta i}-1)}{e^{\theta i}(ae^{\theta i}-c)(ce^{\theta i}-a)}\right|$$

$$|\,(c\cos\theta - a) + ic\sin\theta\,|$$

$$\Rightarrow q = mc\,\frac{\sqrt{(\cos 2\theta - 1)^2 + \sin^2 2\theta}}{\sqrt{\cos^2\theta + \sin^2}\sqrt{(a\cos\theta - c)^2 + a^2\sin^2\theta}}$$

$$\sqrt{(c\cos\theta - a)^2 + c^2\sin^2\theta}$$

$$\text{or } q = \frac{2mc\sin\theta}{a^2 + c^2 - 2ac\cos\theta}.$$

For q to be maximum of minimum $dq/d\theta = 0$, we have

$$\frac{dq}{d\theta} = 2mc\frac{(a^2 + c^2)\cos\theta - 2ac}{(a^2 + c^2) - 2ac\cos\theta)^2} = 0.$$

or $(a^2 + c^2)\cos\theta - 2ac = 0 \Rightarrow \cos\theta = 2ac/(a^2 + c^2)$.]

Now $\theta = 0$ gives minimum velocity as the expression vanishes at this point.

The maximum value of θ will be obtained to the corresponding value of $\cos\theta = 2ac/(a^2 + c^2)$. From (1), we have

$$q = 2mc\frac{\{(a^2 - c^2)(a^2 + c^2)\}}{(a^2 + c^2) - \{4a^2c^2/(a^2 + c^2)\}} = 2mc\frac{(a^2 - c^2)}{(a^2 + c^2) - 4a^2c^2}$$

$$\text{or } q = \frac{2mc}{c^2 - a^2} = \frac{2mOS}{OS^2 - a^2}.$$

The velocity will be along the direction of the tangent to the boundary and is equal to the velocity of slip as the boundary of the circular disc is a stream line. **Proved.**

Example 23: *A source S and a sink T of equal strengths m are situated with in the space bounded by a circle whose centre is O. If S and T are at equal distance from O on opposite sides of it and on the same diameter AOB, shew that the velocity of the liquid at any point P is*

$$2m.\frac{OS^2 + OA^2}{OS}.\frac{PA.PB}{PSPS'.PTPT'},$$

where S′ and T′ are the inverse points of S and T with regards to the circle.

Solution: Consider OS = OT = f, let S′ and T′ be the inverse points of S and T regard to the circle that

$OS.\ OS' = OT.\ OT' = a^2$

$OS' = OT' = a^2/f.$

The image system of a source of a length + m and a sink of strength – m placed at the point S and T with in the circle consist of

(i) a source of strength + m at S (f, 0)

(ii) a source of strength + m at S'$(a^2/f, 0)$

(iii) a sink of strength – m at the origin O

(iv) a sink of strength – m at f (– f, 0)

(v) a sink of strength – m at T′ $(-a^2/f, 0)$

(vi) a source of strength + m at the origin O.

The source and sink of same strength cancels each other at the origin. Thus the complex potential w becomes

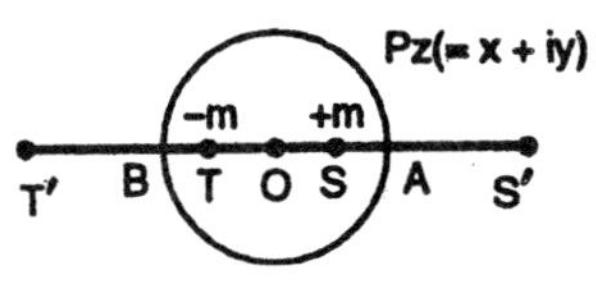

Fig. 3.28

$w = -m \log(z - f) - m \log\{z - (a^2/f)\} + m \log(z + f) + m \log\{z + (a^2/f)\}$

The velocity q at any point P is given by

$$q = \left|\frac{dw}{dz}\right| = m\left|\frac{1}{z-f} - \frac{1}{z-(a^2/f)} + \frac{1}{z+f} + \frac{1}{z+(a^2/f)}\right|$$

$$\text{or } q = m\left|\frac{2f}{z^2-f^2} - \frac{2a^2/f}{z^2-(a^4/f^2)}\right|$$

$$\text{or } q = 2m.\frac{f^2+a^2}{f}\left|\frac{z^2-a^2}{(z^2-f^2)\{z^2-(a^4/f^2)\}}\right|$$

$$\text{or } q = 2m.\frac{f^2+a^2}{f}\left|\frac{(z-a)(z+a)}{(z-f)(z+f)\{z-(a^2/f)\}\{z+(a^2/f)\}}\right|$$

$$\text{or } q = 2m.\frac{f^2+a^2}{f}\left|\frac{|z-a|\;|z+a|}{|z-f||z+f||z-(a^2/f)||z+(a^2/f)|}\right|$$

$$\text{or } q = 2m.\frac{OS^2+OA^2}{OS}.\frac{PA.PB}{PSPS'.PTPT'}.$$ **Proved.**

Example 24: *Find the velocity potential when there is a source and an equal sink inside a circular cavity and show that one of the stream lines is an arc of the circle which passes through the source and sink and cuts orthogonally the boundary of the cavity.*

Solution: Let the source of strength (+ m) and a sink of strength (– m) be placed at the point A and B with in the circular cavity with configuration does not alter by placing a source of strength + m and sink of strength – m and a sink of strength – m at O.

Now a source of strength + m at A and a sink of strength – m at O will gives rise to a source of strength + m at an inverse point A′. Also, a sink of strength – m at B and a source strength + m at the origin O will gives rise to a sink of strength – m at an inverse point B¢. Therefore the image system consists of

(i) a source of strength + m at A (c, 0)

(ii) a source of strength m at an inverse point A′ (a^2/c, 0)

(iii) a sink of strength m at B (b, a) i.e., $z = be^{ai}$

(iv) a sink of strength m at an inverse point B′ (a^2/b, a)

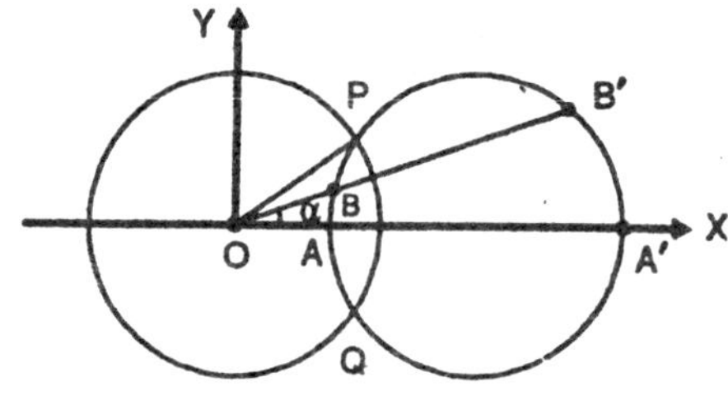

Fig. 3.29

i.e., $z = (a^2/b)e^{ai}$

Hence the complex potential w is given by

$$w = -\cdot m \log [z - c_1 - m \log [z -(a^2/c)] + m \log (z - be^{ai})$$

$$+ m \log [z - (a^2/b) e^{ai}]$$

$$\phi + iy = m \log \left[\frac{c}{b}\cdot\frac{(z-be^{ai})(bz-a^2e^{ai})}{(z-c)(cz-a^2)}\right]$$

Since the points A, A′, B and B′ are concyclic as OA OA′ = a^2, OB. OB′ = b^2, thus the circle through these four points meet the original circle P and Q. Therefore OP is a tangent at P to the circle through P and Q, and the two circle cut orthogonally. The circle A′ B′PQ passes through A and B, hence it must be a streamline. In other words, the streamlines are circles for a source and an equal sink, so the circle through the above four points is a streamline and P and Q are known as stagnation points.

Example 25: *In a region by a fixed quadrantal arc and its radii deduce the motion due to a source and an equal sink situated at the ends of one of the bounding radii, show that the streamline leaving either end (the source or the sink) at an angle a with radius is*

$$r^2 \sin (\alpha + \theta) = a^2 \sin (\alpha - \theta).$$

Solution: The image system due to a source of strength + m at P′ consists of

(i) a source of strength + m at P

(ii) a source of strength + m at P

(iii) a sink of strength – m at O.

The complex potential w for the motion of the fluid element at any point z becomes

$$w = -m \log (z - a) - m \log (z + a) + m \log z$$

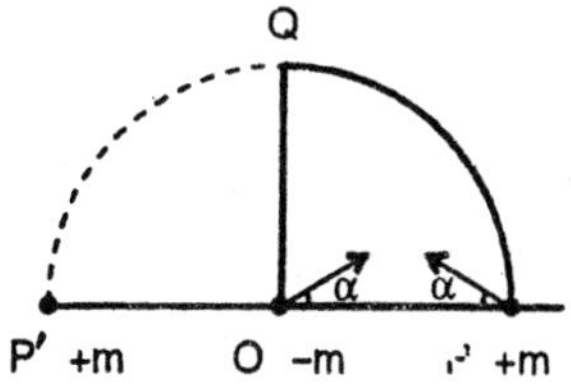

$$\text{or } w = -m \log \left(\frac{z^2 - a^2}{z}\right)$$

$$\text{or } z - \frac{a^2}{z} = \exp.\left\{-\frac{1}{m}(\phi + i\psi)\right\}$$

Fig. 3.30

$$\text{or } r(\cos\theta + i\sin\theta) - \frac{a^2}{r}(\cos\theta - i\sin\theta)$$

$$= e^{-\phi/m}\left(\cos\frac{\psi}{m} - i\sin\frac{\psi}{m}\right)$$

Separating into real and imaginary parts, we have

$$\left(r - \frac{a^2}{r}\right)\cos\theta = e^{-\phi/m}\cos\frac{\psi}{m},$$

$$\text{and } \left(r + \frac{a^2}{r}\right)\sin\theta = -e^{-\phi/m}\sin\frac{\psi}{m},$$

$$\text{or } \tan\frac{\psi}{m} = -\frac{r^2 + a^2}{r^2 - a^2}\tan\theta$$

At $r = a$, $\theta = \frac{\pi}{2}$; $\psi = -\frac{1}{2}m\pi$ and $\theta = 0$ gives $\psi = 0$

Thus OP is the stream line $\psi = 0$, OQ and the arc PQ are the stream line $\psi = -(\pi m/2)$. This determines the motion within the quadrant.

Again $\psi = -m(\theta_1 + \theta_2 - \theta)$.

For a point very near to O, we have

$\theta_2 = 0$, $\theta_1 = \pi$ and $\theta = a$ then $\psi' = -m(\pi - a)$,

represents the stream line which leaves the point O making an angle α with OP.

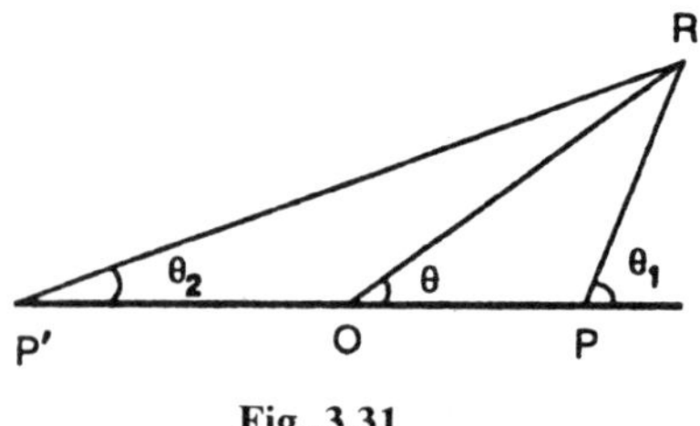

Fig. 3.31

For the stream line which leaves P making an angle α with PO, $\theta_1 = \pi - \alpha$, $\theta_2 = 0$ and $\theta = 0 \Rightarrow \psi = -m(\pi - a)$.

Hence y = – m (p –a) gives the streamlines that make an angle P¢ a and O and at P, therefore

$$\tan(\pi - \alpha) = \frac{r^2 + a^2}{r^2 - a^2}\tan\theta$$

$$\text{or } -\tan\alpha = \frac{r^2 + a^2}{r^2 - a^2}\tan\theta$$

or $(r^2 + a^2)\sin\theta\cos\alpha = -(r^2 - a^2)\sin\alpha\cos\theta$

or $r^2\sin(\alpha + \theta) = a^2\sin(a - \theta)$. **Proved.**

Example 26: *In the parts of an infinite plane bounded by a circular quadrant AB and the productions of the radii OA and OB, there is a two-dimensional motion due to the production of liquid at A, and its absorption at B, at the uniform rate m. Find the velocity potential of the motion ; and shew that the fluid which issues from A in the direction making an angle m with OA follows the path whose polar equation is*

$$r = a\,\sin^{1/2} 2q\,[\cot\mu + \sqrt{\{\cot^2\mu + \operatorname{cosec}^2 2\theta\}}]^{1/2},$$

the positive sign being taken for all square-roots.

Solution : Since there is a production of liquid at the point A of the quadrant, the image system of the source at A with regard to the circular boundary consists of:

(i) a source of strength + (m/2π) at A,

(ii) a source of strength + (m/2π) at A (since the point A is an inverse point of itself),

(iii) a sink of strength – (m/2π) at the origin O.

The image system of the source at A with regard to the circular boundary and the radii OA and OB will give rise to :

(i) two source of equal strength + (m/2π) at A *i.e.*, a source of strength + (m/π) at A,

(ii) two sources of equal strength + (m/2π) at A′ *i.e.*, a source of strength + (m/π) at A′. (This is the image of the source at A).

(iii) a sink of strength – (m/π) at the origin O.

Similarly there is an absorption of liquid at the point B of the quadrant, the image system of the sink at B with regard of the circular boundary consists of:

(i) a sink of strength – (m/2π) at B,

(ii) a sink of strength – (m/2π)at B, (since the point B is an inverse point of itself),

(iii) A source of strength + (m/2π) at the origin O.

The image system of the sink at B with regard to the circular boundary and the radii OA and OB will give rise to

(i) two sinks of equal strength – (m/2π) at B i.e., a sink of strength – (m/2π) at B

(ii) two sinks of equal strength – (m/2π) at B′ i.e., a sink of strength – (m/π) at B′. (This is the image of the sink at B)

(iii) a source of strength + (m/2π) at the origin O.

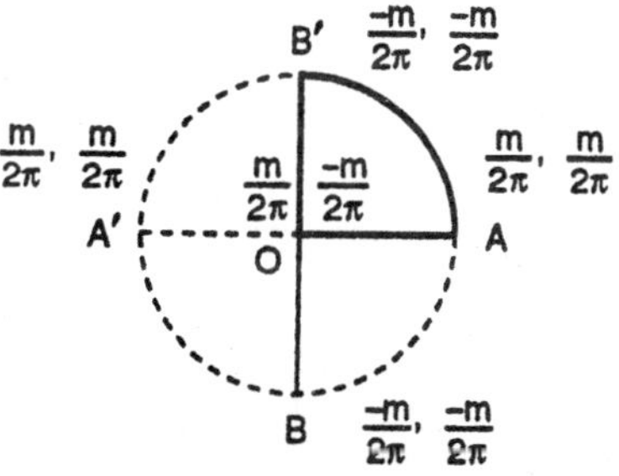

Fig. 3.32

A source of strength + (m/2π) and a sink of strength – (m/2π) at the origin O neutralize each other. Hence the image system, finally, consists of

I. a source of strength + (m/π) at A,

II. a source of strength + (m/π) at A′,

III. a sink of strength – (m/π) at B,

IV. a sink of strength – (m/π) at B′.

The complex potential w at P due to the given system becomes

$$w = -\frac{m}{\pi}\log\ (z-a) - \frac{m}{\pi}\log\ (z+a) + \frac{m}{\pi}\log(z-ai) + \frac{m}{\pi}\log(z+ai)$$

$$\text{or } w = -\frac{m}{\pi}\log(z^2-a^2) + \frac{m}{\pi}\log(z^2+a^2) \qquad ...(1)$$

To determine the velocity potential of the motion, equating the real part from both the sides, we have

$$\phi = -\frac{m}{\pi}\log|z-a| - \frac{m}{\pi}\log|z+a| + \frac{m}{\pi}\log|z-ai| + \frac{m}{\pi}\log\ |z+ai|$$

$$\text{or } \phi = \frac{m}{\pi}\log AP - \frac{m}{\pi}\log A'P + \frac{m}{\pi}\log BP + \frac{m}{\pi}\log B'\ P.$$

$$\text{or } \phi = \frac{m}{\pi}\ [\log BP\ B'\ P - \log APA'P] = \frac{m}{\pi}\log\left(\frac{BPB'P}{APA'P}\right).$$ **Ans.**

Since P $(z = x + iy)$ be any point in the fluid, substituting $z = re^{\theta i}$ in (1), we have

$$\phi + i\psi = -\frac{m}{\pi}\log\ (r^2e^{2\theta i} - a^2) + \frac{m}{\pi}\ \log\ (r^2e^{2\theta i} + a^2)$$

$$\text{or } \phi + i\psi = -\frac{m}{\pi}\ \log\ (r^2\cos 2\theta - a^2) + ir^2 \sin 2\theta) + \frac{m}{\pi}\log\ \{(r^2\cos 2\theta + at^2) + ir^2\sin 2\theta\}$$

$$\text{or } \psi = \frac{m}{\pi}\tan^{-1}\left(\frac{r^2\sin 2\theta}{r^2\cos 2\theta - a^2}\right) + \frac{m}{\pi}\tan^{-1}\left(\frac{r^2\sin 2\theta}{r^2\cos 2\theta + a^2}\right)$$

$$\text{or } \psi = \frac{m}{\pi}\tan^{-1}\left[\frac{\dfrac{r^2\sin 2\theta}{r^2\cos 2\theta + a^2} - \dfrac{r^2\sin 2\theta}{r^2\cos 2\theta - a^2}}{1 + \dfrac{r^2\sin 2\theta}{r^2\cos 2\theta + a^2}\cdot\dfrac{r^2\sin 2\theta}{r^2\cos 2\theta - a^2}}\right]$$

$$\text{or } \psi = -\frac{m}{\pi}\tan^{-1}\left(\frac{2a^2r^2\sin 2\theta}{r^4 - a^4}\right) \qquad ...(2)$$

The streamline that leaves A at an inclination m with OA, is

$\psi = -(m/\pi)\,\mu.$

From (2), we have

$$-\frac{m}{\pi}\tan^{-1}\frac{2a^2r^2\sin 2\theta}{r^4-a^4} = \frac{m}{\pi}\mu$$

or $r^4 - 2a^2r^2 \sin 2\theta \cot^2 \mu - a^4 = 0.$

or $r^2 = a^2 \sin 2\theta \cot \mu \pm \sqrt{(a^4\sin^2 2\theta\cot^2\mu + a^4)}$.

Neglecting the negative sign before the radical because it provides a relation $r^2 < 0$, so we have

$r = a\,(\sin 2\theta)^{1/2}\,[\cot \mu + \sqrt{\{(\cot^2\mu + \operatorname{cosec}^2 2\theta)\}}]^{1/2}$ **Proved.**

Example 27: *Within a rigid boundary in the form of the circle*

$(x + a)^2 + (y - 4\alpha)^2 = 8\alpha^2,$

there is liquid motion due to a doublet of strength m at the point (0, 3α) with its axis along of Y. Show that the velocity potential is

$$\mu\left\{\frac{4(x-3\alpha)}{(x-3\alpha)^2+y^2}+\frac{y-3\alpha}{x^2+(y-2\alpha)^2}\right\}.$$

Solution: The equation to the rigid boundary is given by

$(x + \alpha)^2 + (y - 4\alpha)^2 = 8\alpha^2,$

whose centre is $(-a, 4\alpha)$ and radius is $2\sqrt{2a}$.

The inverse point of P $(0, 3\alpha)$ with regard to the circle of centre C is Q $(3\alpha, 0)$. The distance between two points C and Q is

$$CQ = \sqrt{\{(-\alpha-3\alpha)^2+(4\alpha-0)^2\}} = 4\sqrt{2\alpha}$$

Again, by the definition of the inverse point, we have

$CP.\ CQ = CP\,(CP + PQ) = 8\alpha^2$

or $\alpha\sqrt{2(a\sqrt{2+PQ})} = 8\alpha^2$

or $PQ = 3\sqrt{2\alpha} = 3\alpha\sec\frac{\pi}{4}.$

The gradient of the line PCis

$$\tan\psi = \frac{3\alpha-4\alpha}{0+\alpha} = -1 = \tan\frac{3\pi}{4}$$

$$\Rightarrow \psi = \frac{3\pi}{4},$$

which shows that the line PC makes an angle $\pi/4$ with the vertical axis OY.

Now the image of the doublet of strength μ at P with regard to a circle is a doublet of strength μ' (let) at an inverse point Q. This axis of Q will make an angle $\pi/4$ with PQ and will, therefore, be parallel to X-axis.

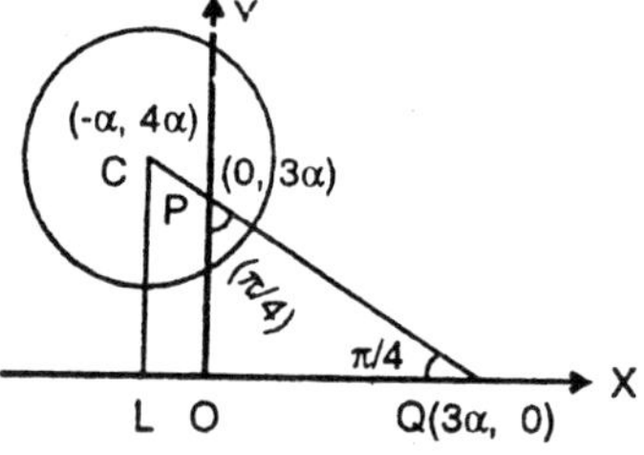

Fig. 3.33

$$m¢ = m\frac{(\text{Radius})^2}{(\text{CP})^2}$$

$$= m.\frac{8a^2}{2a^2} = 4\mu . \qquad ...(1)$$

The complex potential function is given by

$$w = \frac{\mu e^{\pi i/2}}{z-3i\alpha} + \frac{\mu' e^{0i}}{z-3\alpha} = \frac{\mu e^{\pi i/2}}{z-3i\alpha} + \frac{4\mu e^{0i}}{z-3\alpha}$$

$$\text{or } \phi + i\psi = \mu\left[i\frac{1}{x+i(y-3\alpha)} + \frac{4}{(x-3\alpha)+iy}\right]$$

$$\text{or } \phi + i\psi = \mu\left[\frac{i\{x-i(y-3\alpha)\}}{x^2+(y-3\alpha)^2} + \frac{4\{(x-3a)-jy\}}{(x-3\alpha)+iy}\right]$$

Thus the velocity potential is obtained as

$$\text{or } \phi = \mu\left[4\frac{(x-3\alpha)}{(x-3\alpha)^2+y^2} + \frac{(y-3a)}{x^2+(y-3\alpha)^2}\right] \text{ **Proved.**}$$

Example 28(a): *Show that the velocity potential*

$$\phi = \frac{1}{2}\log\frac{(x+a)^2+y^2}{(x-a)^2+y^2},$$

gives a possible motion. Determine the form of streamlines and curves of equal speed.

Solution: $\phi = \dfrac{1}{2}\log\dfrac{(x+a)^2+y^2}{(x-a)^2+y^2}$ ...(1)

$$\frac{\partial \phi}{\partial x} = \frac{x+a}{(x+a)^2+y^2} - \frac{x-a}{(x-a)^2+y^2} \qquad \text{...(2)}$$

$$\frac{\partial \phi}{\partial y} = \frac{y}{(x+a)^2+y^2} - \frac{y}{(x-a)^2+y^2}. \qquad \text{...(3)}$$

$$\text{Also } \frac{\partial u}{\partial x} = \frac{\partial}{\partial x}\left\{\frac{\partial \phi}{\partial x}\right\} = \frac{\partial}{\partial x}\left\{\frac{x-a}{(x-a)^2+y^2} - \frac{x+a}{(x+a)^2+y^2}\right\}$$

$$\text{or } \frac{\partial u}{\partial x} = \frac{y^2-(x-a)^2}{[(x-a)^2+y^2]^2} - \frac{y^2-(x-a)^2}{[(x-a)^2+y^2]^2}$$

$$\text{and } \frac{\partial v}{\partial y} = \frac{\partial}{\partial y}\left(\frac{\partial \phi}{\partial y}\right) = \frac{\partial}{\partial x}\left\{\frac{y}{(x-a)^2+y^2} - \frac{y}{(x+a)^2+y^2}\right\}$$

$$\text{or } \frac{\partial v}{\partial y} = \frac{(x-a)^2-y^2}{[(x-a)^2+y^2]^2} - \frac{(x+a)^2-y^2}{[(x+a)^2+y^2]^2}$$

$\Rightarrow \dfrac{\partial u}{\partial x} + \dfrac{\partial v}{\partial y} = 0$ which satisfies the equation of continuity.

Hence the relation (1) gives a possible liquid motion.

Since the velocity potential f and the stream function y satisfies the property of conjugate functions, so we have

$$\frac{\partial \phi}{\partial x} = \frac{\partial \psi}{\partial y} \text{ and } \frac{\partial \phi}{\partial y} = -\frac{\partial \psi}{\partial x} \qquad \text{...(4)}$$

$$\text{or } \frac{\partial \psi}{\partial y} = \frac{x+a}{(x+a)^2+y^2} - \frac{x-a}{(x-a)^2+y^2}$$

$$\text{or } y = \tan^{-1}\frac{y}{x+a} - \tan^{-1}\frac{y}{x-a} + f(x). \qquad \text{...(5)}$$

To determine f (x), we shall differentiate (5) with regard to x,

$$\frac{\partial \psi}{\partial x} = -\frac{y}{(x+a)^2+y^2} + \frac{y}{(x-a)^2+y^2} + f'(x)$$

$$\therefore \frac{\partial \psi}{\partial x} = -\frac{\partial \phi}{\partial y} \Rightarrow f'(x) = 0 \text{ Þ } f(x) = \text{const.}$$

$$\text{or } \psi = \tan^{-1}\frac{y}{x+a} - \tan^{-1}\frac{y}{x-a} = \tan{-1}\left(\frac{2ay}{x^2+y^2-a^2}\right).$$

The lines of flow can be obtained by ψ = const.

i.e. $\tan^{-1}\left\{\dfrac{2ay}{x^2+y^2-a^2}\right\} = \text{const.}$

or $\dfrac{2ay}{a^2-x^2-y^2} = \tan\mu = A = \text{const.}$

The constant A = 0 gives the stream line y = 0 i.e., a real axis and A =∞ gives the stream line $x^2 + y^2 = a^2$, i.e., a circle. **Ans.**

$$\text{Again } w = \phi + i\,\psi = \frac{1}{2}\log\{(x+a)^2+y^2\} - \frac{1}{2}\log\{(x-a)^2+y^2\}$$
$$+\, i\tan^{-1}\frac{y}{x+a} - i\tan^{-1}\frac{y}{x-a}.$$

$$\text{or } w = \left[\frac{1}{2}\log\{(x+a)^2+y^2\} + i\tan^{-1}\frac{y}{x+a}\right]$$
$$-\left[\frac{1}{2}\log\{(x-a)^2+y^2\} + \tan^{-1}\frac{y}{x-a}\right]$$

or w = log {(x + a) + iy} – log {(x – a) + iy }

or w = log { (z + a) – log (z – a)}; z = x + iy

$$\text{So } q = \left|\frac{dw}{dz}\right| = \left|\frac{1}{z+a} - \frac{1}{z-a}\right| = \frac{2a}{|z+a|\,|z-a|} = \frac{2a}{rr'},$$

where | z + a | = r′, |z – a| = r, r′ and r are the distances from (– a, 0) and (a, 0).

Hence the curves of equal speeds are given by.

$\dfrac{2a}{rr'}$ = const., rr′ = const., which are known as Cassini Ovals.

Example 28(b): *If the fluid fill the region of spaces on the positive side of X-axis, which is a rigid boundary, and if there be a source + m at the point (0, a) and an equal sink at (0, b) and if the pressure on the negative side of the boundary, be the same as the pressure of the fluid at infinity. Shew that the resultant pressure on the boundary is*

$$\pi\rho m^2\,\frac{(a-b)^2}{ab(a+b)},$$

where ρ is density of the fluid.

Solution: The image system due to a source of strength + m placed at the point A (0, a) and a sink of strength – m placed at the point B (0, b) on the same side consists of

(i) a source of strength + m at the point A (0, a) i.e., at a distance z = ai from O

(ii) a source of strength + m at the other side of the surface at A¢ (0, a) *i.e.*, at a distance z = – ai from O

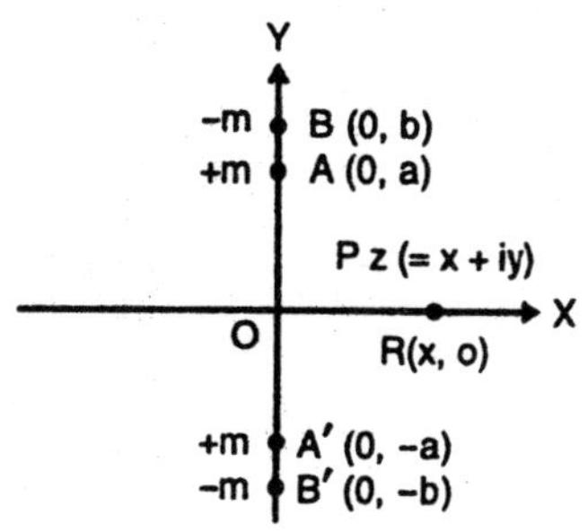

Fig. 3.34

(iii) a sink of strength – m at the other side of the surface at B′ (0, – b) *i.e.*, at a distance z = – bi from O.

Thus the complex potential is given by

$$w = -m \log (z - ai) - m \log (z + ai) + m \log (z - bi) + m \log (z + bi)$$

or $w = -m \log (z^2 + a^2) + m \log (z^2 + b^2)$

or $\dfrac{dw}{dz} = -m \dfrac{2z}{z^2 + a^2} + m \dfrac{2z}{z^2 + b^2}$

or $q = \left|\dfrac{dw}{dz}\right| = 2m \left|\dfrac{z(a^2 - b^2)}{(z^2 + a^2)(z^2 + b^2)}\right|$ ...(1)

Consider a point R (x, 0) on the X-axis. Let q be the velocity at the point R, then substituting z = x in (1), we have

$$q = 2m (a^2 - b^2)\left|\frac{x}{(x^2 + a^2)(x^2 + b^2)}\right|$$

$$q = 2m (a^2 - b^2) \frac{x}{(x^2 + a^2)(x^2 + b^2)}$$

Let p be the pressure at any point, then by Bernoulli's theorem, we have

$$\frac{p}{\rho} = C - \frac{1}{2}q^2.$$

Since $p = p_0$, $q = 0$; $C = p_0/\rho$

or $\frac{p}{\rho} = \frac{p_0}{\rho} - \frac{1}{2}q^2 \Rightarrow p_0 - p = \frac{1}{2}\rho q^2.$

Thus the resultant pressure on the boundary is given by

$$P = \int_{-\infty}^{\infty}(p_0 - p)dx = x\frac{1}{2}\rho\int_{-\infty}^{\infty} q^2 dx$$

$$\Rightarrow P = 4m^2\rho\,(a^2 - b^2)^2 \int_0^{\infty} \frac{x^2 dx}{(x^2 + a^2)^2(x^2 + b^2)^2}$$

$$\Rightarrow P = 4m^2\rho \int_0^{\infty}\left[\frac{a^2 + b^2}{b^2 - a^2}\left\{\frac{1}{x^2 + a^2} - \frac{1}{x^2 + b^2}\right\}\right.$$

$$\left. - \frac{a^2}{(x^2 + a^2)^2} - \frac{b^2}{(x^2 + b^2)^2}\right]dx$$

$$\Rightarrow P = 4m^2\rho\left[\frac{a^2 + b^2}{b^2 - a^2}\left(\frac{\pi}{2a} - \frac{\pi}{2b}\right) - \frac{\pi}{4a} - \frac{\pi}{4b}\right]$$

$$\Rightarrow P = \pi\rho m^2 \frac{(a - b)^2}{ab(a + b)}.$$ **Proved.**

Example 29: *Prove that for liquid circulating irrotationally in part of the plane between two non-interesting circles the curves of constant velocity are Cassini's Ovals.*

Solution: Let the line of centres of the circles be CC′, and A (–a, 0) and B (a, 0) be the inverse points with regard to both the circles.

Consider a point P on one of the circles, that

PA = r and PB = r_1.

Since CA. CB = CP^2.

From the similar triangles CPA and CPB, we have

$$\frac{PA}{PB} = \frac{CP}{CB} = \text{constant}$$

$$\Rightarrow \frac{r}{r_1} = \text{constant}.$$

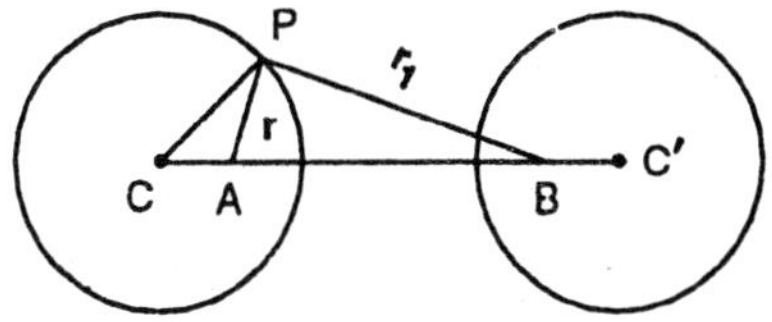

Fig. 3.35

Thus the equations of the two circle are $r/r_1 = C1$ and $r/r_1 = C_2$ where C1 and C_2 are constants.

Since these two circles are the two stream lines, so that the stream function ψ must be of the form (r/r_1), but $f(r/r_1)$ is a plane harmonic then

$$\psi = c \log (r/r_1).$$

Let θ is the conjugate harmonic for log r they y contains log function. Thus the complex potential must be of the form

$$w = \psi - i\phi = c \log (r/r_1) + ic (\theta + \theta_1)$$

$$\text{or } w = c [\log r + i\theta] - c [\log r_1 - i\theta_1]$$

$$\text{or } w = c \log \left(\frac{re^{\theta i}}{r_1 e^{\theta} 1^{i}}\right) = c \log \left(\frac{z+a}{z-a}\right).$$

$$\text{or } q = \left|\frac{dw}{dz}\right| = c\left|\frac{1}{z+a} - \frac{1}{z-a}\right| = \frac{2ac}{|z+a|\,|z-a|} = \frac{2ac}{rr_1}.$$

$$q = \text{const.} \Rightarrow \frac{2ac}{rr_1} = \text{const.}$$

Hence the curves of equal velocities are given by

$\Rightarrow rr_1 =$ constant which is known as *Cassinil's Ovals*. **Proved.**

CONFORMAL REPRESENTATION

If the two curves C, C′ in the domain D intersect at the point $P(x_0, y_0)$ at an angle α, then if the two corresponding curves Γ_1, Γ_2 in the domain D′ intersect the point $P'(u_0, v_0)$ corresponding to P at the same angle α, τηε transformation is said to be *isogonal*. If the sense of the rotation as well as the magnitude of the angle is preserved then the transformation is said to be *conformal.*

Let a bi-uniform or one-one mapping of a region of the Z-plane on a region of the ζ-plane be defined by the transformation

$$\zeta = f(z),$$

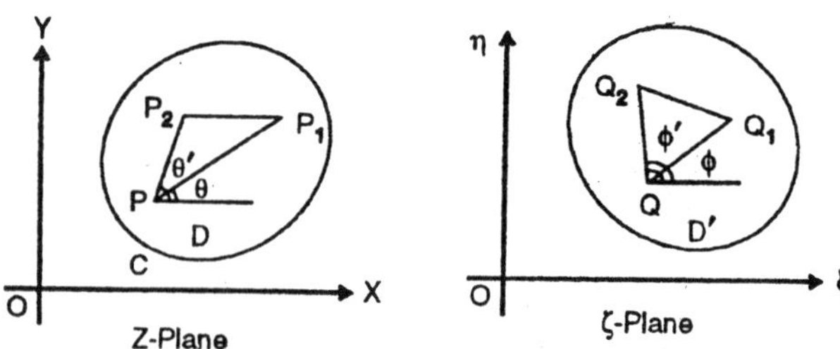

Fig. 3.36

where f (z) be a finite, single valued and differentiable function of a complex variable z (= x + iy) in some domain D enclosed by a contour C in the z-plane. Any point P with in D is transformed uniquely into a definite point Q of the ζ - plane in some domain D′ enclosing by a contour Γ. The necessary condition for such a mapping of Z-plane into -ζ plane is that the derivative f¢ (z) must be finite and non-zero at any point on or within D. Any point P with in D is transformed uniquely into a definite point Q of the ζ-plane, which is represented as P → Q.

Let P (z), P1 (z1), P2 (z2) be three neighbouring points in the domain D. Under transformation P → Q (ζ), $P_1 \to Q_1(\zeta_1)$, $P_2 \to Q_2\ (\zeta_2)$ then ζ- planes where $\zeta_1 = f\ (z_1)$, $\zeta 2 = f\ (z_2)$. Then

$$\frac{\zeta_1 - \zeta}{z_1 - z} = \frac{f(z_1) - f(z)}{z_1 - z} + 0(\,|z_1 - z|\,),$$

$$\frac{\zeta_2 - \zeta}{z_2 - z} = \frac{f(z_2) - f(z)}{z_2 - z} + 0(\,|z_2 - z|\,),$$

where $|z_1 - z|$ and $|z_2 - z|$ is small.

In the limit, if P_1, P_2 approach P and Q_1, Q_2 approach Q, the we have

$$\frac{\zeta_1 - \zeta}{z_1 - z} = f'(z) \text{ and } \frac{\zeta_2 - \zeta}{z_2 - z} = f'(z), \text{ very nearly.}$$

$$\text{Therefore } \frac{\zeta_1 - \zeta}{z_1 - z} = \frac{\zeta_2 - \zeta}{z_2 - z} = f'(z) = \frac{d\zeta}{dz}$$

$$\Rightarrow \frac{QQ_1 e^{i\phi}}{PP_1 e^{i\theta}} = \frac{QQ_2 e^{i\phi'}}{PP_2 e^{i\theta'}} \Rightarrow \frac{QQ_1}{PP_1} e^{i(\phi-\theta)} = \frac{QQ^2}{PP_2} e^{i(\phi-\theta')}$$

$$\Rightarrow \phi - \theta = \phi' - \theta' \Rightarrow \phi' - \phi = \theta$$

$$\Rightarrow \angle Q_2QQ_1 = \angle P_2PP_1$$

or $\dfrac{QQ_1}{PP_1} = \dfrac{QQ_2}{PP_2} = |\, f'(z) \,| = \left|\dfrac{d\zeta}{dz}\right|.$

It follows that geometrically the triangle PP_1P_2 and QQ_1Q_2 are similar. Therefore an infinitesimal triangle in Z-plane maps into a similar infinitesimal triangle in ζ-plane.

If ξ, η denote any functions of x and y, then corresponding to every value of (x + iy) there must be one or more definite values of (ξ + iη); but the ratio of the differential of this function to that of (x + iy) is

$$\frac{\delta\zeta}{\delta z} = \frac{\delta(\xi + i\eta)}{\delta(x + iy)} = \frac{\left(\dfrac{\partial\xi}{\partial x}\delta x + \dfrac{\partial\xi}{\partial y}\delta y\right) + i\left(\dfrac{\partial\eta}{\partial x}\delta x + \dfrac{\partial\eta}{\partial y}\delta y\right)}{\delta x + i\delta y}$$

$$\frac{\delta\zeta}{\delta z} = \frac{\delta\xi + i\xi\eta}{\delta x + i\delta y} = \frac{\left(\dfrac{\partial\xi}{\partial x} + i\dfrac{\partial\eta}{\partial x}\right)\delta x + \left(\dfrac{\partial\xi}{\partial y} + i\dfrac{\partial\eta}{\partial y}\right)\delta y}{\delta x + i\delta y}$$

depends, in general, on the ratio δx/δy. We must have

$$\frac{\partial\xi}{\partial y} + i\frac{\partial\eta}{\partial y} = i\left(\frac{\partial\xi}{\partial x} + i\frac{\partial\eta}{\partial x}\right) \Rightarrow \frac{\partial\xi}{\partial x} = \frac{\partial\eta}{\partial y} \text{ and } \frac{\partial\xi}{\partial y} = -\frac{\partial\eta}{\partial x}.$$

Also $\dfrac{d\zeta}{dz} = \dfrac{\partial\xi}{\partial x} + i\dfrac{\partial\xi}{\partial y}.$

The ratio of corresponding small areas is given by

$$\frac{\Delta Q_1QQ_2}{\Delta P_1PP_2} = \frac{\frac{1}{2}QQ_1QQ_2 \sin Q_1QQ_2}{\frac{1}{2}PP_1PP_2 \sin P_1PP_2} = \frac{QQ_1}{PP_1}\cdot\frac{QQ_2}{PP_2}$$

$$= |\, f'(z) \,|^2 = f'(z) \times \bar{f}'(\bar{z}) = \frac{d\xi}{dz} \times \frac{d\bar{\xi}}{d\bar{z}},$$

where $\bar{f}(\bar{z})$ is the conjugate complex of f (z).

Here $|\, f'(z) \,|^2 = \left|\dfrac{d\zeta}{dz}\right|^2 = \left(\dfrac{\partial\xi}{\partial x}\right)^2 + \left(\dfrac{\partial\xi}{\partial y}\right)^2,$

except at the singular points of the transformation where either speed is zero.

Let w = F (ζ) is the complex potential of a liquid motion in the ζ-plane then a corresponding motion can be determined in the Z-plane by arranging that w takes the same complex value at corresponding points P and Q. So w = F {f (z)}. Since the complex potential w at P is equal to the complex potential at Q, it follows that $\psi_P = \psi_Q$, *i.e.*,a streamline in D′ maps into a streamline along the corresponding curve in D. If D′ is a stream line then D will also be a streamline.

APPLICATION TO FLUID DYNAMICS

1. Let there be a doublet at a point P in the Z-plane then there must be an equal doublet at the corresponding point Q in x-plane.

Also $QQ_1 = hPP_1$.

Again $\mu' = mQQ_1$ and $\mu = mPP_1 \Rightarrow \mu' = h\mu$.

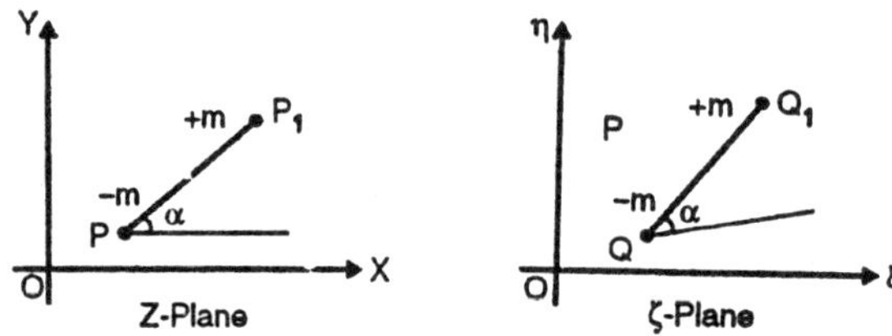

Fig. 3.37

It follows that if the axis PP_1 of the doublet at P makes an angle a with a given direction in Z-plane then the axis QQ_1 will also makes an angle a with the corresponding direction in ζ- plane.

II. Assuming the motion $\zeta = z^{1/n}$. ...(1)

Consider a source of strength + m at P. Let (r, θ) be the polar coordinates of the branch point P in Z-plane and (R,ϕ) be the coordinates of the corresponding point Q in ζ-plane.

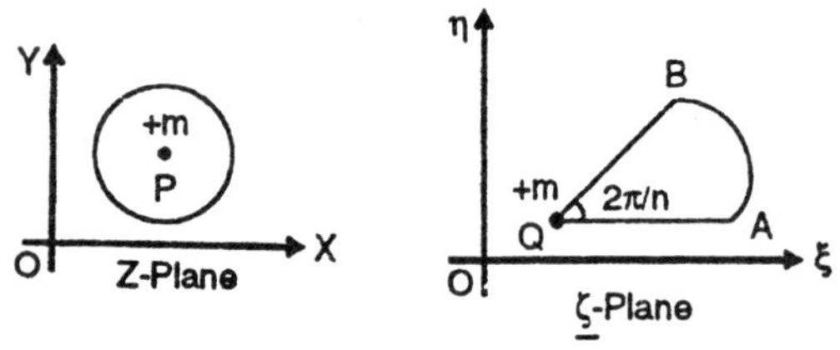

Fig. 3.38

From (1), we have

$Re^{i\phi} = (re^{\theta i})^{1/n} = r^{1/n}\, e^{(\theta/n)i} \Rightarrow R = r^{1/n}$ and $\phi = \theta/n$.

It follows that a complete circle round P in the Z-plane corresponds to an arc AB (= $2\pi/n$) in the ζ-plane. Since the flow across the circle is equal to that across the arc AB.

$$2pm = (2\pi/n)\, m' \Rightarrow m' = nm.$$

Therefore corresponding to a source of strength + m at P there will be a source of strength + m′ (=nm) at Q.

SOLVED EXAMPLES

Example 1: *Use the method of image to prove that if there be a source m at the point z_0 in a fluid bounded by the lines $\theta = 0$ and $\theta = \pi/3$, τηε solution is*

$$\phi + i\psi = -m \log \{(z^3 - z_0^3)\,(z^3 - z_0{}^{8})\},$$

where $z_0 = x_0 + iy_0$ and $z'_0 = x_0 - iy_0$.

Solution: Changing the motion from Z-plane to t-plane by the transformation $t = z^3$, where $t = Re^{\phi i}$ and $z = re^{\theta i}$.

or $Re^{\phi i} = r^3 e^{\theta i} \Rightarrow R\ r^3$ and $\phi = 3\theta$.

Thus the boundaries $\theta = 0$ and $\theta = \pi/3$ in Z-plane transform to $\phi = 0$ and $\phi = \pi$ in t-plane.

A source of strength + m is placed at the point z0 in Z-plane bounded by the line $\theta = 0$ and $\theta = \pi/3$ corresponds by transformation to the points $t_0 = z^3_0$ bounded by the real axis $\phi = 0$ and $\phi = \pi$ in t-plane. The image system consists of

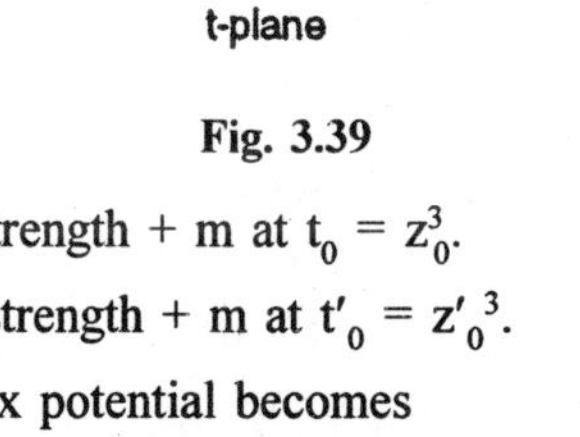

Fig. 3.39

(i) a source of strength + m at $t_0 = z_0^3$.

(ii) a source of strength + m at $t'_0 = z'_0{}^3$.

Thus the complex potential becomes

$$w = m \log (t - z_0^3) - m \log (t - z_0{}'^3)$$

or $w = -m \log (z^3 - z_0^3) - m \log (z^3 - z_0'^3)$

or $\phi + i\psi = -m \log \{(z^3 - z_0^3)(z^3 - z_0'^3)\}$. **Proved.**

Example 2: *Between fixed boundaries $\theta = \pi/4$ and $\theta = -\pi/4$, there is a two dimensional liquid motion due to a source of strength m at the point $r = a$, $\theta = 0$ and an equal sink at the point $r = b$, $\theta = 0$. Use the method of images to show that the stream function is*

$$-m \tan^{-1}\left\{\frac{r^4(a^4 - b^4)\sin 4\theta}{r^8 - r^4(a^4 + b^4)\cos 4\theta + a^4 b^4}\right\}.$$

Shew also that the velocity at (r, θ) is

$$\frac{4m(a^4 - b^4)r^3}{(r^8 - 2a^4 r^4 \cos 4\theta + a^4)^{1/2}(r^8 - 2b^4 r^4 \cos 4\theta + b^8)^{1/2}}$$

Solution: Let the transformation be $t = z^2$, where $t = Re^{\phi i}$ and $z = re^{\theta i}$.

or $Re^{\phi i} = r^2 e^{2\theta i} \Rightarrow R = r^2$ and $\phi = 2\theta$.

The boundaries $\theta = \pi/4$ and $\theta = -\pi/4$ in Z-plane transform to an imaginary axis $\phi = \pi/2$ and $\phi = -\pi/2$ in t-plane.

The source of strength + m at ($r = a$, $\theta = 0$) and a sink of strength – m at ($r = b$, $\theta = 0$) in Z-plane corresponds by the transformation, $t = z^2$ to a source of strength + m at ($R = a^2$, $\phi = 0$) and a sink of strength – m at ($R = b^2$, $\phi = 0$) in t- plane. The image system consists of

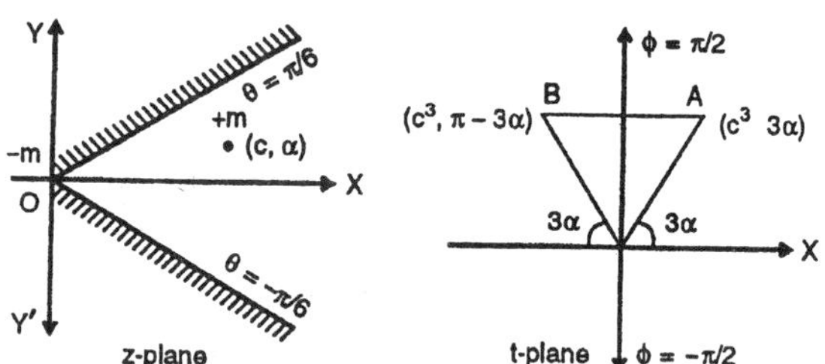

Fig. 3.40

(i) a source of strength + m at $(a^2, 0)$

(ii) a source of strength + m at $(-a^2, 0)$.

(iii) a sink of strength – m at $(b^2, 0)$.

(iv) a sink of strength – m at $(-b^2, 0)$.

Thus the complex potential becomes

$w = -m \log(t - a^2) - m \log(t + a^2) + m \log(t - b^2) + m \log(t + b^2)$

or $w = -m \log(z^2 - a^2) - m \log(z^2 + a^2) + m \log(z^2 - b^2) + m \log(z^2 + b^2)$

or $w = -m \log(z^4 + a^4) + m \log(z^4 + b^4)$ ...(1)

or $\dfrac{dw}{dz} = -m \dfrac{4z^3}{z^4 - a^4} + m \dfrac{4z^3}{z^4 - b^4}$

or $\dfrac{dw}{dz} = -4m\left[\dfrac{z^3}{z^4 - a^4} - \dfrac{z^3}{z^4 - b^4}\right] = -\dfrac{4m(a^4 - b^4)z^3}{(z^4 - a^4)(z^4 - b^4)}$

Then $q = \left|\dfrac{dw}{dz}\right| = 4m(a^4 - b^4)\left|\dfrac{z^3}{(z^4 - a^4)(z^4 - b^4)}\right|.$ ...(2)

Let P ($z = x + iy = re^{\theta i}$) be a point in the Z-plane then by substituting $z = re^{\theta i}$ in (2), we have

$$q = 4m(a^4 - b^4)\left|\frac{r^3 e^{3\theta i}}{(r^4 e^{\theta i} - a^4)(r^4 e^{4\theta i} - b^4)}\right|$$

$$q = 4m(a^4 - b^4)\left|\frac{r^3 e^{3\theta i}}{(r^4 \cos 4\theta - a^4) + i(r^4 \sin 4\theta)] \times [r^4 \cos 4\theta - b^4) + i(r^4 \sin 4\theta)]}\right|$$

$$q = \frac{4m(a^4 - b^4)r^3}{(r^8 - 2a^4r^4 \cos 4\theta + a^8)^{1/2}(r^8 - 2b^4r^4 \cos 4\theta + b^8)^{1/2}}$$

Proved.

Substituting $z = re^{\theta i}$ in (1), we have

$\phi + i\psi = -m \log(r^4 e^{4\theta i} - a^4) + m \log(r^4 e^{4\theta i} - b^4)$

or $\phi + i\psi = -m \log\{r^4 \cos 4\theta - a^4) + i(r^4 \sin 4\theta)\}$
$+ m \log\{(r^4 \cos 4\theta - b^4) + i(r^4 \sin 4\theta)\}$

or $\psi = -m \tan^{-1} \dfrac{r^4 \sin 4\theta}{r^4 \cos 4\theta - a^4} + m \tan^{-1} \dfrac{r^4 \sin 4\theta}{r^4 \cos 4\theta - b^4}$

$$\text{or } \psi = -m \tan^{-1}\left[\frac{\dfrac{r^4 \sin 4\theta}{r^4 \cos 4\theta - a^4} - \dfrac{r^4 \sin 4\theta}{r^4 \cos 4\theta - b^4}}{1 + \dfrac{r^4 \sin 4\theta}{r^4 \cos 4\theta - a^4}\,\dfrac{r^4 \sin 4\theta}{r^4 \cos 4\theta - b^4}}\right]$$

$$\text{or } \psi = -m \tan^{-1}\left[\frac{r^4(a^4 - b^4)\sin 4\theta}{r^8 - r^4(a^4 + b^4)\cos 4\theta + a^4 b^4}\right]$$ **Proved.**

Example 3: *Between the fixed boundaries $\theta = \pi/6$ and $\theta = -\pi/6$, there is a two-dimensional liquid motion due to a source at the point $(r = c,\ \theta = \alpha)$ and a sink at the origin, absorbing, water at the same rate as the source produces it. Find the stream function, and shew that one of the stream lines is a part of the curve*

$r^3 \sin 3\alpha = c^3 \sin 3\theta$.

Solution: Let the transformation be $t = z^3$,

where t $Re^{\phi i}$, $z = re^{\theta i}$.

or $Re^{\phi i} = r^3 e^{3\theta i}$,

$\therefore$ $R = r^3$ and $\phi = 3\theta$.

Now the boundaries $\theta = \pi/6$ and $\theta = -\pi/6$ in Z-plane transforms into $\phi = \pi/2$ and $\phi = -\pi/2$ (i.e., an imaginary axis) in t-plane.

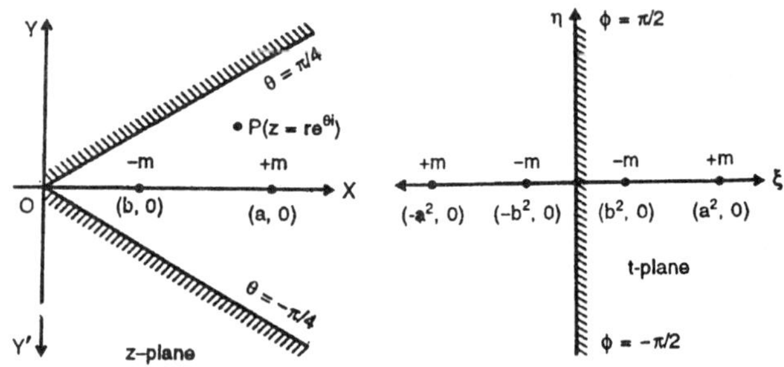

Fig. 3.41

A source of strength + m at the point $(r = c,\ \theta = \alpha)$ in Z-plane transforms to a source of same strength at the point $(R = c^3,\ f = 3\alpha)$ in t-plane. Also a sink of same strength – m at the origin in Z-plane corresponds to a sink of same strength at origin in t-plane. The image system consists of

(i) a source of strength + m at $\{c^3, 3\alpha\}$

(ii) a source of strength + m at $\{c^3, (\pi - 3\alpha)\}$

(iii) a source of strength – m at an origin.

Thus the complex potential becomes

$w = 2m \log t - m \log (t - c^3 e^{3\alpha i}) - m \log (t - c^3 e(\pi - 3\alpha)\ i)$

$w = 2m \log t - m \log (t - c^3 e^{3\alpha i}) - m \log (t + c^3 e^{-3\alpha i})$

$w = 2m \log z^3 - m \log (z^3 - c^3 e^{3\alpha i}) - m \log \ (z^3 + c^3 e - 3^{\alpha i})$,

Let P ($z = re^{\theta i}$) be a point in Z-plane, then

$w = 6m \log (re^{\theta i}) - m \log (r^3 e^{3\theta i} - c^3 e^{\alpha i}) - m \log (r^3 e^{3\theta i} + c^3 e - 3\alpha i)$

or $\phi + i\psi = 6m (\log r + \theta i) - m \log [(r^3 e^{3\theta i} - c^3 e^3 \alpha i)(r^3 e^{3\theta i} + c^3 e - 3\alpha i)]$

or $\phi + i\psi = 6m (\log r + \theta i)$

$- m \log [(c^6 \cos 6\theta + 2c^3 r^3 \sin 3\theta \sin 3\alpha - c^6)$

$+ i (r^6 \sin \theta - 2c^3 r^3 \sin 3\alpha \cos 3\theta)]$

Equating the imaginary parts, we have

$$\psi = 6m\theta - m \tan^{-1}\left\{\frac{r^6 \sin 6\theta - 2c^3 r^3 \sin 3\alpha \cos 3\theta}{r^6 \cos 6\theta + 2c^3 r^3 \sin 3\theta \sin 3\alpha - c^6}\right\}$$

Now one of the stream lines is y = 0, which gives

$$6m\theta - m \tan^{-1}\left\{\frac{r^6 \sin 6\theta - 2c^3 r^3 \sin 3\alpha \cos 3\theta}{r^6 \cos 6\theta + 2c^3 r^3 \sin 3\theta \sin 3\alpha - c^6}\right\} = 0$$

$$\text{or} \quad \frac{\sin 6\theta}{\cos 6\theta} = \frac{r^6 \sin 6\theta - 2c^3 r^3 \sin 3\alpha \cos 3\theta}{r^6 \cos 6\theta + 2c^3 r^3 \sin 3\theta \sin 3\alpha - c^6}$$

or $r^6 \sin 6\theta \cos 6\theta + 2c^3 r^3 \sin 3\theta \sin 3\alpha \sin 6\theta - c^6 \sin 6\theta$

$= r^6 \sin 6\theta \cos 6\theta - 2c^3 r^3 \sin 3\alpha \cos 3\theta \cos 6\theta$

or $c^6 \sin 6\theta = 2r^3 c^3 \sin 3\alpha [\cos 6\theta \cos 3\theta + \sin 6\theta \sin 3\theta]$

or $c^6 \sin 6\theta = 2r^3 c^3 \sin 3\alpha \cos (6\theta - 3\theta)$

or $2c^6 \sin 3\theta \cos 3\theta = 2r^3 c^3 \sin 3a \cos 3\theta$

or $2c^3 \cos 3\theta\ [r^3 \sin 3\alpha - c^3 \sin 3\theta] = 0$.

Either $\cos 3\theta = 0$, or $r^3 \sin 3\alpha - c^3 \sin 3\theta = 0$

If $\cos 3\theta = 0$, then $3\theta = \pm \pi/2 \Rightarrow \theta = \pm \pi/6$, which are the given boundaries.

and $r^3 \sin 3\alpha - c^3 \sin 3\theta = 0$

or $r^3 \sin 3\alpha = c^3 \sin 3\theta$,

which is the required equation of curve. **Proved.**

Example 4: *A source of fluid situated in space of two dimensions, is of such strength that $2\pi\rho m$ represents the mass of fluid of density ρ emitted per unit of time. Show that the force necessary to hold a circular disc at rest in the plane of the source is*

$$\frac{f2\pi\rho m^2 a^2}{r(r^2 - a^2)},$$

where a is the radious of the disc and r is the distance of the source from its centre. In what direction is the disc urged by the pressure?

Solution: Let A′ be an inverse point of A situated at a distance r from the centre with regard to the circular disc, such that

$OA. OA' = a^2 \Rightarrow OA' = a^2/r.$

Since a source of strength + m is situated at the point A, so theimage system give rise to a source of strength + m at an inverse point A′ and a sink of strength – m at the origin O.

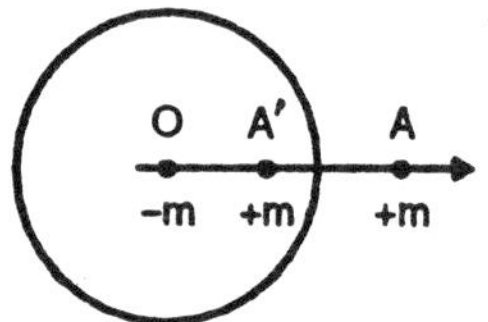

Fig. 3.42

Thus the complex potential due to the image system, is given by

$$w = -m \log (z - r) - m \log \{z - (a^2/r)\} + m \log z$$

$$\Rightarrow \frac{dw}{dz} = -m \frac{1}{z-r} - m\frac{1}{(a^2/r)} + m\frac{1}{z}. \qquad (1)$$

$$\Rightarrow \left(\frac{dw}{dz}\right)^2$$

$$= m^2 \left[\frac{1}{(z-r)^2} + \frac{1}{\{z-(a^2/r)\}^2} + \frac{1}{z^2} + \frac{2}{(z-r)\{z-(a^2/r)\}} - \frac{2}{z(z-r)} - \frac{2}{z\{z-(a^2/r)\}}\right]$$

$$\Rightarrow \left(\frac{dw}{dz}\right)^2$$

$$= m^2\left[\frac{1}{(z-r)^2} + \frac{1}{\{z-(a^2/r)\}^2} + \frac{1}{z^2} + \frac{2}{(r-r)\{z-(a^2/r)\}}\right.$$

$$+\frac{2}{\{z-(a^2/r)\}\{(a^2/r)-r\}}+\frac{2}{(a^2/r)z}-\frac{2}{(a^2/r)\{z-(a2/r)\}}$$

$$+\frac{2}{r(z-r)}+\frac{2}{zr}\Bigg].$$

The poles inside the circular disc are $z=0$ and $z=a^2/r$, so that sum of the residues can be obtained by equating the coefficient of $1/z$ and $1/\{z-(a^2/r)\}$.

The sum of the residues at $z=0$ is

$$= m^2\left[\frac{2}{r}+\frac{2}{a^2/r}\right]=2m^2\left[\frac{1}{r}+\frac{r}{a^2}\right]$$

and the sum of the residues at $z=a^2/r$ is

$$= m^2\left[\frac{2}{(a^2/r)-r}-\frac{2}{(a^2/r)}\right]= m^2\left[\frac{2}{a^2-r^2}-\frac{2r}{a^2}\right]$$

Thus the total sum of the residues is

$$= 2m^2\left[\frac{1}{r}+\frac{r}{a^2}+\frac{r}{a^2-r^2}-\frac{r}{a^2}\right].$$

$$= 2m^2\left[\frac{a^2-r^2+r^2}{r(a^2-r^2)}\right]=\frac{2m^2a^2}{r(r^2-a^2)}.$$

Let X and Y are the components of the forces that act on the disc, then by Blasius theorem, we have

$$X-iY=\frac{1}{2}i\rho.\int_C\left(\frac{dw}{dz}\right)^2 dz$$

$$X-iY=\frac{1}{2}i\rho\left\{2\pi i\times \text{Sumof the residues of}\left(\frac{dw}{dz}\right)^2 \text{with in the contour}\right\}$$

$$X-iY=\frac{1}{2}i\rho\ 2\pi i\left[-\frac{2m^2a^2}{r(r^2-a^2)}\right]$$

$$\text{or } X-iY=\frac{2\pi\rho m^2a^2}{r(r^2-a^2)}\Rightarrow X=\frac{2\pi\rho m^2a^2}{r(r^2-a^2)} \text{ and } Y=0.$$

Thus the circular disc is attracted towards the sources and the pressure will be greater on the opposite side of the disc to that of the source.

Proved.

Example 5: *With in a circular boundary of radius a there is a two dimensional liquid motion due to a source producing liquid at the rate m, at a distance f from the centre and an equal sink at the centre. Find the velocity potential, and show that the resultant of the pressure on the boundary is*

$$\frac{\rho m^2 f^3}{2a^2\pi(a^2 - f^2)},$$

where ρ is the density. Deduce, as a limit, the velocity potential due to the doublet at a centre.

Solution: Let S be an inverse point of S′ = f) with regard to the circular boundary, such that

OS. OS′ = a^2 ⇒ OS = (a^2/f).

The image of the source of strength $(m/2\pi)$ at S′ gives rise to an equal source at an inverse point S and an equal source at the origin O. Also, the image of the sink of strength – $(m/2\pi)$ at origin gives rise to a sink at an inverse point of O which is at infinity and an equal source at the origin. The source and the sink of equal strength at the origin O cancels each other. Hence the image system consists of

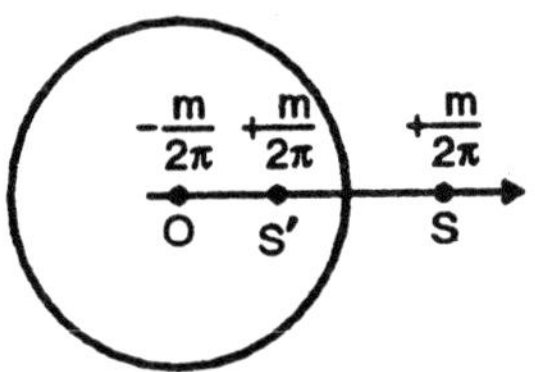

Fig. 3.43

(i) a source of strength + $\frac{m}{2\pi}$ at S′ (f, 0)

(ii) a source of strength + $\frac{m}{2\pi}$ at S$\left(\frac{a^2}{f}, 0\right)$.

(iii) a sink of strength – $\frac{m}{2\pi}$ at O (0, 0).

Thus the complex potential w is given by

$$w = -\frac{m}{2\pi}\log(z - f) - \frac{m}{2\pi}\log\left(z - \frac{a^2}{f}\right) + \frac{m}{2\pi}\log z. \qquad ...(1)$$

$$\text{or } \phi + i\psi = \frac{m}{2\pi}\left\{\log z - \log(z - f) - \log\left(z - \frac{a^2}{f}\right)\right\}$$

Equating the real parts from both the sides, we have

$$f = \frac{m}{2\pi}\log OP - \frac{m}{2\pi}\log S'P - \frac{m}{2\pi}\log SP$$

$$\text{or } f = -\frac{m}{2\pi}\log\frac{S'P.SP}{OP}. \qquad ...(2) \qquad \textbf{Ans.}$$

Again, differentiating (1), with regard to z, we have

$$\frac{dw}{dz} = \frac{m}{2\pi}\left[\frac{1}{z} - \frac{1}{z-f} - \frac{1}{z-(a^2/f)}\right].$$

$$\Rightarrow \left(\frac{dw}{dz}\right)^2 = \frac{m^2}{4\pi^2}\left[\frac{1}{z^2} + \frac{1}{(z-f)^2} + \frac{1}{(z-a^2/f)^2} - \frac{2}{z(z-f)} - \frac{2}{z\{z-(a^2/f)\}} + \frac{2}{(z-f)\{z-(a^2/f)\}}\right]$$

$$\Rightarrow \left(\frac{dw}{dz}\right)^2 = \frac{m^2}{4\pi^2}\left[\frac{1}{z^2} + \frac{1}{(z-f)^2} + \frac{1}{\{z-(a^2/f)\}^2} - \frac{2}{f(z-f)} + \frac{2}{zf} - \frac{2}{(a^2/f)\{z-(a^2/f)\}} + \frac{2}{z(a^2/f)} + \frac{2}{(z-f)\{f-(a^2/f)\}} + \frac{2}{\{z-(a^2/f)\}\{(a^2/f)-f\}}\right]$$

The poles inside the circular boundary are z = 0 and z = f, so the residue at z = 0 is

$$= \left(\frac{2}{f} + \frac{2}{(a^2/f)}\right) = 2\left(\frac{1}{f} + \frac{f}{a^2}\right),$$

and the residue at z = f

$$= \frac{2}{f-a^2/f} - \frac{2}{f} = \frac{2f}{f^2-a^2} - \frac{2}{f}.$$

Hence the sum of the residues is

$$= \frac{m^2}{4\pi^2}\left[\frac{2}{f} + \frac{2f}{a^2} + \frac{2f}{f^2-a^2} - \frac{2}{f}\right] = \frac{m^2}{4\pi^2 a^2}\frac{2f^3}{(f^2-a^2)}$$

Let X and Y are the components, parallel to the axes of the force that acts on the circular boundary, then by Blasius's theorem, we have

$$X - iY = \frac{1}{2} i\rho \int_C \left(\frac{dw}{dz}\right)^2 dz = \frac{1}{2} i\rho \{2\pi i \times \text{ Sum of the residues of}$$

$(dw/dz)^2$ with in the contour}.

$$X - iY = \frac{1}{2} i\rho \cdot 2\pi i \frac{m^2}{4\pi^2} \frac{2f^3}{a^2(f^2 - a^2)} = \frac{\rho m^2 f^3}{2\pi a^2 (a^2 - f^2)}$$

$$\Rightarrow X = \frac{\rho m^2 f^3}{2a^2 \pi (a^2 - f^2)} \text{ and } Y = 0. \quad \textbf{Proved.}$$

Now we shall determine the velocity potential due to a doublet at the centre of the circular boundary.

The combination forms a doublet if f tends to zero or (a^2/f) tends to infinity, such that

$$(m/2\pi)\ f = \mu. \qquad ...(3)$$

From the relation (1), we have

$$w = -\frac{m}{2\pi} \log\left(1 - \frac{f}{z}\right) - \frac{m}{2\pi} \log\left(1 - \frac{z}{a^2/f}\right) + \text{Const.}$$

$$\text{or } w = -\frac{m}{2\pi}\left[-\frac{f}{z} - \frac{f^2}{2z^2} - \ldots\right] - \frac{m}{2\pi}\left[-\frac{z}{a^2/f} - \frac{z^2}{2(a^2/f)^2} - \ldots\right]$$

$$\text{or } w = -\frac{m}{2\pi}\left[\frac{f}{z} + \frac{z}{a^2/f}\right]$$

Using the relation (3), we have

$$w = \frac{\mu}{z} + \frac{\mu z}{a^2} \Rightarrow \phi + i\psi = \frac{\mu}{r} e^{-\theta i} + \frac{\mu}{a^2} r e^{\theta i}$$

$$\Rightarrow \phi + i\psi = \frac{\pi}{r}(\cos\theta - i\sin\theta) + \frac{\mu}{a^2}(\cos\theta + i\sin\theta)$$

$$\Rightarrow \phi = \frac{\mu}{r}\cos\theta + \frac{\mu r}{a^2}\cos\theta$$

$$= \mu\left(\frac{1}{r} + \frac{r}{a^2}\right)\cos\theta. \quad \textbf{Ans.}$$

Example 6: *Prove that the equation of motion is satisfied for an inviscid, incompressible, steady flow with negligible body force whose velocity components are given by*

$$q_r = U\left(1-\frac{A^3}{r^3}\right)\cos\theta,\ q_\theta = -U\left(1+\frac{A^3}{2r^3}\right)\sin\theta,\ q_\phi = 0,$$

where A is constant. Find the resultant velocity when $r \to \infty$.

Solution: The equations of motion for an inviscid, incompressible and steady flow with negligible external force, in spherical polar coordinates, are given as

$$q_r\frac{\partial p_r}{\partial r}+\frac{q\theta}{r}\frac{\partial q_r}{\partial \theta}+\frac{q\phi}{r\sin\theta}-\frac{\partial q_r}{\partial \phi}-\frac{q_\theta^2+q_\phi^2}{r} = -\frac{1}{\rho}\frac{\partial p}{\partial r},$$

$$q_r\frac{\partial q_\theta}{\partial r}+\frac{q_\theta}{r}\frac{\partial q_\theta}{\partial \theta}+\frac{q\phi}{r\sin\theta}\frac{\partial q\theta}{\partial \phi}+\frac{q_r q_\theta}{r}-\frac{q_\phi^2\cot\theta}{r} = -\frac{1}{\rho}\frac{\partial p}{\partial \theta},$$

$$q_r\frac{\partial q_\phi}{\partial r}+\frac{q_\theta}{r}\frac{\partial q\phi}{\partial \theta}+\frac{q\phi}{r\sin\theta}\frac{\partial q\phi}{\partial \phi}+\frac{q_\phi q_r}{r}-\frac{q_\theta q_\phi\cot\theta}{r}$$

$$= -\frac{1}{\rho}\frac{1}{r\sin\theta}\frac{\partial p}{\partial \phi} \qquad ...(1, 2, 3)$$

Here $q_r = U\left(1-\frac{A^3}{r^3}\right)\cos\theta,\ q_\theta = -U\left(1+\frac{A^3}{2r^3}\right)$

$\sin\theta,\ q_\phi = 0,$ ...(4)

From the relation (4), equations (1,2,3) reduce to

$$U\left(1-\frac{A^3}{r^3}\right)\cos\theta\left(\frac{3UA^3}{r^4}\right)\cos\theta+\frac{U}{r}\left(1+\frac{A^3}{2r^3}\right)\sin\theta$$

$$\times U\left(1-\frac{A^3}{r^3}\right)\sin\theta-\frac{U^2}{r}\left(1+\frac{A^3}{2r^3}\right)\sin^2\theta = -\frac{1}{\rho}\frac{\partial p}{\partial r},$$

$$U\left(1-\frac{A^3}{r^3}\right)\cos\theta\left(\frac{3UA^3}{r^4}\right)\sin\theta+\frac{U}{r}\left(1+\frac{A^3}{2r^3}\right)\sin\theta$$

$$\times U\left(1+\frac{A^3}{2r^3}\right)\cos\theta-\frac{U^2}{r}\left(1-\frac{A^3}{r^3}\right)\left(1+\frac{A^3}{2r^3}\right)\sin\theta\cos\theta$$

$$= -\frac{1}{\rho}\frac{\partial p}{r\partial \theta},$$

$$0 = \frac{1}{\rho}\frac{1}{r\sin\theta}\frac{\partial p}{r\phi}. \qquad ...(5, 6, 7)$$

Equation (7) shows that the pressure p is independent of ϕ, therefore $p = p(r, \theta)$. On simplifying the equation (5) and (6), we have

$$\frac{3U^2A^3}{r^4}\left(1-\frac{A^3}{r^3}\right)\cos^2\theta+\frac{U^2}{r}\left(1-\frac{A^3}{2r^3}-\frac{A^6}{2r^6}\right)\sin^2\theta$$

$$-\frac{U^2}{r}\left(1+\frac{A^3}{r^3}+\frac{A^6}{4r^6}\right)\sin^2\theta = -\frac{1}{\rho}\frac{\partial p}{\partial r},$$

or $$\frac{3U^2A^3}{r^4}\left(1-\frac{A^3}{r^3}\right)\cos^2\theta+\frac{U^2}{r}\left(-\frac{3A^3}{2r^3}-\frac{3A^6}{4r^6}\right)\sin^2\theta = \frac{1}{\rho}\frac{\partial p}{\partial r},$$

or $$\frac{3U^2A^3}{r^4}\left(1-\frac{A^3}{r^3}\right)\cos^2\theta-\frac{3U^2A^3}{2r^4}\left(1+\frac{A^3}{2r^3}\right)\sin^2\theta$$

$$= -\frac{1}{\rho}\frac{\partial p}{\partial r}. \quad \text{...(8)}$$

and $$\frac{3U^2A^3}{2r^4}\left(1-\frac{A^3}{r^3}\right)\sin\theta\cos\theta+\frac{U^2}{r}\left(1+\frac{A^3}{r^3}+\frac{A^6}{4r^6}\right)\sin\theta\cos\theta$$

$$-\frac{U^2}{r}\left(1-\frac{A^3}{2r^3}-\frac{A^6}{2r^6}\right)\sin\theta\cos\theta \; -\frac{1}{\rho}\frac{\partial p}{\partial\theta}.$$

or $$\frac{3U^2A^3}{2r4}\left(1-\frac{A^3}{r^3}\right)\sin\theta\cos\theta+\frac{3U^2A^3}{2r^4}\left(1+\frac{A^3}{2r^3}\right)\sin\theta\cos\theta =$$

$$-\frac{1}{\rho}\frac{\partial p}{r\partial\theta}. \quad \text{...(9)}$$

Differentiating equation (8) with regard to θ, we have

$$-\frac{6U^2A^3}{r^4}\left(1-\frac{A^3}{r^3}\right)\cos\theta\sin\theta \; - \; \frac{3U^2A^3}{r^4}\left(1+\frac{A^3}{2r^3}\right)\sin\theta\cos\theta$$

$$= -\frac{1}{\rho}\frac{\partial^2 p}{\partial r\partial\theta}$$

or $$\left(-\frac{9U^2A^3}{r^4}+\frac{9U^2A^6}{2r^7}\right)\sin\theta\cos\theta = -\frac{1}{\rho}\frac{\partial^2 p}{\partial r\partial\theta} \; - \quad \text{...(10)}$$

Differentiating equation (9) with regard to r, we have

$$\frac{3U^2A^3}{2}\left(-\frac{3}{r^4}+\frac{6A^3}{r^7}\right)\cos\theta\sin\theta$$

$$+\frac{3U^2A^3}{2}\left(-\frac{3}{r^4}+\frac{3A^3}{r^7}\right)\sin\theta\cos\theta = -\frac{1}{\rho}\frac{\partial^2 p}{\partial r\partial\theta}$$

or $$\frac{9}{2}\frac{U^2A^3}{2}\left(1-\frac{2A^3}{r^3}\right)\cos\theta\sin\theta$$

$$-\frac{9}{2}\frac{U^2A^3}{2}\left(1+\frac{A^3}{r^3}\right)\sin\theta\cos\theta = -\frac{1}{\rho}\frac{\partial^2 p}{\partial r\partial\theta}$$

or $$\left(-\frac{9U^2A^3}{r^4}+\frac{9U^2A^6}{2r^7}\right)\sin\theta\cos\theta = -\frac{1}{\rho}\frac{\partial^2 p}{\partial r\partial\theta} - \quad ...(11)$$

Equation (10) and (11) are identical. Hence, the equation of motion is satisfied.

When $r \to \infty$, the resultant velocity is equal to U. **Proved.**

Example 7: *Prove that the velocity components*

$$q_r\ (r,\theta) = -U\left(1-\frac{a^2}{r^2}\right)\cos\theta, q_\theta(r,\theta) = U\left(1+\frac{a^2}{r^2}\right)\sin\theta,$$

satisfy the equation of motion for a two-dimensional inviscid incompressible flow. Find the pressure associated with this velocity field. U and a are constants.

Solution: The equations of motion for a two-dimensional steady, inviscid, incompressible flow in the absence of external force, in spherical polar coordinates, are given by

$$q_r\frac{\partial q_r}{\partial r}+\frac{q_\theta}{r}\frac{\partial q_r}{\partial\theta}-\frac{q_\theta^2}{r} = -\frac{1}{\rho}\frac{\partial p}{\partial r},$$

$$q_r\frac{\partial q_\theta}{\partial r}+\frac{q_\theta}{r}\frac{\partial q_\theta}{\partial\theta}+\frac{q_r q_\theta}{r} = -\frac{1}{\rho}\frac{\partial p}{r\partial\theta},$$

$$0 = -\frac{1}{\rho}\frac{1}{r\sin\theta}\frac{\partial p}{\partial\phi}. \quad ...(1, 2, 3)$$

Here $q_r\ (r, \theta) = -U\left(1-\frac{a^2}{r^2}\right)\cos\theta,$

$$q_\theta\ (r,\ \theta) = U\left(1+\frac{a^2}{r^2}\right)\sin\theta. \qquad ...(4)$$

From (3), it follows that the pressure p is independent of ϕ *i.e.*, p = p (r, θ).

Using the relation (4) into (1) and (2), we have

$$U\left(1-\frac{a^2}{r^2}\right)\cos\theta+\left(\frac{2Ua^2}{r^3}\right)\cos\theta+\frac{U}{r}\left(1+\frac{a^2}{r^2}\right)\sin\theta$$

$$\times\ U\left(1-\frac{a^2}{r^2}\right)\sin\theta-\frac{U^2}{r}\left(1+\frac{a^4}{r^2}\right)^2\sin^2\theta = -\frac{1}{\rho}\frac{\partial p}{\partial r},$$

or
$$\frac{2U^2a^2}{r^3}\left(1-\frac{a^2}{r^2}\right)\cos^2\theta-\frac{2U^2a^2}{r^3}\left(1+\frac{a^2}{r^2}\right)\sin^2\theta$$

$$= -\frac{1}{\rho}\frac{\partial p}{\partial r}. \qquad ...(5)$$

and
$$\frac{2U^2a^2}{r^3}\left(1-\frac{a^2}{r^2}\right)\sin\theta\cos\theta+\frac{U^2}{r}\left(1+\frac{a^2}{r^2}\right)\sin\theta\cos\theta$$

$$-\frac{U^2}{r}\left(1-\frac{a^2}{r^4}\right)\sin\theta\cos\theta = -\frac{1}{\rho}\frac{\partial p}{r\partial\theta}$$

or
$$\frac{2U^2a^2}{r^3}\left(1-\frac{a^2}{r^2}\right)\sin\theta\cos\theta+\frac{2U^2a^2}{r}\left(1+\frac{a^2}{r^2}\right)\sin\theta\cos\theta$$

$$= -\frac{1}{\rho}\frac{\partial p}{r\partial\theta}$$

$$\frac{4U^2a^2}{r^3}\sin\theta\cos\theta = -\frac{1}{\rho}\frac{\partial p}{r\partial\theta}. \qquad ...(6)$$

Differentiating (5) with regard to θ, we have

$$-\frac{4U^2a^2}{r^3}\left(1-\frac{a^2}{r^2}\right)\cos\theta\sin\theta\frac{4U^2a^2}{r^2}\left(1+\frac{a^2}{r^2}\right)\sin\theta\cos\theta$$

$$= -\frac{1}{\rho}\frac{\partial^2 p}{\partial r\partial\theta}$$

$$\frac{8U^2a^2}{r^3}\cos\theta\sin\theta = \frac{1}{\rho}\frac{\partial^2 p}{\partial r\partial\theta}. \qquad ...(7)$$

Differentiating (6) with regard to r, we have

$$\frac{8U^2a^2}{r^3}\cos\theta\sin\theta = \frac{1}{\rho}\frac{\partial^2 p}{\partial r\partial\theta}. \qquad ...(8)$$

Equation (7) and (8) are identical. Hence, the equation of motion are satisfied.

Again, p is a function or r and θ, we have

$$dp = \left(\frac{\partial p}{\partial r}\right) dr + \left(\frac{\partial p}{\partial \theta}\right) d\theta$$

$$\text{or } dp = -\,2r\ U2a2 \left[\left(\frac{1}{r^3}-\frac{a^2}{r^5}\right)\cos^2\theta - \left(\frac{1}{r^3}+\frac{a^2}{r^5}\right)\sin^2\theta\right] dr$$

$$-\ \frac{4\rho U^2a^2}{r^2}(\sin\theta\cos\theta)\, d\theta$$

By integrating, we have

$$p = -2\rho U^2a^2\left[\left(-\frac{1}{2r^2}+\frac{a^2}{4r^4}\right)\cos^2\theta - \left(-\frac{1}{2r^2}-\frac{a^2}{4r^4}\right)\sin^2\theta\right] + C$$

$$\text{or } p = \frac{4\rho U^2a^2}{r}(\cos^2\theta - \sin^2\theta) + C,$$

where C is an integration constant.

$$\text{or } p = 2\rho U^2a^2$$

$$\left[\left(\frac{1}{2r^r}-\frac{a^2}{4r^4}-\frac{2}{r^2}\right)\cos^2\theta - \left(\frac{1}{2r^2}-\frac{a^2}{4r^4}-\frac{2}{r^2}\right)\sin^2\theta\right] + C$$

$$\text{or } p = 2\rho U^2a^2\left[-\frac{3}{2r^2}\cos 2\theta - \frac{a^2}{4r^4}\right] + C$$

$$\text{or } p = -\,\rho U^2a^2\left[-\frac{3}{r^2}\cos 2\theta + \frac{a^2}{2r^4}\right] + C,$$

Which gives the required pressure distribution.

Example 8: *An elastic fluid, the weight of which is neglected, obeying Boyle's law is in motion in a uniform straight tube; show that on the hypothesis of parallel sections the velocity at any time t at a distance r from a fixed point in the tube is defined by the equation*

$$\frac{\partial^2 v}{\partial t^2}+\frac{\partial}{\partial r}\left(2v\frac{\partial v}{\partial t}+v^2\frac{\partial v}{\partial r}\right)=k\,\frac{\partial^2 v}{\partial r^2}.$$

Solution: Since the fluid obeys Boyle's law then

$$p = k\rho. \qquad ...(1)$$

The equation of continuity and the equation of motion is given by

$$\frac{\partial \rho}{\partial t}+\frac{\partial}{\partial r}(\rho v)=0, \qquad ...(2)$$

and

$$\frac{\partial v}{\partial t}+v\frac{\partial v}{\partial r}=-\frac{1}{\rho}\frac{\partial p}{\partial r} \qquad ...(3)$$

Using the relation (1), we have

$$\frac{\partial v}{\partial t}+v\frac{\partial v}{\partial r}=-\frac{k}{\rho}\frac{\partial \rho}{\partial r}. \qquad ...(4)$$

Differentiating (4) partially with regard to t, we have

$$\frac{\partial^2 v}{\partial t^2}+\frac{\partial}{\partial t}\left(v\frac{\partial v}{\partial r}+\frac{k}{\rho}\frac{\partial \rho}{\partial r}\right)=0,$$

$$\Rightarrow \quad \frac{\partial^2 v}{\partial t^2}+\frac{\partial}{\partial r}\left(v\frac{\partial v}{\partial t}+\frac{k}{\rho}\frac{\partial \rho}{\partial t}\right)=0$$

$$\Rightarrow \quad \frac{\partial^2 v}{\partial t^2}+\frac{\partial}{\partial r}\left\{v\frac{\partial v^*}{\partial t}+\frac{k}{\rho}\left(-\frac{\partial}{\partial r}(\rho v)\right)^t\right\}=0$$

$$\Rightarrow \quad \frac{\partial^2 v}{\partial t^2}+\frac{\partial}{\partial r}\left\{v\frac{\partial v}{\partial t}+\frac{k}{\rho}\left(\rho\frac{\partial}{\partial r}+v\frac{\partial \rho}{\partial r}\right)^t\right\}=0$$

$$\Rightarrow \quad \frac{\partial^2 v}{\partial t^2}+\frac{\partial}{\partial r}\left\{v\frac{\partial v}{\partial t}-k\frac{\partial v}{\partial r}-\frac{k}{\rho}\frac{\partial \rho}{\partial r}v\right\}=0$$

$$\Rightarrow \frac{\partial^2 v}{\partial t^2}+\frac{\partial}{\partial r}\left\{v\frac{\partial v}{\partial t}-k\frac{\partial v}{\partial r}+\left(\frac{\partial v}{\partial t}+v\frac{\partial v}{\partial r}\right)v\right\}=0$$

$$\Rightarrow \qquad \frac{\partial^2 v}{\partial t^2} + \frac{\partial}{\partial r}\left(2v\frac{\partial v}{\partial t} + v^2\frac{\partial v}{\partial r} - k\frac{\partial v}{\partial r}\right) = 0$$

$$\Rightarrow \qquad \frac{\partial^2 v}{\partial t^2} + \frac{\partial}{\partial r}\left(2v\frac{\partial v}{\partial t} + v^2\frac{\partial v}{\partial r}\right) = k\,\frac{\partial^2 v}{\partial r^2}.$$

Proved.

Example 9: *A line source is in the presence of an infinite plane on which is placed a semi-circular cylindrical boss, the direction of the source is parallel to the axis of the boss, the source is at distance c from the plane and the axis of the boss, whose radius is a. Show that the radius to the point on the boss at which the velocity is a maximum makes an angle θ with the radius to the source where*

$$\theta = \cos^{-1}\left\{\frac{a^2 + c^2}{\sqrt{\{2(a^4 + c^4)\}}}\right\}$$

Solution: Let R be the inverse point of P (OP = c) with regard to the circular boundary, then

OR. OP = $a^2 \Rightarrow$ or = (a^2/c).

The image of a source of strength + m at P will give rise to an equal source of strength + m at an inverse point R and a sink of strength – m at an origin with regard to the cylindrical boundary.

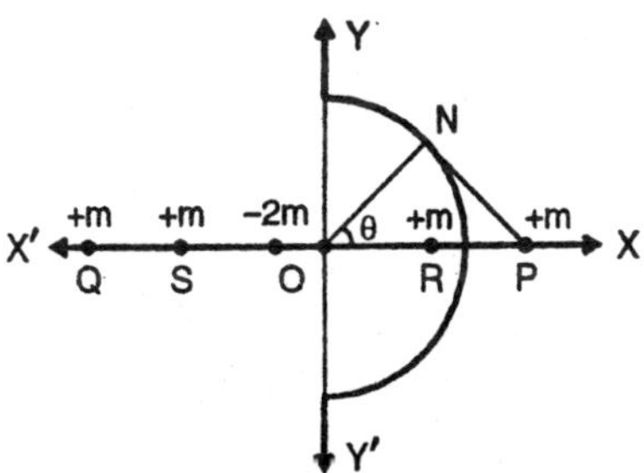

Fig. 3.44

Again, the image of a source of strength + m at P in the line x = 0 will give rise to a source of strength + m at Q (z = – c), a source of strength + m at S (z = – a^2/c) and a sink of strength – m at the origin.

The complex potential w becomes

$w = 2m \log z - m \log (z - c) - m \log (z + c)$

$- m \log \{z - (a^2/c)\} - m \log \{z + (a^2/c)\}$ ...(1)

or $w = 2m \log z - m \log (z^2 - c^2) - m \log \{z^2 - (a^4/c^2)\}$

or $\dfrac{dw}{dz} = 2m\dfrac{1}{z} - m\dfrac{2z}{z^2 - c^2} - m\dfrac{2z}{z^2 - (a^4/c^2)}$

or $\dfrac{dw}{dz} = -2m \dfrac{(z^4 - a^4)}{z(z^2 - c^2)\{z^2 - (a^4/c^2)\}}$

The velocity at any point N ($z = ae^{\theta i}$) on the circular boundary is

$$q = \left|\frac{dw}{dz}\right| = 2m \frac{\left|z^4 - a^4\right|}{\left|z(z^2 - c^2)\{z^2 - (a^2/c^2)\}\right|}$$

or $q = \left|\dfrac{dw}{dz}\right| = \dfrac{2ma^4\left|e^4\theta i - 1\right|}{\left|ae^{\theta i}(a^2e^{2\theta i} - c^2)\{a^2e^{2\theta i} - (a^4/c^2)\}\right|}$

or $q = \dfrac{2ma^4c^2}{a^3} \dfrac{|(\cos 4\theta - 1) + i\sin 4\theta|}{|(\cos\theta + i\sin\theta)\{(a^2\cos 2\theta - c^2) + ia^2\sin 2\theta)\}\{(c^2\cos 2\theta - a^2) + ic^2\sin 2\theta|}$

or $q = \dfrac{4mac^2 \sin 2\theta}{a^4 + c^4 - 2a^2c^2\cos 2\theta}$

or $\dfrac{4mac^2}{q} = \dfrac{a^4 + c^4 - 2a^2c^2\cos 2\theta}{\sin 2\theta}$

or $m = (a^4 + c^4) \operatorname{cosec} 2\theta - 2a^2c^2 \cot 2\theta$, where $\mu = 4mac^2/q$

If q is maximum then (1/q) must be minimum i. e.

$\dfrac{d\mu}{d\theta} = -2(a^4 + c^4) \operatorname{cosec} 2\theta \cot 2\theta + 4a^2c^2 \operatorname{cosec}^2 2\theta,$ ...(2)

and $\dfrac{d^2\mu}{d\theta^2} = 4(a^4 + c^4) \operatorname{cosec} \theta (\operatorname{cosec}^2 2\theta + \cot^2 2\theta)$

$- 8a^2c^2 \operatorname{cosec}^2 2\theta \cot 2\theta$

$= 4 \operatorname{cosec} \theta \{(a^2 \operatorname{cosec} \theta - c^2 \cot \theta)^2 + a^4 \cot^2 \theta + c^4 \operatorname{cosec}^2 \theta\}.$

Since θ " $\pi/2$ then $\dfrac{d^2\mu}{d\theta^2} > 0$, so that μ should be minimum then q (i.e.

velocity will be maximum.

From (2) $\frac{d\mu}{d\theta} = 0$, we have

$$(a^4 + c^4) \operatorname{cosec} 2\theta \cot 2\theta = 2a^2c^2 \operatorname{cosec}^2 2\theta$$

or $\cos 2\theta = \frac{2a^2c^2}{a^4 + c^4} \Rightarrow \cos^2 \theta = \frac{(a^2 + c^2)^2}{2(a^4 + c^4)}$

or $\theta = \cos^{-1}\left\{\frac{a^2 + c^2}{\sqrt{2(a^4 + c^4)\}}}\right\}$. **Proved.**

Example 10: *An area A is bounded by that part of the X-axis for which $x > a$ and by that branch of $x^2 - y^2 = a^2$ which is in the positive quadrant. There is a two dimensional unit source at (a, 0) which sends out liquid uniformly in all direction. Show by means of the transformation $w = \log (z^2 - a^2)$ that in steady motion the stream lines of the liquid within the area A are portions of rectangular hyperbolas. Determine the stream lines corresponding to $\psi = 0$, $\pi/4$ and $\pi/2$. If ρ_1 and ρ_2 are the distance of a point P within the fluid from the points (± a, 0). Shew that the velocity of fluid at P is measured by $\frac{2OP}{\rho_1\rho_2}$, O being the origin.*

Solution: The complex potential is given by

$$w = \log (z^2 - a^2) \qquad ...(2)$$

or $\phi + i\psi = \log \{ (x + iy)^2 - a^2\} = \log \{(x^2 - y^2 - a^2) + 2ixy\}$

or $\psi = \tan^{-1} \frac{2xy}{x^2 - y^2 - a^2} \Rightarrow \tan \psi = \frac{2xy}{x^2 - y^2 - a^2}$ ...(3)

The stream lines are given by y = const.

i.e., $\frac{2xy}{x^2 - y^2 - a^2} = \text{const.} = k \text{ (let)}$

If k is zero, the stream lines will be x = 0 and y = 0.

If k is infinite, the stream lines will be $x^2 - y^2 = a^2$.

Thus the liquid flows in the area A bound by x = 0, y = 0, and $x^2 - y^2 = a^2$ *i.e.*, the portion of a rectangular hyperbola in the positive quadrant.

Again, the complex potential (1) can be written as

$$w = \log (z - a) + \log (z + a),$$

which shows that the image of a unit source at the point (a, 0) consists a unit source at the point (– a, 0) with regard to Y-axis. **Proved.**

The velocity of the fluid at any point is

$$q = \left|\frac{dw}{dz}\right| = \left|\frac{1}{z-a} + \frac{1}{z+a}\right| = \frac{|2z|}{|z-a|\,|z+a|} = \frac{2OP}{\rho_1\rho_2}. \quad \textbf{Proved.}$$

The stream lines corresponding to $\psi = 0$ and $\psi = \pi/2$ are x = 0, y = 0 and $x^2 - y^2 = a^2$.

Also, the stream lines corresponding to $\psi = \pi/4$ are

$$\frac{2xy}{x^2 - y - a^2} = \tan \pi/4 = 1$$

or $x^2 - y^2 - a^2 = 2xy$. **Ans.**

Example 11: *The space on one side of an infinite plane wall y = 0, is filled with inviscid, incompressible fluid, moving at infinity with velocity U in the direction of the axis of X. The motion of the fluid is wholly two dimensional, in the (x, y) plane. A doublet of strength m is at a distance a from the wall and points in the negative direction of the axis of X. Shew that if μ is less than $4a^2U$, the pressure of the fluid on the wall is a maximum at points distance $\sqrt{3}a$ from O, the foot of the perpendicular from the doublet on the wall, and is minimum at O.*

If μ is equal to $4a^2U$, find the points where the velocity of the fluid is zero, and shew that the stream lines include the circle

$$x^2 + (y - a)^2 = 4a^2,$$

where the origin is taken at O.

Solution: We know that the complex potential for a doublet of strength μ at $z = z_0$ inclined at an angle α to the real axis is given by

$$w = e^{i\alpha}./(z - z_0)$$

The doublet points in the negative direction of the X-axis will make an angle π with it. Since the doublet of strength m is placed at a distance a from the real axis then the image system consists of an equal doublet on the other side of that axis. The complex potential for the system is

$$w = \frac{\mu e^{i\pi}}{z - ia} + \frac{\mu e^{i\pi}}{z + ia} - Uz = -\frac{2\mu z}{z^2 + a^2} - Uz \qquad ...(1)$$

$$\text{or} \quad \frac{dw}{dz} = -\frac{2\mu(a^2 - z^2)}{(a^2 + z^2)^2} - U \qquad ...(2)$$

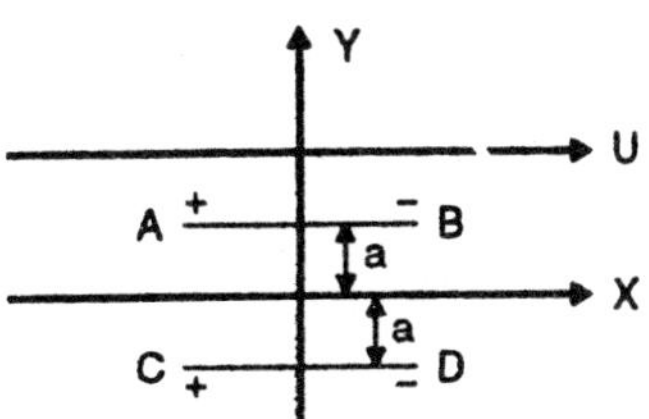

Fig. 3.45

Let q be the velocity on the wall then

$$q = \left|\frac{dw}{dz}\right| = \left|\frac{2\mu(a^2 - z^2)}{(a^2 + z^2)} + U\right|$$

Consider p be any point on the wall. Substituting z = x, we have

$$q = \left[2\mu\frac{a^2 - x^2}{(a^2 + x^2)^2} + U\right]$$

By Bernoulli's theorem, we get $\frac{p}{\rho} + \frac{1}{2}q^2 = \text{const.}$

As $p = \Pi$, $q = U$; const. $= \frac{\Pi}{\rho} + \frac{1}{2}U^2$

$$\Rightarrow \frac{p}{\rho} + \frac{1}{2}q^2 = \frac{\Pi}{\rho} + \frac{1}{2}U^2 \Rightarrow \quad \Pi - p = \frac{1}{2}\rho(q^2 - U^2)$$

$$\Rightarrow \Pi - p = \frac{1}{2}\rho\left\{\left(2\mu\frac{(a^2 - x^2)}{(a^2 + x^2)^2} + U\right)^2 - U^2\right\}$$

$$\Rightarrow \Pi - p = 2\mu\rho\left[\mu\frac{(a^2 - x^2)}{(a^2 + x^2)^4} + U\frac{a^2 - x^2}{(a^2 + x^2)^2}\right]$$

The pressure will be maximum or minimum if $\frac{dp}{dx} = 0$

$$\text{i.e. } 2\mu^2\left\{\frac{4x(a^2 - x^2)}{(a^2 + x^2)^4} + \frac{8x(a^2 - x^2)^2}{(a^2 + x^2)^5}\right\}$$

$$+ 2\mu U\left\{\frac{2x}{(a^2 + x^2)^2} + \frac{4x(a^2 - x^2)^2}{(a^2 + x^2)^3}\right\} = 0$$

or $4\mu x\ [2\mu\ (a^2 - x^2) + U\ (a^2 + x^2)^2]\dfrac{3a^2 - x^2}{(x^2 + a^2)^5} = 0$

Either $x\ (3a^2 - x^2) = 0$ or $[2\mu\ (a^2 - x^2) + U\ (a^2 + x^2)] = 0$.

If x (3a2 – x2) = 0 then $x = 0,\ \sqrt{3a}\ ,\ -\sqrt{3a}$.

Now on the wall (y = 0) at $x = +\ \sqrt{3a}$

$$\frac{dw}{dx} = -U - \frac{2\mu(a^2 - 3a^2)}{(a^2 + 3a^2)^2} = \frac{\mu}{4a^2} - U .$$

If $\mu < 4a^2U$, the value of $(d^2p/dx^2)x = a\sqrt{3}$ is negative, then the pressure of the fluid at the wall is maximum. Also if $(d^2p/dx^2)_x = 0$ is positive then the pressure at the wall is minimum, and if $m = 4a^2U$, then $(dw/dz) = 0$.

From (2), we have

$$8a^2U\frac{a^2 - z^2}{(a^2 + z^2)^2} + U = 0$$

or $z^4 - 6a^2z^2 + 9a^4 = 0 \Rightarrow z = \pm\ a\sqrt{3}$.

The stagnation points are $(a\sqrt{3}\ , 0)\ (-\ a\sqrt{3}\ ,0)$.

Again, to determine the stream function, equating the imaginary parts from (1) and substituting $\mu = 4a^2U$, we have

$$\psi = \frac{2\mu(x^2 + y^2 - a^2)}{(x^2 + y^2)^2 + 2a^2(x^2 - y^2) + a^4} - Uy$$

$$\text{or}\quad \psi = \frac{8a^2Uy(x^2 + y^2 - a^2)}{(x^2 + y^2)^2 + 2a^2(x^2 - y^2) + a^4} - Uy$$

The stream lines are given by $\psi = 0$

i.e., $-\ y\ \{(x^2 + y^2)^2 + 2a^2\ (x^2 - y^2) + a^4\} + 8a^2y\ (x^2 + y^2 - a^2) = 0$

or $(x^2 + y^2)^2 - 6a^2x^2 - 10\ a^2y^2 + 9a^4 = 0$.

or $(x^2 + y^2 - 2ay - 3a^2)\ (x^2 + y^2 + 2ay - 3a^2) = 0$.

Obviously which includes the circle

$x^2 + y^2 - 2ay - 3a^2 = 0$,

or $x^2 + (y - a)^2 = 4a^2$. **Proved.**

EXERCISE

1. Prove that in two dimensions:
   (i) There is always a stream function whether the motion is irrotational or rotational.
   (ii) If the speed is everywhere the same, the stream lines are straight.
2. Parallel line sources (perp. to XY – plane) of equal strength m are parallel at the points z = nia where n = ...–2, – 1, 0, 1, 2, ... Prove that the complex potential is w = – m log sin h (πz/a).

   Hence, shew that the complex potential for two dimensional doublets (lines doublets), with their axes parallel to the X-axis, of strength μ at the same points is given by

   w = m coth (πz/a).

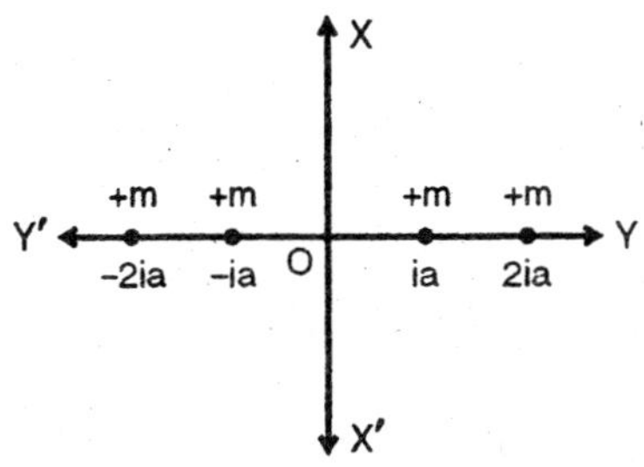

Fig. 3.46

Hint : Sources are given at the points z = 0, ia, – ia, 2ia, –2ia, ... and so on. The complex potential is given by

w = – m log z – m log (z – ai) – m log (z + ai) – m log (z – 2ai) – m log (z + 2ai) ...

or $w = -m \log z (z^2 + a^2)(z^2 + 2^2 a^2)(z^2 + 3^2 a^2) \ldots (z^2 + n^2 a^2)$

$$\text{or } w = -m \log \left\{ \frac{\pi}{a} z \left(1 + \frac{z^2}{a^2}\right)\left(1 + \frac{z^2}{2^2 a^2}\right) \ldots \right\}$$

$$- m \log \left\{ \frac{a}{\pi} a^2 (2^2 a^2)(3^2 a^2) \right\}$$

$$\text{or } w = -m \sin h \frac{\pi}{a} z + \text{const.}$$

The complex potential w_1 for the doublets at the same points is

$$w_1 = -\frac{\partial w}{\partial z} = \frac{m\pi}{a}\coth\left(\frac{\pi z}{a}\right) = \mu \coth\left(\frac{\pi z}{a}\right), \text{ where } \mu = \frac{m\pi}{a}.$$

3. With in a circular boundary of radius a there is a source of strength m at a distance f from the centre and an equal sink at the centre. Determine the resultant thrust on the boundary.

4. In two- dimensional irrotational fluid motion show that, if the stream lines are confocal ellipses.

$$\frac{x^2}{a^2+\lambda} + \frac{y^2}{b^2+\lambda} = 1.$$

$\psi = A \log\{\sqrt{(a^2+\lambda)} + \sqrt{(b^2+\lambda)}\} + B$ and the velocity at any point is inversely proportional to the square root of the rectangle under the focal radii of the point.

Hint : Let the conformal transformation be $z = c \cos w$

or $x + iy = c \cos (f + iy) = c \cos f \cos h\, y + ic \sin f \sin h\, y$.

By eliminating f, we have

$$\frac{x^2}{c^2 \cos h^2 \psi} + \frac{y^2}{c^2 \sinh^2 \psi} = 1. \qquad ...(1)$$

Stream lines will be obtained by ψ = Const. which are confocal ellipses. Comparing (1) with original equation of confocal ellipses, we have

$c^2 \cos h^2 \psi = a^2 + \lambda$, $c^2 \sin h^2 \psi = b^2 + \lambda$; $a^2 - b^2 = c^2$, $ae = c$.

Also $\sqrt{(a^2+\lambda)} + \sqrt{(b^2+\lambda)} = c\,(\cos h\, \psi + \sin h\, \psi) = ce^{\psi}$

or $\psi = \log\{\sqrt{(a^2+\lambda)} + \sqrt{(b^2+\lambda)}\} - \log$ `c.

5. Shew that the force per unit length exerted on a circular cylinder, radius a, due to a source of strength m, at a distance c from the axis is

$$\frac{2\pi\rho m^2 a^2}{c(c^2-a^2)}.$$

6. Prove that in two dimensional liquid motion due to any number of sources at different points on a circle, the circle is a stream line provided that there is no boundary and that the algebraic sum of the strengths of the sources is zero.

7. If a homogeneous liquid is acted on by repulsive force the origin, the magnitude of which at a distance r from the origin is μr per unit mass. Shew that it is possible for the liquid to move steadily, without being contained by any boundaries, in the space between one branch of the hyperbola $x^2 - y^2 = a^2$ and the asymptotes. Also, find the velocity potential.

8. A source and a sink, each of strength μ, exist in an infinite liquid on opposite sides of, and at equal distances c from the centre of rigid sphere of radius a. Shew that the velocity potential V may be expressed in the form.